Novel Membrane Technologies for Traditional Industrial Processes

Novel Membrane Technologies for Traditional Industrial Processes

Special Issue Editors

Pei Li
Yan Wang
Lan-Ying Jiang
Tai-Shung Chung

MDPI • Basel • Beijing • Wuhan • Barcelona • Belgrade

MDPI

Special Issue Editors

Pei Li
Beijing University of Chemical Technology
China

Yan Wang
Huazhong University of Science and Technology
China

Lan-Ying Jiang
Central South University
China

Tai-Shung Chung
National University of Singapore
Singapore

Editorial Office
MDPI
St. Alban-Anlage 66
4052 Basel, Switzerland

This is a reprint of articles from the Special Issue published online in the open access journal *Processes* (ISSN 2227-9717) from 2018 to 2019 (available at: https://www.mdpi.com/journal/processes/special_issues/membrane_processes)

For citation purposes, cite each article independently as indicated on the article page online and as indicated below:

LastName, A.A.; LastName, B.B.; LastName, C.C. Article Title. *Journal Name* **Year**, *Article Number*, Page Range.

ISBN 978-3-03897-790-2 (Pbk)
ISBN 978-3-03897-791-9 (PDF)

Contents

About the Special Issue Editors

Pei Li received his Bachelor's and Master's degrees in Chemical Engineering from Tsinghua University (Beijing China) in 2001 and 2004. He got his Ph.D. degree in the department of Chemical and Environmental Engineering in 2010 from the University of Toledo in OH, USA. Then, he worked in the department of Chemical and Biomolecular Engineering of the National University of Singapore as a Research Fellow for the next 4 years. In 2013, he joined the faculty of the College of Materials Science and Engineering at Beijing University of Chemical Technology as an Associate Professor. He has published over 60 articles in the areas of gas separation membranes, nanofiltration, electrospinning, pervaporation, and forward osmosis.

Yan Wang is a Professor of Chemical Engineering at Huazhong University of Science and Technology in China. Her main research activities are about new membrane materials for separation applications and separation processes including nanofiltration, forward osmosis, reverse osmosis, and pervaporation, as well as membrane fouling. She was enrolled in the "1000 Young Talent Program of China" in 2013. She also serves as the subject editor of Chemical Engineering Research and Design. She has co-authored over 50 refereed journal papers and 4 book chapters.

Lanying Jiang is a Professor at the School of Metallurgy and Environment at Central South University, China. Her current and upcoming efforts are concentrated on hollow fibers membrane formation, nano-scale mixed matrix membranes, polymer blending, and membrane surface modification, with particular emphasis on their applications in gas purification, solvent dehydration, and organic/organic separations in bioenergy, chemical, petrochemical, and metallurgical industries. She has edited one book (Elsevier) and participated in writing 4 book/encyclopedia chapters.

Tai-Shung Chung is a Professor and Provost Chair of Department of Chemical and Biomolecular Engineering, Singapore. He is the Editor or Editorial/Advisory Board Member for 20 journals. He was an inventor of Hyflux Kristal^{TM} 600 ultrafiltration membranes. He is the author of 1 book, 24 book chapters, more than 650 journal papers, more than 70 patents (including 45 US patents), and 350 conference papers. He is the Academician of the Asia Pacific Academy of Materials (APAM) (2015–present) and Fellow (Academician) of the Academy of Engineering Singapore (2012–present). He received the Distinction Award in Water Reuse and Conservation, the International Desalination Association (IDA) in 2016), Singapore President's Technology Award in 2015, IES (Institution of Engineers, Singapore) Prestigious Engineering Achievement Award, and the Hyflux-SNIC (Singapore National Institute of Chemistry) Award in Environmental Chemistry in 2010.

Preface to "Novel Membrane Technologies for Traditional Industrial Processes"

Traditional industry is indispensable for human society, but is challenged by high mass and energy consumption. Among the technologies addressing these problems, membrane separation exhibits great prospect. In recent years, its application has been extended to more diverse fields in traditional industry, as novel ideas in material and engineering have contributed to enhanced membrane performance and process efficiency. This Special Issue engaged researchers from 17 universities and institutes to discuss and illustrate some of the latest advancements in gas separation, biorefinery, forward osmosis, membrane bioreactors, and nanofiltration. It is hoped that more excellent work on membrane technology can be delivered in the future through Processes, encouraging its wider application.

Pei Li, Yan Wang, Lan-Ying Jiang, Tai-Shung Chung
Special Issue Editors

processes

MDPI

Editorial

Special Issue on "Novel Membrane Technologies for Traditional Industrial Processes"

Lan Ying Jiang [1],*, Pei Li [2] and Yan Wang [3]

1 School of Metallurgy and Environment, Central South University, Changsha 410083, China
2 College of Material Science and Engineering, Beijing University of Chemical Technology, Beijing 100029, China; lipei@mail.buct.edu.cn
3 School of Chemistry and Chemical Engineering, Huazhong University of Science and Technology, Wuhan 430073, China; wangyan@hust.edu.cn
* Correspondence: jianglanyingsme@csu.edu.cn

Received: 2 March 2019; Accepted: 4 March 2019; Published: 7 March 2019

1. Background

Traditional industries span multiple sectors, such as coal, iron and steel, textile, machinery, chemical engineering, shipbuilding, and construction materials. They are indispensable to the social economy but consume significant quantities of raw materials and energy and release myriads of waste. With rising population and expansion of urbanization, there is a higher demand for the supply of goods and services, the problem may be intensified. Raising awareness about sustainable development has led to an increased pressure on technological upgrades to enhance the capacity of traditional industries in conducting "responsible consumption and production" [1].

Membrane technology, in general, refers to membrane separation that covers a spectrum of processes involving various organic and inorganic membranes as selective mediums. Investigation into combining membrane technology with traditional industrial processes is abundant and has identified its potential and versatility in tackling problems concerning mass and energy efficiency [2,3]. The key strength of membrane separation processes as a whole as compared with others is due to process continuity, environmental friendliness, and low footprint [4]. For several specific membrane technologies, for example, pervaporation (PV) and gas separation (GS), energy efficiency is an extra bonus [5]. Given the above points, a widespread implementation of membrane separation in the traditional industrial sector is expected to bring about more social and economic benefits. The truth, however, is that the overall level for industrial adoption is, thus far, limited.

It is argued by some that "innovation management" is not efficient. The consequence comprises the "awareness gap" and delayed transfer of advanced technology to low- and medium-tech industries [6]. A more important factor for scientists and engineers to consider is the technical problems arising in the process of the combination. On the one hand, traditional industrial processes commonly consume multifarious raw materials that makes the mass flows more or less complex and, hence, gives rise to the impossibility of solely relying on a single membrane unit or a simple process [7]. To realize workable solutions, upstream and/or downstream operations with sound design are needed, which requires multidisciplinary coordination and a long lead time. On the other hand, the drawbacks of membrane technology constitute a setback for its attractiveness. The separation properties of some commercial membranes are not sufficiently high. And decline in separation efficiency often occurs during service [8]. Several critical issues for practically operating a membrane system, for example, the capital and operational costs, are closely correlated with membrane separation performance. Another point to be noted is the diversity in separation mechanisms and operation modes among the varied membrane processes. Therefore, the determinant for process efficiency is not merely membrane property. As for forward osmosis (FO), for instance, an energy efficient manner to handle draw solution (DS) will greatly facilitate its spread [9].

Obviously, the scope of the problems require a multi-dimensional effort for further actualizing the boon of membrane technology. Novel scientific and engineering approaches need to be developed to identify effective solutions for the challenges. This encourages us to compile this special issue of *Novel Membrane Technologies for Traditional Industrial Processes*. It is now available online at: https: //www.mdpi.com/journal/processes/special_issues/membrane_processes.

2. Content in This Special Issue

Nine articles are selected for this special issue. In the form of research or review, they present us with the latest progress addressing several aspects critical for developing efficient membrane systems. The papers are grouped into five categories:

(i) Gas separation; (ii) Biorefinery; (iii) Forward osmosis; (iv) Membrane bioreactor; (v) Membranes with aquaporins.

2.1. Gas Separation (GS)

A critical step for the effective operation of Integrated Gasification Combined Cycle (IGCC) is air separation for gasification. Membrane separation is an energy-efficient way, but lacks membranes with high selectivity. In Yang et al.'s work [10], on-stream catalytic cracking deposition (CCD) was applied to modify MFI-type zeolite membranes. Selectivity of the modified membrane with an alumina-containing ZSM-5 surface (~1.95) was slightly lower than that (~2.25) of modified membrane without a ZSM-5 surface, but the former membrane exhibited a permeance 15 times higher. The channel narrowing during CCD was constrained within the thin layer near the outer surface, due to the "gate" effect in the presence of a ZSM-5 surface. This helped avoid back diffusion of less permeable N_2. Attention was also directed to intercrystalline spaces for further improving selectivity.

The spectrum of utilization of natural gas has broadened in past decade. Seeburg et al. carried out low-temperature hydrogen production from natural gas via steam reforming with 47 vol% successfully and H_2 was generated at 350 °C [11]. The novelty of the work for obtaining this result is associated with two points. The first was the production of a natural gas stream enriched with liquefied petroleum gas (LPG) alkanes. *n*-butane enrichment in a process using zeolite membranes of type MFI (Mobile Five) was 37.6 and 22.6 vol% in permeate. Another is a highly active Rh catalyst. It was formed in situ on Al_2O_3 during the reaction of C1–C5 alkanes with steam. The active component in the resultant $Rh1/Al_2O_3$ compound was Rh in nanoparticles (1–3 nm). The reforming was performed with real natural gas.

A problem raised before applying natural gas for any purpose is the presence of acid gas (e.g., CO_2, H_2S). They will reduce the caloric value of fuel and cause corrosion of pipelines. For polymeric membranes used in natural gas treatment to remove CO_2, a severe challenge is performance degradation due to CO_2-induced plasticization. Zhang et al. composed a review of the advanced methods for suppressing plasticization of polyimide membranes [12]. These include thermally-induced crosslinking, chemical crosslinking, physical crosslinking, ultraviolet-radiation crosslinking, and blending.

2.2. Biorefinery

Utilization of microalgae solely for obtaining biofuels is not an economically viable process. To solve the problem, several other aspects should be considered, such as inclusive CO_2 capture, water quality improvement, procurement of commodities, and products with high-added value. The work by Lorente et al. showed us an innovative downstream route for bringing down production cost [13]. Acid-catalyzed steam explosion was employed to disrupt cells, generate hydrolysis of carbohydrates and partial hydrolysis of proteins. Dynamic filtration produced a permeate containing water and monosaccharides and a retentate containing the lipids and proteins. The steam explosion operation cost varied between 0.005 $/kg and 0.014 $/kg of the microalgae dry sample, depending on the cost of fuel. Membrane filtration cost was about 0.12 $/kg of microalgae (in dry state).

The challenge for the commercial application of lignin lies with its heterogeneity in structure and the low-purity of lignin delivered in traditional fractionation. Concerning the second challenge, several researches paid attention to the supported ionic liquid membrane (SILM). The selected paper in this special issue made a more comprehensive study to identify the suitable materials for effective lignin extraction and purification [14]. Among the combination from five different membrane supports and nine ionic liquids (ILs), the one with IL of [BMIM][DBP] embedded in polytetrafluoroethylene (PTFE) demonstrated selective transport for target solutes.

2.3. Forward Osmosis (FO)

FO is an emerging and versatile membrane separation that derives its driving force from a highly concentrated draw solution (DS). DS regeneration step will consume energy, which renders the necessity to find methods that could reduce energy consumption. Long et al. conducted a review analyzing the types of energy applied in DS regeneration [15]. Some DS (e.g., natural sugar, fertilizer) can be directly used without recovery, constituting the most effective way of saving energy. Chemical reactions involving precipitation and redissolving capitalize on chemical energy. The biggest family may be related to the utilization of thermal energy. Technologies making use of magnetic energy via magnetic particles and solar energy via light-sensitive hydrogels were also introduced. The most common one is electricity. This work gives a clear guideline for research into DS in the future.

Thin-film composite (TFC) membranes are widely used in FO. Membrane performance deterioration due to degradation and the polyamide (PA) skin layer breaking off often occurs. The former is due to disinfectants cleaning, and the latter results from the different materials for the selective layer and substrate. Wang et al. proposed a new strategy to overcome the problem [16]. Dopamine (DA) was the only amine in the aqueous phase to react with trimesoyl chloride (TMC) during interfacial polymerization (IP). In the reaction, PDA particles formed by self-assembly led to significant π–π and hydrogen-bonding interactions between the active and support layers. Additionally, the ester bonds had higher resistance to active chlorine than amide bonds of conventional PA layers.

2.4. Membrane Bioreactor (MBR)

This review by Argurio focuses on photocatalytic membrane reactors (PMR), a combination of photocatalytic and membrane processes [17]. PMR is a relatively new membrane technology, and the pioneering work was performed by Anderson and coworkers using TiO_2-based membranes. This review is comprehensive as it introduces system configurations based on catalyst confinement, the influence of operation parameters, and materials selection and membrane formation. Within these parts, the advantages and limits of PMR were analyzed. The final section is selected case studies.

2.5. Membranes with Aquaporins

Aquaporins which are highly permeable for water molecules and resistant to contaminants and small molecules are receiving great attention from membranologists, particularly for water treatment. An important challenge for their utilization in membrane separation is associated with maintaining their integrity and performance during developing membranes with embedded and aligned aquaporins, while keeping support porous enough. The steps key to the solution designed by Wagh et al. included [18]: (1) The connection of Cysteine modified aquaporins (Aqp-SH) to the polymeric support covalently, which results in an aligned pattern of the proteins; (2) which results in the remaining sites on the support reacting with polyvinyl alcohol with long alkyl chains (PVA-alkyl) to seal the gaps without aquaporin molecules. The membranes with aquaporins exhibited much higher and more stable rejection towards inorganic salts.

Conflicts of Interest: The authors declare no conflicts of interest.

References

1. Leal Filho, W.; Marisa Azul, A.; Brandli, L.; Gokcin Ozuyar, P.; Wall, T. (Eds.) *Responsible Consumption and Production*; Encyclopedia of the UN Sustainable Development Goals; Springer: Basel, Switzerland, 2019.
2. Figoli, A.; Criscuoli, A. (Eds.) *Sustainable Membrane Technology for Water and Wastewater Treatment*; Chemistry and Sustainable Technology; Springer: Singapore, 2017.
3. Jiang, L.Y.; Li, N. (Eds.) *Membrane-Based Separation in Metallurgy: Principles and Applications*; Elsevier: Amsterdam, The Netherlands, 2017.
4. Wilcox, J. Membrane Technology. In *Carbon Capture*; Springer: New York, NY, USA, 2012.
5. Shao, P.; Huang, R.Y.M. Review: Polymeric membrane pervaporation. *J. Membr. Sci.* **2007**, *287*, 162–179. [CrossRef]
6. McKelvey, M.; Ljungberg, D. How public policy can stimulate the capabilities of firms to innovate in a traditional industry through academic engagement: The case of the Swedish food industry. *Res. Dev. Manag.* **2017**, *47*, 535–544. [CrossRef]
7. Zhang, Q.X. *Science and Engineering for Metallurgical Separation*; Science Press: Beijing, China, 2004.
8. Ismail, A.F.; Rahman, M.A.; Othman, M.H.D.; Matsuura, T. (Eds.) *Membrane Separation Principles and Applications: From Material Selection to Mechanisms and Industrial Uses*; Hand Books of Separation Science; Elsevier: Amsterdam, The Netherlands, 2018.
9. Shaffer, D.L.; Werber, J.R.; Jaramillo, H.; Lin, S.; Elimelech, M. Forward osmosis: Where are we now? *Desalination* **2015**, *356*, 271–284. [CrossRef]
10. Yang, S.; Arvanitis, A.; Cao, Z.; Sun, X.; Dong, J. Synthesis of Silicalite Membrane with an Aluminum-Containing Surface for Controlled Modification of Zeolitic Pore Entries for Enhanced Gas Separation. *Processes* **2018**, *6*, 13. [CrossRef]
11. Seeburg, D.; Liu, D.; Dragomirova, R.; Atia, H.; Pohl, M.; Amani, H.; Georgi, G.; Kreft, S.; Wohlrab, S. Low-Temperature Steam Reforming of Natural Gas after LPG-Enrichment with MFI Membranes. *Processes* **2018**, *6*, 263. [CrossRef]
12. Zhang, M.; Deng, L.; Xiang, D.; Cao, B.; Hosseini, S.; Li, P. Approaches to Suppress CO_2-Induced Plasticization of Polyimide Membranes in Gas Separation Applications. *Processes* **2019**, *7*, 51. [CrossRef]
13. Lorente, E.; Haponska, M.; Clavero, E.; Torras, C.; Salvadó, J. Steam Explosion and Vibrating Membrane Filtration to Improve the Processing Cost of Microalgae Cell Disruption and Fractionation. *Processes* **2018**, *6*, 28. [CrossRef]
14. Abejón, R.; Rabadán, J.; Lanza, S.; Abejón, A.; Garea, A.; Irabien, A. Supported Ionic Liquid Membranes for Separation of Lignin Aqueous Solutions. *Processes* **2018**, *6*, 143. [CrossRef]
15. Long, Q.; Jia, Y.; Li, J.; Yang, J.; Liu, F.; Zheng, J.; Yu, B. Recent Advance on Draw Solutes Development in Forward Osmosis. *Processes* **2018**, *6*, 165. [CrossRef]
16. Wang, Y.; Fang, Z.; Xie, C.; Zhao, S.; Ng, D.; Xie, Z. Dopamine Incorporated Forward Osmosis Membranes with High Structural Stability and Chlorine Resistance. *Processes* **2018**, *6*, 151. [CrossRef]
17. Argurio, P.; Fontananova, E.; Molinari, R.; Drioli, E. Photocatalytic Membranes in Photocatalytic Membrane Reactors. *Processes* **2018**, *6*, 162. [CrossRef]
18. Wagh, P.; Zhang, X.; Blood, R.; Kekenes-Huskey, P.; Rajapaksha, P.; Wei, Y.; Escobar, I. Increasing Salt Rejection of Polybenzimidazole Nanofiltration Membranes via the Addition of Immobilized and Aligned Aquaporins. *Processes* **2019**, *7*, 76. [CrossRef]

processes

MDPI

Article

Synthesis of Silicalite Membrane with an Aluminum-Containing Surface for Controlled Modification of Zeolitic Pore Entries for Enhanced Gas Separation

Shaowei Yang [†], Antonios Arvanitis [ORCID], Zishu Cao, Xinhui Sun and Junhang Dong *

Department Chemical and Environmental Engineering, University of Cincinnati, Cincinnati, OH 45221, USA; shaowei.yang@chbe.gatech.edu (S.Y.); arvanias@mail.uc.edu (A.A.); caozu@mail.uc.edu (Z.C.); sunxh@mail.uc.edu (X.S.)
* Correspondence: junhang.dong@uc.edu
† Current address: Georgia Institute of Technology, Atlanta, GA 30332, USA.

Received: 15 January 2018; Accepted: 31 January 2018; Published: 5 February 2018

Abstract: The separation of small molecule gases by membrane technologies can help performance enhancement and process intensification for emerging advanced fossil energy systems with CO_2 capture capacity. This paper reports the demonstration of controlled modification of zeolitic channel size for the MFI-type zeolite membranes to enhance the separation of small molecule gases such as O_2 and N_2. Pure-silica MFI-type zeolite membranes were synthesized on porous α-alumina disc substrates with and without an aluminum-containing thin skin on the outer surface of zeolite membrane. The membranes were subsequently modified by on-stream catalytic cracking deposition (CCD) of molecular silica to reduce the effective openings of the zeolitic channels. Such a pore modification caused the transition of gas permeation from the N_2-selective gaseous diffusion mechanism in the pristine membrane to the O_2-selective activated diffusion mechanism in the modified membrane. The experimental results indicated that the pore modification could be effectively limited within the aluminum-containing surface of the MFI zeolite membrane to minimize the mass transport resistance for O_2 permeation while maintaining its selectivity. The implications of pore modification on the size-exclusion-enabled gas selectivity were discussed based on the kinetic molecular theory. In light of the theoretical analysis, experimental investigation was performed to further enhance the membrane separation selectivity by chemical liquid deposition of silica into the undesirable intercrystalline spaces.

Keywords: zeolite membrane; pore modification; gas separation

---

1. Introduction

The energy-efficient membrane gas separation technology can play significant roles in performance enhancement and process intensification for the advanced fossil energy systems with CO_2 capture capacity. For the emerging coal- and natural gas-based IGCC power production with CO_2 capture and sequestration (CCS), cost effective air separation unit (ASU) for the gasifier and H_2/CO_2 separation units for pre-combustion CO_2 capture are critical to commercial success. However, both the O_2 and CO_2 gas separations currently remain challenging that hinders the realization of the IGCC-CCS technology. Traditional industrial O_2 separation by cryogenic distillation or pressure swing adsorption (PSA) is highly energy intensive and involves large capital investment. Membrane separation has shown promises as energy-efficient alternative for O_2 separation in the past few decades [1]. However, because of the very close kinetic diameters (d_k) of the O_2 ($d_k = 0.346$ nm) and N_2 ($d_k = 0.364$ nm) molecules and similarly weak adsorbing behaviors of both gases in common membrane materials, low temperature O_2 separation membranes reported in the literature have been of relatively poor performance with limited selectivity and low permeance. The microporous zeolite membranes, because of their extraordinary stability and molecular sieving effects enabled by their uniform sub-nanometer pore sizes, are promising for separating many small molecule gases. The kinetic molecular theory of gas transport in zeolitic pores suggests that, in order to achieve O_2/N_2 selectivity by non-adsorptive diffusion mechanism, the zeolite pore size must situate in between or very near the kinetic diameters of O_2 and N_2 [2,3]. Therefore, in order to achieve size-selectivity between O_2 and N_2 molecules, the zeolite membranes must possess pore diameter (d_p) very close to the kinetic size of N_2, i.e., <0.4 nm.

Membranes of zeolites with pore openings defined by 8-membered rings are promising candidates for O_2 separation because their d_p are around 0.4 nm. The highly siliceous DDR zeolite (d_p ~0.4 nm) membrane has been reported to exhibit O_2/N_2 permselectivity around 2 but the O_2 permeance was only in the order of 10^{-9}–10^{-8} mol/m^2·s·Pa at 298 K [4–6]. Since O_2 and N_2 are both weakly adsorbing with similarly small adsorption amounts in the DDR zeolite [5], the very low O_2 permeance in the DDR zeolite membrane is a result of the large resistance for O_2 transport in the critically sized zeolitic pores over the entire membrane thickness. The membrane of NaA zeolite (d_p ~0.41 nm) was once reported to have surprisingly high O_2/N_2 separation factor of ~7 and O_2 permeance of 2.6×10^{-7} mol/m^2·s·Pa at room temperature [7]. This performance though was observed under vacuum pressure and its reproducibility is yet to be confirmed. In general, for zeolite membranes of very large aluminum contents, such as the A-type membranes, the gas permeation properties are often severely affected by the presence of trace water vapor due to its extremely ionic and hydrophilic surface [8]. In addition, zeolite membranes of such high-aluminum contents are known to experience gradual degradation when subjected to dehydration in vacuum over extended time [9].

Zeolites with pore openings formed by 10-membered rings have effective d_p of around 0.6 nm, which are much less resistive to small gas transport but become incapable of achieving size-selectivity or molecular sieving effect between the closely sized molecules like O_2, N_2, H_2 and CO_2. However, such relatively large pore size allows for depositing molecular modifiers to the internal pore wall to fine tune the effective channel openings. In the literature, highly siliceous MFI-type zeolite membranes, which have nearly cylindrical zeolitic channels formed by 10-membered rings with a diameter of ~0.56 nm, were successfully modified by depositing mono-silica species on their internal pore surfaces. The effective d_p of the internally modified zeolitic channels was estimated to be around 0.36 nm based on the observation of obvious permeance cut-off effects between H_2 and N_2 or CO_2 [3,10,11]. The modification of the MFI-type zeolite membranes by deposition of mono-silica on the internal pore wall using methyldiethoxysilane (MDES) as precursor was first reported by Masuda et al. [10]. The MDES molecule has a linear geometry with nominal dimension of 0.41 nm × 0.91 nm, which can penetrate into the pristine MFI zeolitic channels of 0.56 nm in diameter. The modification was conducted by pre-loading MDES in the zeolitic channels at low temperature followed by calcination to thermally decompose the MDES molecules and deposit mono-silica inside the zeolite channels [10]. After modification, the H_2/N_2 separation factor increased from 1.5~4.5 to 90~140 at 383 K but the H_2 permeance decreased by an order of magnitude to only 2×10^{-8} mol/m^2·s·Pa. The drastic loss of H_2 permeance was attributed to the deposition of mono-silica over the entire channel length that caused enormous resistance to molecular diffusion. An on-stream catalytic cracking deposition (CCD) method was developed in our lab to avoid excessive silica deposition over the channel length [12]. In the on-stream CCD modification process, the MDES precursor is carried in a feed stream during membrane gas permeation operation at around 450 °C where the CCD process can occur at catalytically active sites such as the alumina tetrahedron sites in channel surface without inducing non-catalytic MDES thermal decomposition [3,13,14]. Thus, in principle, after CCD deposition of silica at a single site, this mono-silica deposit creates a narrow "gate" to prevent MDES molecules from entering the rest of channel length. Consequently, the deposition of mono-silica is effectively limited in a small segment of the zeolitic channel while large part of the channel remains in its original size (~0.56 nm) for fast diffusion [3,15]. The on-stream CCD method alleviated the mass transport resistance increase while achieving selectivity improvement. However, the supported zeolite membrane had largely random distribution of catalytic sites and the locations of CCD depositions were consequentially uncontrolled along the membrane thickness, which is a multiple of individual crystal size in the film. It is conceivable that when the modified site, which could be thought of as a "gate," locates deep down the channel length, the less permeable molecules will have to back diffuse from the gated point to the feed stream during gas mixture permeation. Such a back-diffusion process would inevitably hinder the permeation of the more permeable component.

Here we report the synthesis and on-stream CCD modification of an MFI-type zeolite membrane, which has a pure-silica (i.e., silicalite) base layer with an outer surface of aluminum-containing framework (i.e., ZSM-5), in attempt to effectively control the pore modification within a small depth near the membrane outer surface. The hypothesis is that such a surface "gated" channel structure can improve the gas permeance by avoiding the back diffusion of the less permeable gases without sacrificing permeation selectivity. The modified MFI-type zeolite membranes are examined by permeation for a number of small molecule gases including O_2 and N_2.

2. Experimental

2.1. Materials and Chemicals

The porous α-alumina disc substrates were made of α-alumina powders with an average diameter of 0.46 μm (SG16, Alcoa, TN, USA). The alumina disc was 2-mm thick and 2.6-cm in diameter with porosity and average pore diameter (d_p) of 27–30% and 0.1 μm, respectively. The edge of the disc was sealed by glass and the active membrane area was ~2.5 cm^2 after excluding the glass sealed edge area. The surface of the disc used for membrane coating was polished by 800 mesh sandpaper and washed and dried in an oven at 313 K for overnight before membrane synthesis. The following chemicals and materials were used in the present work: sodium hydroxide (99.99% trace metal basis, pellet, Sigma-Aldrich, USA), fumed silica (0.007 μm, Aldrich, USA), tetrapropylammonium hydroxide (TPAOH, 1 M, Aldrich, USA), sodium aluminate (anhydrous, Al content as Al_2O_3: 50–56 wt. %, Riedel-de Haën, Germany), HNO_3 solution (1.0 M, Fluka, USA) and methyldiethoxysilane (MDES, 96%, Aldrich, USA). All chemicals were used as received. Gases used for membrane permeation tests included H_2 (99.999%), He (99.999%), N_2 (99.999%), O_2 (99.6%), CO_2 (99.99%), CH_4 (99.999%), i-C_4H_{10} (99.5%) and SF_6 (99%). All gases were obtained from Wright Brothers Inc. (Cincinnati, OH, USA) and used as received.

2.2. Synthesis of MFI Zeolite Membranes

Two kinds of MFI-type zeolite membranes were synthesized by the in situ hydrothermal crystallization method under different conditions. The first was a silicalite membrane without an aluminum-containing zeolite surface and the second was a silicalite membrane with an aluminum-containing MFI-zeolite (ZSM-5) outer surface obtained through a two-stage synthesis procedure.

The silicalite membrane without a ZSM-5 surface was obtained from an alumina-free synthesis precursor, which was prepared by mixing 0.35 g NaOH, 5 g SiO_2 and 25 mL 1 M TPAOH solution under vigorous stirring at 353 K. The TPAOH was used as the structure directing agent (SDA). The synthesis solution was aged for four hours at room temperature and then transferred into the synthesis autoclave where the disc substrate was placed horizontally at the Teflon-lined bottom with the polished side facing upward. About 25 mL of the synthesis solution was poured into the autoclave alone the autoclave wall to immerse the alumina disc. The hydrothermal synthesis was conducted at 453 K for 5 h and then the disc membrane was recovered and washed with DI water until the water pH was close to 7. The cleaned membrane disc was dried in an oven at 313 K and then calcined in air at 773 K for 6 h to remove the SDA from the zeolitic pores. This membrane is denoted as M1Si hereafter.

The membrane with a silicalite base layer and a ZSM-5 outer surface was obtained by a two-stage synthesis procedure where the liquid phase composition was varied for the second stage. A specially designed synthesis autoclave, as schematically shown in Figure 1, was employed for the continuous two-stage synthesis. The zeolite precursor solution and hydrothermal reaction conditions for the first stage of synthesis were identical to those used for synthesizing the M1Si. However, the hydrothermal treatment was not terminated in 5 h; and instead, 1.2 mL of sodium aluminate solution (21 wt. % NaAlO$_2$) was injected into the liquid phase by a high-pressure precision syringe pump (KD Scientific, KDS-410, Holliston, MA, USA) and the hydrothermal treatment continued for another hour at 453 K. The reaction was then terminated and the membrane was recovered. The thus synthesized membrane is denoted as M2Al hereafter. Membrane M2Al underwent the same washing, drying and calcining processes as used for preparation of M1Si.

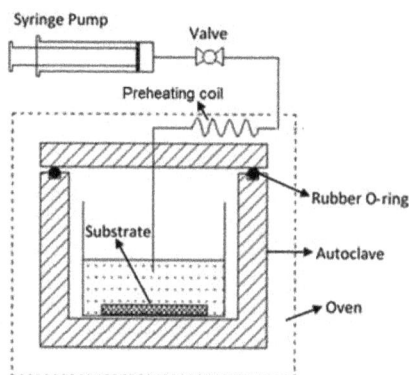

Figure 1. Schematic diagram of the setup used for synthesizing M2Al.

To confirm the formation of a ZSM-5 thin skin on the outer surface of the silicalite crystals by the above described synthesis process, silicalite crystals were synthesized with and without the second stage treatment in NaAlO$_2$ containing solution following the exact procedures used for the syntheses of the two kinds of membranes. The pure-silica MFI-type zeolite (i.e., silicalite) crystals were synthesized using the same aluminum-free precursor and reaction temperature (453 K) and duration (5 h). The thus obtained silicalite particles are denoted as P1Si. In synthesis of the silicalite crystals with a ZSM-5 outer surface, 1.2 mL 21wt. % NaAlO$_2$ solution was introduced into the above silicalite synthesis solution (~25 mL) after 5 h of reaction and the hydrothermal treatment continued for another hour. The thus obtained zeolite particles, which have a ZSM-5 outer surface, are denoted as P2Al.

Both zeolite particulate samples were extensively washed to ensure that any aluminum ions adsorbed on the external surface were completely removed. The washing process including the following consecutive steps: (1) repeated process of zeolite particle dispersion in DI water followed by filtration and rinsing until the filtered water reached pH of ~7; (2) surface cleaning by treating the particles in a 0.1 M NaOH solution under stirring and subsequent ultrasonication for 1 h; (3) washing the NaOH solution-treated zeolite particles by DI water until pH of the washing water reached pH of ~7; (4) treating the cleaned particles in 0.1 M HNO$_3$ solution under stirring and subsequent ultrasonication for 1 h; and (5) final cleaning by rinsing the zeolite particles with DI water until the filtered water reached pH of ~7. This rigorous surface cleaning process was repeated twice to ensure the complete removal of any aluminum ions and silicate species adsorbed to the external surface of the zeolites. The zeolite particles were then dried and calcined at 450 °C for SDA removal.

2.3. Material Characterizations

The zeolite membranes were examined by X-ray diffraction (XRD, PANalytical, X'Pert Pro MPD, The Netherlands) to confirm the crystal phase and purity. Scanning electron microscopy (SEM, FEI, XL30, Philips, USA) was used to observe the membrane morphology and determine the individual crystallite size and membrane thickness. The energy dispersive X-ray spectroscopy (EDS, EDAX, PV9761/70, NJ, USA) was employed to estimate the elemental compositions at different locations of the zeolite crystals in combination with the transmission electron microscopy (TEM; FEI CM20, Philips, USA).

2.4. Membrane Modification

Figure 2 shows the schematic diagram of the experimental system for CCD modification of the membranes that was essentially the same as that described in our previous publication [12]. This apparatus was also used for membrane gas mixture separation measurements. The disc-shaped zeolite membrane was mounted in a stainless-steel permeation cell using soft graphite gasket seals (Mercer Gasket & Shim, NJ, USA). The membrane permeation cell was placed in a temperature-programmable furnace.

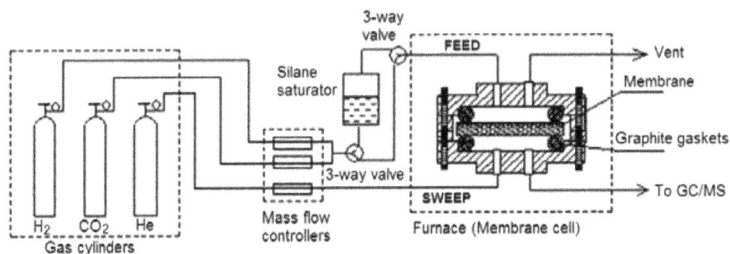

Figure 2. Schematic diagram of the apparatus for membrane modification and gas separation tests.

The CCD modification started with permeation of an equimolar H_2/CO_2 mixture gas fed at a total flow rate of 40 cm^3 (STP)/min when the permeate side was swept by a helium stream at a flow rate of 30 cm^3 (STP)/min. The partial pressure of MDES in the H_2/CO_2 carrier gas was determined to be ~4.3 kPa from the room temperature saturator. The temperature for CCD modification was 723 K at which MDES decomposition occurs catalytically at $[AlO_2^-]$ sites but not on pure silicalite surface, which is noncatalytic. During the gas permeation, the membrane cell was heated up to 723 K at a heating rate of 0.5 K/min and kept at 723 K throughout the modification process. The H_2/CO_2 mixture feed was then switched to bubble through the MDES liquid column to carry MDES vapor before entering the membrane cell. The permeate stream was continuously analyzed by an online gas chromatographer (GC, Agilent 6890N, USA) equipped with a Carboxen 1000 packed column (Supelco, USA) and a thermal conductivity detector (TCD). The H_2/CO_2 separation through the zeolite membrane was continuously monitored during the entire CCD modification process. The feed gas flow was switched back to dry gas feeding route to stop the MDES supply when the online monitored H_2/CO_2 separation factor and gas permeance were stabilized for 2 h.

2.5. Gas Permeation

The membrane was degassed in the permeation cell by vacuuming at 453 K for 12 h prior to testing permeation for each gas. The permeance of pure gases was measured by the transient permeation method in a temperature range from 297 to 773 K with feed pressure of 2 bar using an experimental setup reported a previous publication [16]. The separation of gas mixtures was performed by the conventional steady-state permeation setup (Figure 2) at a feed flow rate of 40 cm^3 (STP)/min and a helium sweep flow rate of 30 cm^3 (STP)/min. The permeance for gas i ($P_{m,i}$), permselectivity (i.e., ideal selectivity) of gas i over gas j ($\alpha^o_{i/j}$) and separation factor of gas i over gas j in mixture ($\alpha_{i/j}$) are defined as follows:

$$P_{m,i} = \frac{Q}{A_m \cdot t \cdot \Delta P_i} \tag{1}$$

$$\alpha^o_{i,j} = P^o_{m,i} / P^o_{m,j} (i \neq j) \tag{2}$$

$$\alpha_{i/j} = \frac{(y_i/y_j)}{(x_i/x_j)} (i \neq j) \tag{3}$$

where Q (mol) is the moles of gas i permeated through the membrane over a time period of t (s); ΔP_i is the partial pressure difference (i.e., $\Delta P_i = P_{i,f} - P_{i,p}$, where $P_{i,f}$ and $P_{i,p}$ are the partial pressures of gas component i in the feed and permeate sides, respectively); $P^o_{m,i}$ and $P^o_{m,j}$ are pure gas permeance of gas i and gas j, respectively; and x and y are molar compositions of the feed and permeate gas mixtures, respectively.

3. Results and Discussion

3.1. Material Characterizations

The supported zeolite membranes were verified to be of pure MFI-type zeolite phase by the XRD patterns shown in Figure 3. The low intensity of zeolite peaks observed in contrast to the strong peaks of α-alumina substrate was resulted from the thinness of the zeolite layer. Figure 4 shows the SEM images of the two zeolite membranes, i.e., M1Si and M2Al. The zeolite crystals in the two membrane surfaces had very similar shapes of typical siliceous MFI zeolite crystals with individual crystallite size of ~0.3 μm. The SEM pictures of the membrane cross-section also showed zeolite layer thicknesses of ~1 μm for both M1Si and M2Al membranes.

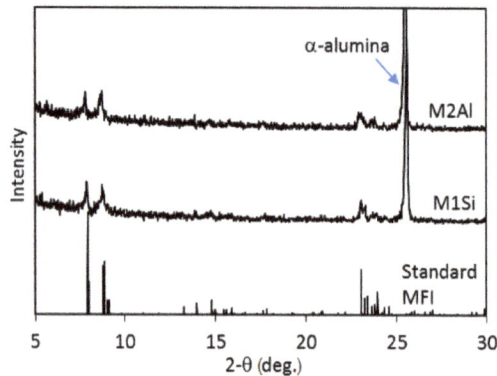

Figure 3. XRD patterns of the MFI membranes M1Si and M2Al together with the standard pattern of silicalite powders [17].

It is well-known that silicalite membranes obtained by in situ crystallization on α-alumina supports in the highly alkaline Al-free precursor used in this work have nearly pure silicalite outer surface but inevitably contain framework aluminum at the zeolite/support interface region [18]. However, because of the thinness of the zeolite layers on the alumina substrates, determining the Si/Al ratios in the zeolite membrane surfaces by EDS elemental analysis is rather challenging. Thus, the incorporation of Al into the framework of silicalite surface was investigated by comparing the two kinds of zeolite particles, i.e., P1Si and P2Al, which were obtained under conditions identical to the syntheses of M1Si and M2Al, respectively. Each zeolite crystal was examined by EDS survey at two locations, namely the center and corner surfaces of the crystal as shown in Figure 5. The EDS examination used an X-ray beam size of 20 nm in diameter and an operation voltage of 200 kV under which the beam penetration depth was expected to well exceed the zeolite crystal thickness, which was around 200 nm. Crystal P1Si had no detectable Al content (i.e., Si/Al $\rightarrow \infty$) at both locations because no alumina source existed in the precursor throughout the synthesis process. The crystal P2Al had Si/Al ratios of 150 ± 18 at the center surface and 40 ± 15 at the corner that confirmed Al incorporation into the framework of zeolite surface. The measured Si/Al ratio was much lower at the corner than at the center because the X-ray beam sampled more surface materials from the side surfaces at the corner than did from the top surface at the center. Since the EDS results are averaged over the depth penetrated by the interrogating X-ray, the Si/Al ratio in the P2Al outer surface must be <40.

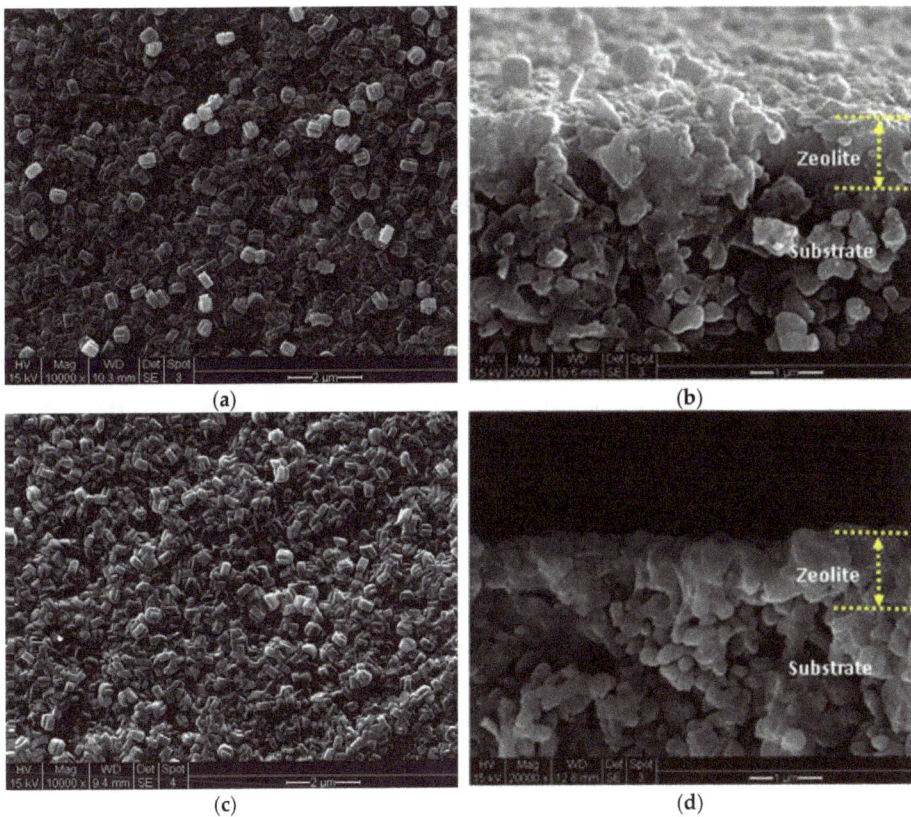

Figure 4. SEM pictures of M1Si and M2Al: (**a**) surface of M1Si and (**b**) cross-section of M1Si; (**c**) surface of M2Al and (**d**) cross-section of M2Al.

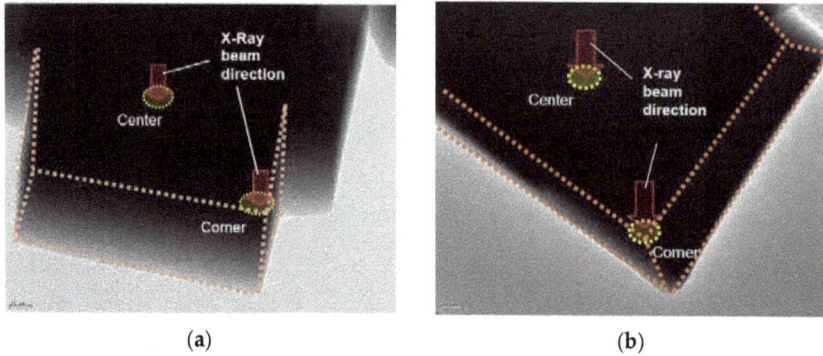

(a)   (b)

Figure 5. TEM pictures and illustration of EDS sampling locations for elemental analyses for the zeolite crystals: (a) P1Si and (b) P2Al.

3.2. Membrane Modification

The two membranes, i.e., M1Si and M2Al, had similar H_2/CO_2 gas permeation properties before modification that made the modification and subsequent performance analyses and comparisons meaningful. Figure 6 presents the evolution of H_2/CO_2 separation performance for membranes M1Si and M2Al during the CCD modification process. Both fresh membranes exhibited H_2/CO_2 separation factors ($\alpha_{H2/CO2}$) of ~0.2 at room temperature because at room temperature CO_2 is preferentially adsorbed into and diffusing through the zeolitic pores to hinder the entry and permeation of the non-adsorbing H_2 that results in CO_2-selective permeation. The integrity of the membranes was also evidenced by their high permselectivity of H_2 over SF_6 (d_k ~0.55 nm) ($\alpha^o_{H2/SF6,K}$) prior to CCD modification, which was 159 for M1Si and 120 for M2Al at 453 K. The $\alpha_{H2/CO2}$ changed to 2.9 and 2.8 (H_2-selective) for M1Si and M2Al, respectively, when temperature increased to 723 K where adsorption of CO_2 becomes negligible and transport of both CO_2 and H_2 is governed by gaseous diffusion mechanism, which is selective towards the lighter H_2 [3,12]. However, these $\alpha_{H2/CO2}$ values are below the theoretical value ($\alpha_{H2/CO2} = \sqrt{M_{w,CO2}/M_{w/H2}}$) of ~4.69 given by the gaseous diffusion mechanism that indicates enlargement of intercrystalline spaces at high temperature because of the mismatch of thermal expansion coefficient between the zeolite layer and the alumina substrate [19].

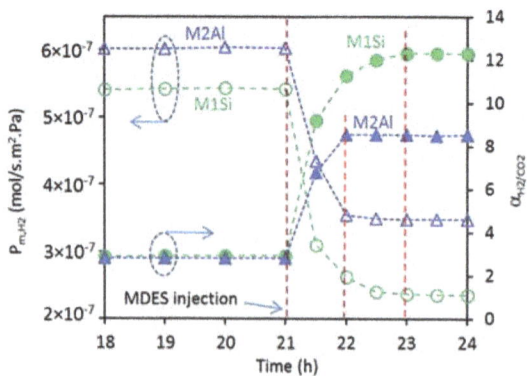

Figure 6. Evolution of the H_2/CO_2 separation factor and H_2 permeance for the two membranes during the CCD modification process.

As shown in Figure 6, after introducing the MDES vapor at 723 K, the permeance of H_2 and CO_2 decreased sharply from 5.42×10^{-7} and 1.87×10^{-7} mol/m^2·s·Pa to 2.36×10^{-7} and 1.92×10^{-8} mol/m^2·s·Pa, respectively, in 2 h for membrane M1Si; and from 6.03×10^{-7} and 2.16×10^{-7} mol/m^2·s·Pa to 3.48×10^{-7} and 4.08×10^{-8} mol/m^2·s·Pa, respectively, in 1 h for membrane M2Al. Meanwhile, $\alpha_{H2/CO2}$ of M1Si and M2Al increased from 2.9 and 2.8 before modification to 12.3 and 8.5 after modification, respectively. The time needed for stabilizing $\alpha_{H2/CO2}$ and H_2 permeance after introducing MDES vapor was noticeably shorter for M2Al (~1.0 h) than for M1Si (~2 h) that might be explained by the different locations of CCD modification in the two membranes. The CCD deposition of mono-silica in the zeolitic channels of M2Al membrane presumably occurred at the outer surface that is expected to complete quickly and essentially no further CCD modification could happen in deeper locations down the zeolite channels because the pore entries are narrowed and become inaccessible to the MDES. On the contrary, the M1Si membrane, are known to have a silicalite outer surface but the framework near the zeolite/alumina-substrate interface contains certain amount of Al^{3+} [12,18]. Thus, the CCD modification in the M1Si membrane takes place deep along the thickness in the channels (close to the zeolite/substrate interface) and thus requires longer time due to the difficult diffusion of MDES molecules and debris of catalytic cracking. The modified M1Si and M2Al are denoted as M-M1Si and M-M2Al, respectively, hereafter. The effective control of pore modification within the surface of M2Al has led to its much higher H_2 permeance ($P_{m,H2}$ ~3.46×10^{-7} mol/m^2·s·Pa) than that of the M-M1Si ($P_{m,H2}$ ~2.36×10^{-7} mol/m^2·s·Pa). However, the M-M2Al exhibited an $\alpha_{H2/CO2}$ of ~8.5 which is lower than that of the M-M1Si ($\alpha_{H2/CO2}$ ~12.3), suggesting that more intercrystalline pores exited in M-M2Al, which could also contribute to its high $P_{m,H2}$.

3.3. Single Gas Permeation for O_2 and N_2

The O_2 and N_2 Single gas permeation was performed as a function of temperature for both membranes before and after the CCD modification. The results are presented in Figures 7 and 8. Before modification, M1Si had O_2/N_2 permselectivity ($\alpha^o_{O2/N2}$) of ~0.925–0.970, which was very close to the Knudsen factor ($\alpha^o_{O2/N2,K}$ ~0.935) for gaseous diffusion, while M2Al had $\alpha^o_{O2/N2}$ of ~0.996–1.03, which was slightly greater than the $\alpha^o_{O2/N2,K}$ in the temperature range of 297–773 K. Since the d_k of O_2 (~0.346 nm) and N_2 (0.364 nm) are significantly smaller than the diameter of the pristine MFI zeolitic pores (d_p ~0.56 nm). The $\alpha^o_{O2/N2}$ was slightly larger than the $\alpha^o_{O2/N2,K}$ because both O_2 and N_2 are weakly adsorbing in the MFI-type zeolites with O_2 having slightly higher adsorption amount [20]. The M2Al exhibited slightly higher $\alpha^o_{O2/N2}$ than M1Si that may be attributed to the higher content of extra-framework Na^+ in the former, which enhances the O_2 adsorption. In addition, the extra-framework metal ions can slightly reduce the zeolitic pore size to favor the diffusion of the smaller O_2. Observations of O_2/N_2 permselectivity greater than Knudsen factor on the alumina supported ZSM-5 membranes have been previously reported in the literature ($\alpha^o_{O2/N2}$ ~1.02–1.16) [21]. The permeance of both O_2 and N_2 exhibited "S"-shaped temperature-dependences on both M1Si and M2Al as shown in Figures 7a and 8a, respectively. As temperature increases, the permeances of O_2 and N_2 increase first and then decrease after passing maxima; as temperature further increases, the permeances increase again after passing minima. Such a temperature-dependence of the permeance ($P^o_{m,i}$) is a result of the adsorption-diffusion transport mechanism because increasing temperature reduces adsorption but enhances the transport diffusivity [22]. The $\alpha^o_{O2/N2}$ values of both unmodified membranes were almost unchanged over the entire temperature range as expected.

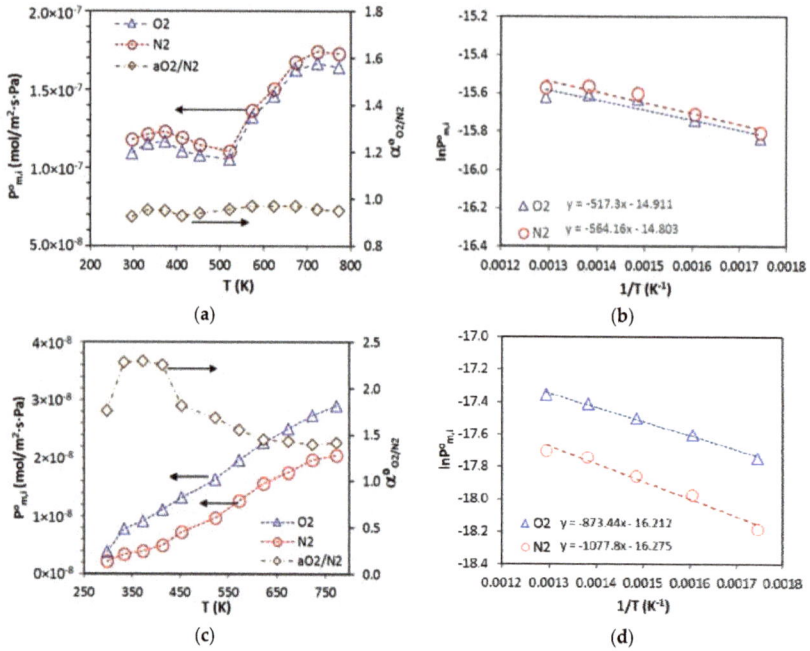

Figure 7. O$_2$ and N$_2$ single gas permeation on M1Si and M-M1Si. (**a,b**) M1Si and (**c,d**) M-M1Si.

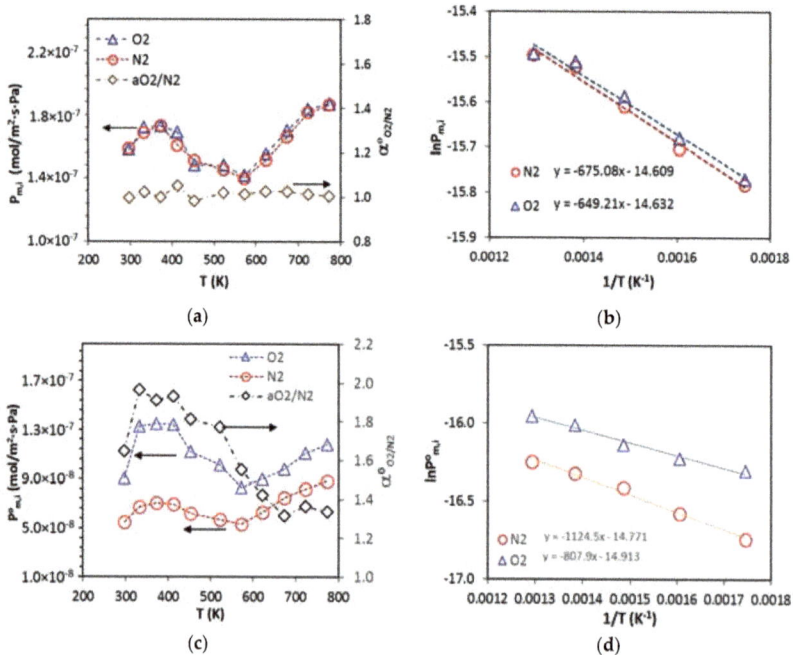

Figure 8. O$_2$ and N$_2$ single gas permeation on M2Al and M-M2Al. (**a,b**) M2Al and (**c,d**) M-M2Al.

In Figures 7b and 8b, the gas permeance of the two unmodified membranes are presented by the Arrhenius plots in the temperature range of 573–773 K, where the adsorption of O_2 and N_2 become negligible. The activation energies ($E_{d,i}$) for diffusion of O_2 and N_2 were similar in the two unmodified membranes, i.e., $E_{d,O2}$ = 4.30 kJ/mol and $E_{d,N2}$ = 4.69 kJ/mol in M1Si; and $E_{d,O2}$ = 5.40 kJ/mol and $E_{d,N2}$ = 5.61 kJ/mol in M2Al. However, the $E_{d,O2}$ and $E_{d,N2}$ are slightly greater in M2Al than in M1Si that again suggest variation of zeolitic channel size due to the difference in amount of extra-framework metal ions between the two membranes.

The gas permeation behavior of the two modified membranes became qualitatively different from their unmodified counter parties. As shown in Figures 7c and 8c, the M-M1Si and M-M2Al were found to be O_2 selective over N_2 with $\alpha^o_{O2/N2}$ much higher than that of the M1Si and M2Al and the selectivity values are opposite to the N_2-selective Knudsen factor for gaseous diffusion mechanism. These indicate that the zeolite channel openings have been effectively narrowed to enable size-selectivity between O_2 and N_2. For the modified membranes, the diffusion activation energy $E_{d,O2}$ and $E_{d,N2}$ were also significantly increased from the values of the unmodified membranes as can be seen in Figures 7d and 8d. The $E_{d,O2}$ and $E_{d,N2}$ were 7.26 kJ/mol and 8.96 kJ/mol, respectively, for M-M1Si; and were 6.72 kJ/mol and 9.35 kJ/mol, respectively, for M-M2Al. Apparently, the CCD modification caused larger increase in $E_{d,N2}$ than $E_{d,O2}$ because N_2 is bigger in size than O_2. In other words, O_2 and N_2 and permeation through the two membranes were governed by gaseous diffusion mechanism before modification, which is selective towards N_2 of smaller mass and became controlled by the activated diffusion mechanism after pore size reduction by the modification, which is selective towards O_2 of smaller size. Interestingly, the temperature-dependences of gas permeance were quite different between the two membranes as shown in Figures 7c and 8c. The M-M1Si exhibited monotonic increase of permeance with increasing temperature that deviates from the "S"-shaped dependence observed before modification. However, the M-M2Al maintained "S"-shaped permeance dependence on temperature, which is similar to that observed on the unmodified membrane. This difference may be caused by the different locations of pore narrowing in the channels of the two modified membranes. In M-M1Si, because the modification presumably occurs near the exiting end of channel and thereby the gas permeation is less influenced by the adsorbing effect since the transport is mainly controlled by the section of reduced channel size.

3.4. O_2/N_2 Mixture Separation

The two modified membranes were also examined for separation of O_2/N_2 mixtures. Figure 9 shows the temperature and pressure dependencies of gas permeance ($P_{m,i}$; i= O_2 and N_2) and separation factor ($\alpha_{O2/N2}$) on M-M1Si for an O_2/N_2 mixture feed with 20/80 molar ratio. It was found that, as feed pressure increased, the $P_{m,O2}$ remained almost unchanged at temperature of 297 K but showed moderate deceases at temperatures of $\geq$333 K; the $P_{m,N2}$, on the other hand, increased slightly as feed pressure increased at 297 K but also decreased moderately at $\geq$333 K. Overall, $P_{m,O2}$ decreased with pressure more rapidly than did $P_{m,N2}$, resulting in decline of $\alpha_{O2/N2}$ with increasing feed pressure. Similar pressure-dependences of gas permeance and $\alpha_{O2/N2}$ were observed for M-M2Al. The decrease in $\alpha_{O2/N2}$ with increasing feed pressure may be explained by considering the increasing contribution of N_2-selective gaseous diffusion through the intercrystalline pores while the transport through the zeolitic pores is less sensitive to pressure as limited by the weak adsorption at zeolite surface. Both $P_{m,O2}$ and $P_{m,N2}$ of the M-M1Si were found to increase monotonically with temperature for all feed pressures due to the activated diffusion mechanism for both gases in the modified membranes. However, due to the difference in magnitude of the temperature dependencies between $P_{m,O2}$ and $P_{m,N2}$, the $\alpha_{O2/N2}$ exhibited bell-shaped temperate-dependence as shown in Figure 9f. The maximum $\alpha_{O2/N2}$ appeared at 373 K for all feed pressures and achieved the best value of ~2.25 with a $P_{m,O2}$ of 7.24×10^{-9} mol/m^2·s·Pa at feed pressure of 1 bar. Thus, for M-M1Si, the gas mixture separation factor $\alpha_{O2/N2}$ was very close to the permselectivity $\alpha^o_{O2/N2}$ (~2.29) but $P_{m,O2}$ was notably lower than the $P^o_{m,O2}$ (9.10×10^{-9} mol/m^2·s·Pa).

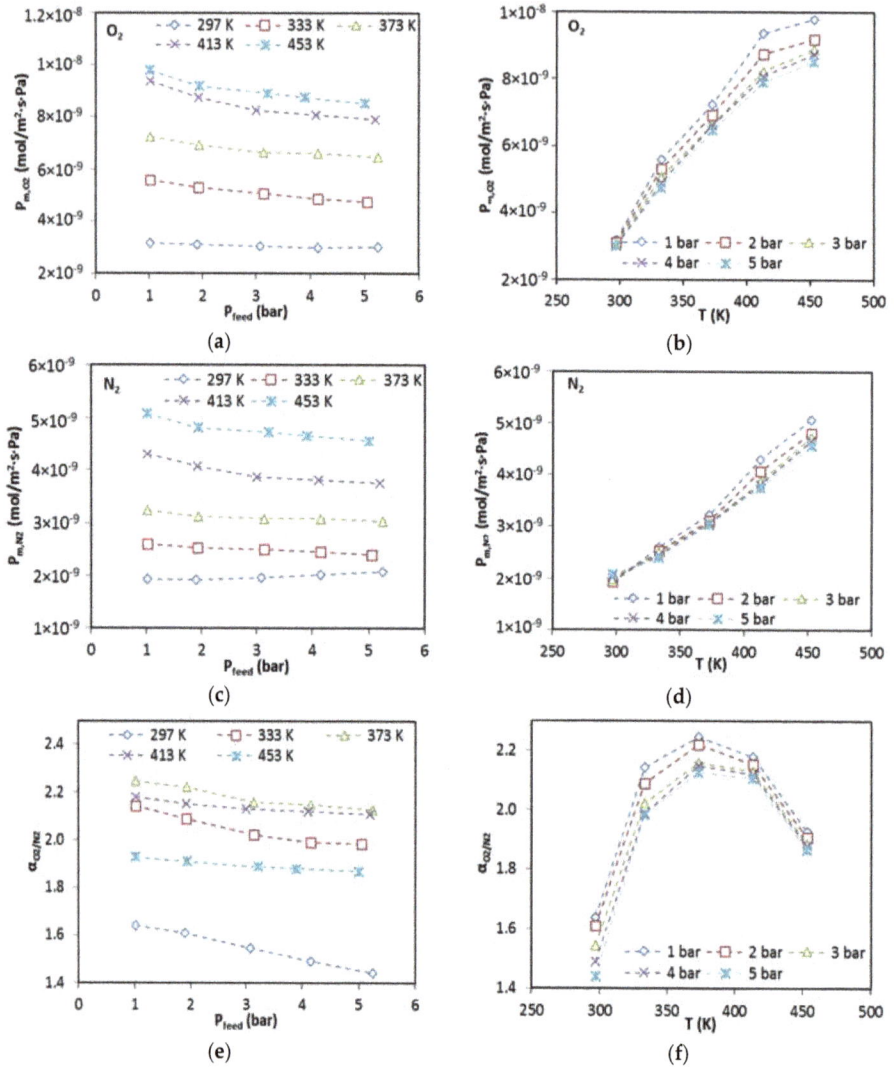

Figure 9. Effects of feed pressure and operating temperature on gas permeance and O_2/N_2 separation factor in the M-M1Si for an O_2/N_2 mixture feed containing 20 mol% O_2. (a) $P_{m,O2}$ as a function of pressure; (b) $P_{m,O2}$ as a function of temperature; (c) $P_{m,N2}$ as a function of pressure; (d) $P_{m,N2}$ as a function of temperature; (e) $\alpha_{O2/N2}$ as a function of pressure; and (f) $\alpha_{O2/N2}$ as a function of temperature.

The results of separation for the 20/80 O_2/N_2 mixture on the M-M2Al membrane are presented in Figure 10. Both $P_{m,O2}$ and $P_{m,N2}$ deceased with feed pressure at all temperatures tested while both $P_{m,O2}$ and $P_{m,N2}$ exhibited bell-shaped dependencies on temperature at each pressure, i.e., increased first and then decreased after reaching maxima at around 333 K. The temperature-dependence of gas permeance on the M-M2Al was again different from that on M-M1Si, which was similar to that observed in single gas permeation. The $\alpha_{O2/N2}$ also exhibited a bell-shaped profile of temperate dependence at each feed pressure as shown in Figure 10f. The $\alpha_{O2/N2}$ achieved maxima at 333 K, which was lower than the temperature (373 K) for maximum $\alpha_{O2/N2}$ on the M-M1Si. At feed pressure of 1 bar, M-M2Al obtained the best $\alpha_{O2/N2}$ of 1.95, which was almost identical to the $\alpha^o_{O2/N2}$ (~1.96) and a $P_{m,O2}$ of 1.2×10^{-7} mol/m^2·s·Pa, which was also very close to its $P^o_{m,O2}$ (1.3×10^{-7} mol/m^2·s·Pa) and far greater than that of the M-M1Si.

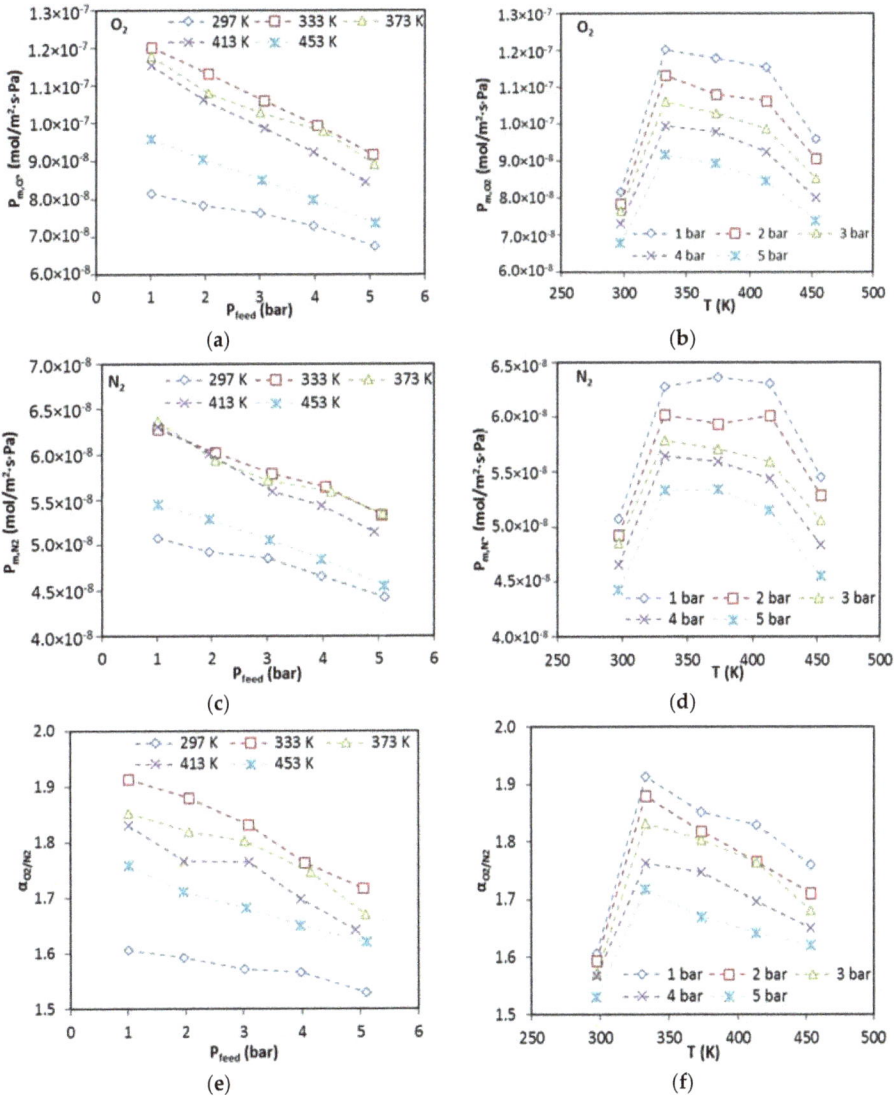

Figure 10. Effects of feed pressure and operating temperature on gas permeance and O$_2$/N$_2$ separation factor in M-M2Al for O$_2$/N$_2$ mixture feed with 20 mol% O$_2$: (**a**) $P_{m,O2}$ as a function of pressure; (**b**) $P_{m,O2}$ as a function of temperature; (**c**) $P_{m,N2}$ as a function of pressure; (**d**) $P_{m,N2}$ as a function of temperature; (**e**) $\alpha_{O2/N2}$ as a function of pressure; (**f**) $\alpha_{O2/N2}$ as a function of temperature.

As presented above, in separation of the O_2/N_2 mixture, both M-M1Si and M-M2Al had $\alpha_{O2/N2}$ values very close to their respective $\alpha^o_{O2/N2}$ that was expected from a selectivity largely determined by size exclusion or steric effects at the reduced pore openings. However, $P_{m,O2}$ in mixture permeation was significantly lower than the $P^o_{m,i}$ for M-M1Si while $P_{m,O2}$ remained similar to the $P^o_{m,i}$ for M-M2Al. Such a difference in permeance change between single gas and mixture permeation for M-M1Si and M-M2Al may be caused by the variation of the modification locations along the zeolite channel and membrane thickness as schematically illustrated in Figure 11. The deposition of silica modifier was near the zeolite/substrate interface in M-M1Si but was at the outer surface in M-M2Al. Thus, for gas mixture feed, the O_2 permeation rate is expected to remain very similar to that of single gas permeation in M-M2Al as depicted in Figure 11c but to be inevitably hindered by the back diffusion of the less permeable N_2 from the narrowed section in M-M1Si as depicted by Figure 11b.

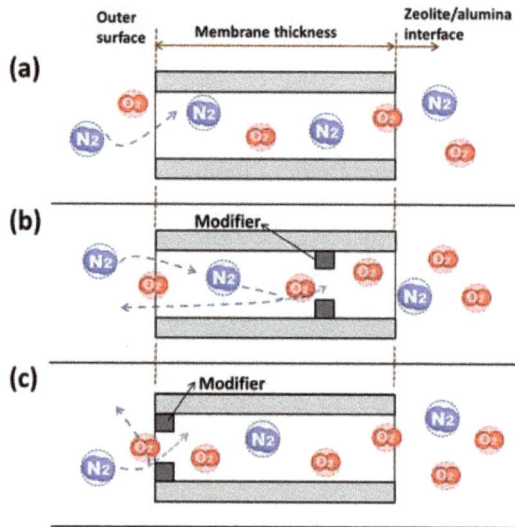

Figure 11. Schematic illustration of the presumed channel structures in the MFI-type zeolite membranes. (a) Unmodified; (b) M-M1Si; and (c) M-M2Al.

3.5. Gas Permeation Selectivity by Size Exclusion

Both M-M1Si and M-M2Al membranes were able to achieve O_2 permselectivity over N_2; however, the M-M2Al with a ZSM-5 surface drastically improved the O_2 permeance as compared to M-M1Si. However, the $\alpha_{O2/N2}$ for both membranes were limited to around 2. The relatively low $\alpha_{O2/N2}$ was attributed primarily to the intercrystalline spaces, which are known to be of nanometer sizes and selective towards the permeation of the lighter N_2. In literature, the effect of pore narrowing in MFI-type zeolite membranes by CCD modification on small gas permeation has been discussed based on the molecular transport model developed by Xiao and Wei [2,3,23]. In the kinetic molecular transport model, gas permeance (P_m) is expressed by Equation (4),

$$P_m = \frac{\varphi}{\delta}(D_o\beta) = \frac{\varphi}{\delta}\cdot\frac{\alpha}{z}\left(\frac{8RT}{\pi M_w}\right)^{1/2}\cdot\beta\cdot\exp\left(-\frac{E_d}{RT}\right)$$ (4)

where D_o is jump diffusivity, which correlates with the transport diffusivity D_c and molecule load q in zeolite, i.e., $D_o = D_c/q$ [24]. Under conditions where ideal gas behavior can be assumed and gas adsorption is negligible, q may be represented by $q = \beta P$ [23,25]. φ is a constant depending on the porosity and tortuosity; δ is the membrane thickness; and α and z are the single jump distance and diffusion coordination number, respectively.

Equation (4) can be used to understand the diffusion mechanisms inside the zeolitic channels according to the magnitude of diffusion activation energy (E_d) obtained from the Arrhenius representation of permeance but cannot correlate the gas selectivity with the size contrast between the molecule and pore opening, i.e., d_k/d_p (=λ), since neither λ nor d_k and d_p is explicitly related in Equation (4). The selectivity for permeating molecules by size discrimination at the channel entrance may be qualitatively analyzed based on kinetic molecular theory and an assumption of elastic collision of hard sphere molecule at pore mouth. At a specific temperature and pressure, the frequency of molecule colliding to the area of channel opening is given the product of gas phase molecular density ($\rho = p/RT$) and molecule travel velocity (v), which is related inversely to the square root of molecular weight ($v \propto M^{-0.5}$). As schematically illustrated in Figure 12, when $r_k < r_p$, collisions that can lead to effective entry of molecules must occur within the center area of channel opening defined by r_e (=$r_p - r_k$).

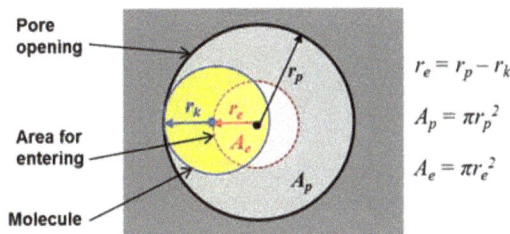

Figure 12. Schematic illustration of molecule entering the zeolite channel entrance.

Therefore, using the original pore radius ($r_p = 0.56$ nm for MFI zeolite) as the basis of consideration, for molecule "i" with $r_k < r_p$, when permeance is limited by the rate of molecule entering the entrance, $P_{m,i}$ depends on the collision frequency in the entire original pore area A_p (= πr_p^2) and probability (ς) of effective collision in A_e, i.e., $\varsigma = A_e/A_p = r_e^2/r_p^2 = (1 - \lambda_i)^2$. Thus, the gas permeation selectivity for molecule i over j by size exclusion effect at the pore opening can be related to the molecular weight and kinetic size and the pore size as

$$\alpha_{i/j}^o \propto \left(\frac{1 - \lambda_i}{1 - \lambda_j}\right)^2 \cdot \frac{\sqrt{M_{w,j}}}{\sqrt{M_{w,i}}} \cdot \exp\left(-\frac{E_{a,i} - E_{a,j}}{RT}\right), \lambda_i = r_{k,i}/r_p \tag{5}$$

where $E_{a,i}$ is the energy barrier for molecule i to enter the zeolitic channel, which is presumably dependent of λ_i. It is well-known that molecules can enter zeolites even when the nominal r_k and r_p are equal (i.e., $\lambda_i = 1$) or r_k is slightly bigger than r_p (i.e., $\lambda_i > 1$). Thus, theoretically quantifying $\alpha_{i/j}$ or $\alpha_{i/j}^o$ is difficult because both the zeolitic pores and molecular size are not exactly known due to their structural flexibility. Nevertheless, relation (5) can still help understanding the size exclusion effects. Figure 13 shows the relation between $[(1 - \lambda_{O2})^2/(1 - \lambda_{N2})^2 \cdot (\sqrt{M_{w,N2}}/\sqrt{M_{w,O2}})]$ and d_p that is predicted for the ideal system of rigid pore opening and hard sphere molecules. Because $E_{a,O2} < E_{a,N2}$, the value of $\exp[-(E_{a,i} - E_{a,j})/RT]$ is always >1; therefore, $\alpha_{O2/N2}^o > [(1 - \lambda_{O2})^2/(1 - \lambda_{N2})^2 \cdot (\sqrt{M_{w,N2}}/\sqrt{M_{w,O2}})]$. Apparently, $[(1 - \lambda_{O2})^2/(1 - \lambda_{N2})^2 \cdot (\sqrt{M_{w,N2}}/\sqrt{M_{w,O2}})]$ and consequently $\alpha_{O2/N2}^o$ remains close to 1.0 when d_p is relatively large (e.g., $d_p > 0.45$ nm) and increases sharply when d_p becomes < 0.4 nm to approach $d_{k,N2}$.

The above simple calculation indicates a conservative estimate of $\alpha_{O2/N2}^o > 4.22$ for $d_p < 0.38$ nm by size-selectivity at the pore entrance, which is significantly greater than the $\alpha_{O2/N2}^o$ of the CCD modified membranes in this study. The O$_2$/N$_2$ selectivity was thus believed to be lowered primarily by gas permeation through the intercrystalline pores where the molecular transport is mainly governed by the gaseous diffusion mechanism and selective towards N_2 [26].

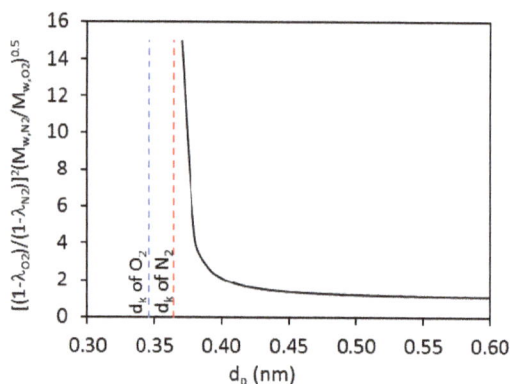

Figure 13. Dependence of $[(1 - \lambda_{O2})^2 / (1 - \lambda_{N2})^2 \cdot (\sqrt{M_{w,N2}} / \sqrt{M_{w,O2}})]$ on d_p.

3.6. Reduction of Intercrystalline Spaces

Preliminary experiment was carried out in attempt to reduce the intercrystalline spaces and improve the O_2/N_2 separation selectivity by liquid phase deposition of silica using tetramethoxysilane (TMOS) as the precursor. This chemical liquid deposition (CLD) method was recently developed and proven very effective for filling the nonzeolitic spaces in the DDR zeolite membrane to enhance the small gas permeation selectivity [6].

An MFI-type membrane was prepared using the same precursor and the same synthesis procedure as described earlier. The membrane was however obtained by two times of hydrothermal treatment that minimized nonzeolitic pores in the membrane as evidenced by its very small $\alpha_{O2/N2}$ ($\approx$0.1) for separation of an equimolar H_2/CO_2 mixture at room temperature. The CCD modification was terminated in 4 h when the online-monitored $\alpha_{O2/N2}$ and $P_{m,H2}$ stabilized at 14 and of 1.2×10^{-7} mol/m²·s·Pa, respectively. The CCD modified membrane was tested for O_2/N_2 separation and then treated by the CLD method for reducing the intercrystalline pores, which was similar to that reported in literature [27,28]. The CLD modification was conducted by the following procedure. First, the alumina substrate of the disc membrane was soaked with water and any liquid water was wiped off the substrate surface; second, the zeolite membrane surface was dried by sweeping with compressed air for about 10 s after the membrane disc was mounted in a permeation cell; third, the liquid TMOS was immediately filled into the cell chamber on the zeolite surface side while the chamber on the substrate side was filled with D.I. water; finally, the membrane cell was kept static at room temperature with the zeolite membrane surface facing upward for 2 h. During the finally step, TMOS hydrolysis and hydrous silica deposition took place inside the intercrystalline spaces by liquid phase counter diffusion of TMOS and water but not in the zeolitic pores because d_k of TMOS is far bigger than the zeolitic pore diameter. After the CLD process, the membrane disc was dried at 333 K for 1 day followed by further drying at 473 K for 6 h and finally calcined at 723 °C for 6 h.

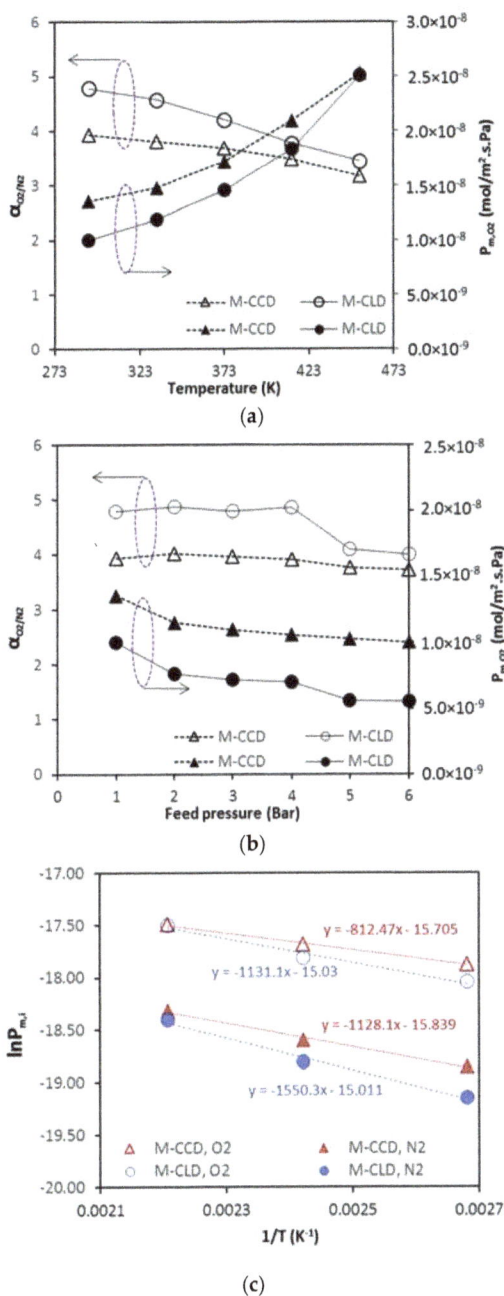

Figure 14. Results of separation of equimolar O_2/N_2 gas mixture on the CCD modified membrane before (M-CCD) and after the CLD treatment (M-CLD). (**a**) $\alpha_{O2/N2}$ and $P_{m,O2}$ as functions of temperature; (**b**) $\alpha_{O2/N2}$ and $P_{m,O2}$ as functions of feed pressure; and (**c**) Arrhenius plots of gas permeance.

The CLD modified zeolite membrane was again tested for separation of an O_2/N_2 equimolar mixture in a temperature range of 293–473 K and a feed pressure range of 1–6 bar. The O_2/N_2 separation results obtained on the CCD-modified membrane before and after the CLD treatment are presented in Figure 14. The modified membrane is named as M-CCD before the CLD treatment and named as M-CLD after CLD treatment. The temperature- and pressure-dependencies of $\alpha_{O2/N2}$ and $P_{m,O2}$ for M-CCD and M-CLD were in general consistent with those for M-M1Si and M-M2Al, both decrease moderately with increasing temperature or pressure. However, M-CCD and M-CLD exhibited significantly high $\alpha_{O2/N2}$ but much lower $P_{m,O2}$ as compared to M-M1Si and M-M2Al because the former was obtained by two time of hydrothermal crystallization that resulted in thicker membrane with less microdefects. The M-CLD had notable improvement in $\alpha_{O2/N2}$ but with large decreases in $P_{m,O2}$ as compared to M-CCD that demonstrated the effective reduction of intercrystalline spaces by the CLD modification. The best $\alpha_{O2/N2}$ on the M-CLD was nearly 5 but the $P_{m,O2}$ was only ~1.0×10^{-8} mol/m^2·s·Pa, which was less than 10% that of the M-M2Al. Figure 14c presents the Arrhenius correlations of gas permeance at relatively high temperatures ($\geq$373 K) for M-CCD and M-CLD. The $E_{d,O2}$ and $E_{d,N2}$ of M-CCD were 6.75 kJ/mol and 9.38 kJ/mol, respectively, which were very similar to the values observed in M-M2Al as expected. The $E_{d,O2}$ and $E_{d,N2}$ of M-CLD increased to 9.40 kJ/mol and 12.89 kJ/mol, respectively; both became significantly greater than the values of M-CCD that further indicated the effective reduction of gaseous diffusion through the intercrystalline spaces by the CLD modification.

4. Conclusions

The CCD modified MFI-type zeolite has been demonstrated to have O_2/N_2 permselectivity and mixture separation factor around 2. The O_2 permeance was drastically enhanced for the modified MFI zeolite membrane by forming of an alumina-containing zeolite membrane surface, which effectively limited the channel narrowing within a small thickness near the outer surface by CCD modification. The modified membrane with a ZSM-5 surface (i.e., M-M2Al) had a moderately lower $\alpha_{O2/N2}$ (~1.95) than the modified membrane M-M1Si ($\alpha_{O2/N2}$~2.25) but the former had a $P_{m,O2}$ of 1.2×10^{-7} mol/m^2·s·Pa, which was 16 times that of the M-M1Si ($P_{m,O2} = 7.24 \times 10^{-9}$ mol/m^2·s·Pa). The theoretical analysis of size exclusion effect by the reduced pore size was carried out based on a simplified system of rigid pore openings and hard sphere molecules. The results suggested that the CCD modified MFI-type zeolite membrane was expected to possess much higher O_2/N_2 selectivity than those obtained on the two CCD modified membranes. The lower-than-expected $\alpha_{O2/N2}$ on the modified membranes was attributed to the inevitably existing intercrystalline spaces, which allow the undesirable gas transport by gaseous diffusion to lower the O_2/N_2 selectivity. Preliminary investigation on CLD treatment of a CCD modified MFI-type zeolite membrane demonstrated that O_2 selectivity of the modified zeolite membrane could be further improved by reducing the nonzeolitic pores by repeated hydrothermal synthesis and the CLD repairing of intercrystalline pores. The CLD-modified membrane achieved an impressive $\alpha_{O2/N2}$ of 4.9 but with a low $P_{m,O2}$ of ~1.0×10^{-8} mol/m^2·s·Pa. To further improve the separation of small gases such as the O_2/N_2 mixture, the challenge of synthesizing high quality silicalite membranes with very small thickness and a ZSM-5 surface need to be addressed so that the membrane can achieve both high selectivity and high permeance after the CCD modification.

Acknowledgments: This research was financially supported by the U.S. Department of Energy/NETL through grant number DE-FE0026435 and the Development Service Agency of Ohio through the Ohio Coal Research and Development program (Grants # OOECDO-D-15-16 and # OOECDO-D-17-13).

Author Contributions: S. Yang and A. Arvanitis designed and performed the experiments and did main part of the data analyses; Z. Cao and X. Sun involved in the synthesis and characterization work; J. Dong directed the research, conceived the concept, and contributed to the data analyses and theoretical work.

Conflicts of Interest: The authors declare no conflict of interest.

References

1. Baker, R.W. Future directions of membrane gas separation technology. *Ind. Eng. Chem. Res.* **2002**, *41*, 1393–1411. [CrossRef]
2. Xiao, J.; Wei, J. Diffusion mechanism of hydrocarbon in zeolites-1. Theory. *Chem. Eng. Sci.* **1992**, *47*, 1123–1141. [CrossRef]
3. Tang, Z.; Nenoff, T.M.; Dong, J. Internal surface modification of MFI-type zeolite membranes for high selectivity and high flux for hydrogen. *Langmuir* **2009**, *25*, 4848–4852. [CrossRef] [PubMed]
4. Tomita, T.; Nakayama, K.; Sakai, H. Gas separation characteristics of DDR type zeolite membrane. *Micropor. Mesopor. Mater.* **2004**, *68*, 71–75. [CrossRef]
5. Van den Bergh, J.; Zhu, W.; Gascon, J.; Moulijn, J.A.; Kapteijn, F. Separation and permeation characteristics of a DD3R zeolite membrane. *J. Membr. Sci.* **2008**, *316*, 35–45. [CrossRef]
6. Yang, S.; Cao, Z.; Arvanitis, A.; Sun, X.; Xu, Z.; Dong, J. DDR-type zeolite membrane synthesis, modification and gas permeation studies. *J. Membr. Sci.* **2016**, *505*, 194–204. [CrossRef]
7. Yin, X.; Zhu, G.; Yang, W.; Li, Y.; Zhu, G.; Xu, R.; Sun, J.; Qiu, S.; Xu, R. Stainless-steel-net-supported zeolite NaA membrane with high permeance and high permselectivity for oxygen over nitrogen. *Adv. Mater.* **2005**, *17*, 2006–2010. [CrossRef]
8. Zhu, W.; Gora, L.; van den Berg, A.W.C.; Kapteijn, F.; Jansen, J.C.; Moulijn, J.A. Water vapour separation from permanent gases by a zeolite-4A membrane. *J. Membr. Sci.* **2005**, *253*, 57–66. [CrossRef]
9. Caro, J.; Albrecht, D.; Noack, M. Why is it so extremely difficult to prepare shape-selective Al-rich zeolite membranes like LTA and FAU for gas separation? *Sep. Purif. Technol.* **2009**, *66*, 143–147. [CrossRef]
10. Masuda, T.; Fukumoto, N.; Kitamura, M.; Mukai, S.R.; Hashimoto, K.; Tanaka, T.; Funabiki, T. Modification of pore size of MFI-type zeolite by catalytic cracking of silane and application to preparation of H_2-separating zeolite membrane. *Micropor. Mesopor. Mater.* **2001**, *48*, 239–245. [CrossRef]
11. Hong, M.; Falconer, J.L.; Noble, R.D. Modification of Zeolite Membranes for H_2 Separation by Catalytic Cracking of Methyldiethoxysilane. *Ind. Eng. Chem. Res.* **2005**, *44*, 4035–4041. [CrossRef]
12. Gu, X.; Tang, Z.; Dong, J. On-stream modification of MFI zeolite membranes for enhancing hydrogen separation at high temperature. *Micropor. Mesopor. Mater.* **2008**, *111*, 441–448. [CrossRef]
13. O'Connor, C.T.; Moller, K.P.; Manstein, H. The Effect of Silanization on the Catalytic and Sorption Properties of Zeolites. *CatTech* **2001**, *5*, 172–182. [CrossRef]
14. Hong, Z.; Wu, Z.; Zhang, Y.; Gu, X. Catalytic cracking deposition of Methyldiethoxysilane for modification of zeolitic pores in MFI/α-Al_2O_3 zeolite membrane with H^+ ion exchange pretreatment. *Ind. Eng. Chem. Res.* **2013**, *52*, 13113–13119. [CrossRef]
15. Wang, H.; Lin, Y.S. Synthesis and modification of ZSM-5/silicalite bilayer membrane with improved hydrogen separation performance. *J. Membr. Sci.* **2012**, *396*, 128–137. [CrossRef]
16. Dong, J.; Lin, Y.S. In situ synthesis of P-type zeolite membranes on porous alpha-alumina supports. *Ind. Eng. Chem. Res.* **1998**, *37*, 2404–2409. [CrossRef]
17. Baerlocher, C.; McCusker, L.B.; Olson, D.H. *Atlas of Zeolite Framework Types*, 6th ed.; Elsevier: Oxford, UK, 2007.
18. Dong, J.; Wegner, K.; Lin, Y.S. Synthesis of submicron silicalite membranes on porous ceramic supports. *J. Membr. Sci.* **1998**, *148*, 233–241. [CrossRef]
19. Dong, J.; Lin, Y.S.; Hu, M.Z.C.; Peascoe, R.A.; Payzant, E.A. Template-removal-associated microstructural development of porous-ceramic-supported MFI zeolite membranes. *Micropor. Mesopor. Mater.* **2000**, *34*, 241–253. [CrossRef]
20. Dunne, J.A.; Rao, M.; Sircar, S.; Gorte, R.J.; Myers, A.L. Calorimetric heats of adsorption and adsorption isotherms. 2. O_2, N_2, Ar, CO_2, CH_4, C_2H_6, and SF_6 on NaX, H-ZSM-5, and Na-ZSM-5 zeolites. *Langmuir* **1996**, *12*, 5896–5904. [CrossRef]
21. Bernal, M.P.; Bardaji, M.; Coronas, J.; Santamaria, J. Facilitated transport of O_2 through alumina-zeolite composite membranes containing a solution with a reducible metal complex. *J. Membr. Sci.* **2002**, *203*, 209–213. [CrossRef]
22. Dong, J.; Lin, Y.S.; Liu, W. Multicomponent hydrogen/hydrocarbon separation by MFI-type zeolite membranes. *AIChE J.* **2000**, *46*, 1957–1966. [CrossRef]
23. Kanezashi, M.; O'Brien-Abraham, J.; Lin, Y.S.; Suzuki, K. Gas permeation through DDR-type zeolite membranes at high temperatures. *AIChE J.* **2008**, *54*, 1478–1486. [CrossRef]

24. Babarao, R.; Jiang, J. Diffusion and separation of CO_2 and CH_4 in silicalite, C-168 schwarzite, and IRMOF-1: A comparative study from molecular dynamics simulation. *Langmuir* **2008**, *24*, 5474–5484. [CrossRef] [PubMed]
25. Gu, Y.; Hacarlioglu, P.; Oyama, S.T. Hydrothermally stable silica-alumina composite membranes for hydrogen separation. *J. Membr. Sci.* **2008**, *310*, 28–37. [CrossRef]
26. Dong, J.; Lin, Y.S.; Kanezashi, M.; Tang, Z. Microporous inorganic membranes for high temperature hydrogen purification. *J. Appl. Phys.* **2008**, *104*, 121301. [CrossRef]
27. Zhang, B.; Wang, C.; Lang, L.; Cui, R.; Liu, X. Selective defect-patching of zeolite membranes using chemical liquid deposition at organic/aqueous interfaces. *Adv. Funct. Mater.* **2008**, *18*, 3434–3443. [CrossRef]
28. Hong, Z.; Zhang, C.; Gu, X.; Jin, W.; Xu, N. A simple method for healing nonzeolitic pores of MFI membranes by hydrolysis of silanes. *J. Membr. Sci.* **2011**, *366*, 427–435. [CrossRef]

processes

MDPI

Article

Low-Temperature Steam Reforming of Natural Gas after LPG-Enrichment with MFI Membranes

Dominik Seeburg [†], Dongjing Liu [†], Radostina Dragomirova, Hanan Atia, Marga-Martina Pohl, Hadis Amani, Gabriele Georgi, Stefanie Kreft and Sebastian Wohlrab *

Leibniz Institute for Catalysis at the University of Rostock, Albert-Einstein-Str. 29a, D-18059 Rostock, Germany; dominik.seeburg@catalysis.de (D.S.); liudongjing19@163.com (D.L.); radidragomirova@gmail.com (R.D.); hanan.atia@catalysis.de (H.A.); marga.pohl@freenet.de (M.-M.P.); hadis.amani@catalysis.de (H.A.); gabi.georgi@web.de (G.G.); stefanie.kreft@catalysis.de (S.K.)
* Correspondence: sebastian.wohlrab@catalysis.de; Tel.: +49-381-128-1328
† These authors contributed equally to this work.

Received: 15 November 2018; Accepted: 8 December 2018; Published: 12 December 2018

Abstract: Low-temperature hydrogen production from natural gas via steam reforming requires novel processing concepts as well as stable catalysts. A process using zeolite membranes of the type MFI (Mobile FIve) was used to enrich natural gas with liquefied petroleum gas (LPG) alkanes (in particular, propane and n-butane), in order to improve the hydrogen production from this mixture at a reduced temperature. For this purpose, a catalyst precursor based on Rh single-sites (1 mol% Rh) on alumina was transformed in situ to a $Rh1/Al_2O_3$ catalyst possessing better performance capabilities compared with commercial catalysts. A wet raw natural gas (57.6 vol% CH_4) was fully reformed at 650 °C, with 1 bar absolute pressure over the $Rh1/Al_2O_3$ at a steam to carbon ratio $S/C = 4$, yielding 74.7% H_2. However, at 350 °C only 21 vol% H_2 was obtained under these conditions. The second mixture, enriched with LPG, was obtained from the raw gas after the membrane process and contained only 25.2 vol% CH_4. From this second mixture, 47 vol% H_2 was generated at 350 °C after steam reforming over the $Rh1/Al_2O_3$ catalyst at $S/C = 4$. At $S/C = 1$ conversion was suppressed for both gas mixtures. Single alkane reforming of $C_2–C_4$ showed different sensitivity for side reactions, e.g., methanation between 350 and 650 °C. These results contribute to ongoing research in the field of low-temperature hydrogen release from natural gas alkanes for fuel cell applications as well as for pre-reforming processes.

Keywords: steam reforming; pre-reforming; alkanes; hydrogen; membrane separation; single-sites

1. Introduction

Exploitation of the world's reserves of natural gas has increased tremendously, while the spectrum of utilization has broadened continuously. Methane, the main component in natural gas, is typically combusted for energy generation, since it is difficult to activate this smallest alkane and form more valuable products [1]. In terms of direct methane activation, several directions can be pursued which are currently only of academic interest [2–6]. Hence, for chemical processing, methane is mainly converted into H_2/CO mixtures (syngas) which provide key components for the production of chemical products [7–10], e.g., for the synthesis of ammonia, methanol, and liquid hydrocarbons [11]. The steam reforming process of methane is typically performed at high temperatures between 750 and 900 °C. At a pressure of $p = 24$ bar and a steam to carbon ratio of $S/C = 3$, nearly 75 vol% H_2 is produced (excluding water) [12].

Wet natural gas, rich in $C_2–C_5$ alkanes, or $C_2–C_5$ alkane fractions from natural gas conditioning, are valuable sources for syngas production. Depending on the alkane composition and the processing technology (i.e., steam reforming, partial oxidation, autothermal reforming, dry reforming), syngas with different H_2/CO ratios can be obtained by using suitable catalysts [13]. Generally, the highest possible amount of H_2 can be achieved by using steam as a reactant for alkanes, due to the inherent hydrogen content of the water. Industrially, pre-reforming of the higher alkanes in wet natural gas can help to reduce the size of a downstream tubular reformer [9]. Moreover, liquid fuels such as butane, alcohols, or diesel are increasingly being considered as easily storable sources of hydrogen which can be applied in fuel cell systems [14–18]. Among those, reforming of bioethanol could contribute to sustainable hydrogen production. However, a high temperature of about 700 °C is required to obtain hydrogen yields above 70% [19].

Steam reforming of alkanes requires a significant energy input [20,21]. For methane conversion, the required $\Delta H°$ is +206 kJ/mol [22], hence, the reaction has to be performed at high temperatures (often higher than 900 °C) to ensure full conversion [23]. In order to decrease the required reaction temperature, low pressure and a relatively high S/C ratio are necessary [14]. Several attempts have been made in order to lower the required energy input for alkane steam reforming [22]. Interestingly, the reforming of higher alkanes can be performed at a lower temperature compared to methane. For instance, complete conversion of *n*-butane can be achieved at 405 °C over Pt-Ni/δ-Al_2O_3 [15]. Schädel et al. [20] systematically tested an industrial Rh catalyst as a washcoat on cordierite honeycomb monoliths for the steam reforming of methane, ethane, propane, butane, and natural gas. The authors showed that methane requires a much higher temperature for conversion compared to the other alkanes. Consequently, steam reforming of a natural gas with a high liquefied petroleum gas (LPG) fraction would offer lower reaction temperatures.

The enrichment of natural gas with LPG alkanes via MFI (Mobile FIve)-membranes has been reported in the past. Selective gas transport across zeolite membranes can be achieved by either selective adsorption or size exclusion [24]. Due to their selective adsorption properties, MFI zeolite membranes are extensively studied for separation of different alkanes [25–29]. From a mechanistic point of view, the higher molecular weight alkanes adsorb preferentially at the membrane surface and thus block the pores for further passage of lighter components. However, a fundamental understanding of transport across the membrane and the accurate analysis of optimal operating parameters are essential for achieving high membrane performances. The separation of natural gas alkanes based on the adsorption behavior of the different alkanes was first demonstrated by Arruebo et al. [30] and detailed parameter studies were performed by our group [31–34].

Unfortunately, stable catalysts working at lower temperatures in the steam reforming of LPG enriched natural gas are rather scarce. Carbon deposition, the most serious problem affecting the stability of catalysts at low reaction temperatures (400–550 °C), was investigated by Angeli et al. [35] who found significant carbon residues on Ni catalysts, while in the presence of Rh, coke was oxidized. Our initial idea was to reduce the size of the active sites in order reduce coking ability. In terms of methane activation, Bao and co-workers developed an oxygen free route towards C_2 and aromatic products over Fe single-sites [36]. Moreover, direct transformation of methane to methanol can be performed over Pd single-sites at a low temperature [37], or over single-site copper species in zeolites [38–41]. Molecularly attached VO_x single-sites on silica have been used as catalysts for the selective oxidation of methane towards formaldehyde [42–44]. However, in terms of steam reforming of methane, Bokhoven and co-workers recently found that the catalytic reaction over single Rh sites on stabilizing supports requires additional nanoparticles to oxidize formed carbon species [45]. In the absence of nanoparticles, reaction rates are lowered due to the formation of carbon species which are strongly bound to the surface.

In this work, we started with single Rh site rhodium catalyst precursors in the steam reforming of alkanes and LPG rich natural gas. The single-sites were transformed during the reaction and the in situ formed catalysts containing Rh nanoparticles surprisingly delivered a much better performance compared to commercial catalysts at low temperature. Therefore, a membrane-based pre-enrichment of LPG can be introduced as a feasible concept to obtain an alkane mixture from which a high amount of hydrogen can be produced at temperatures as low as 350 °C.

2. Materials and Methods

2.1. Catalyst Preparation

Rh1/Al$_2$O$_3$ with 1.0 mol% Rh was prepared by impregnation of Al$_2$O$_3$. First, 15 g Disperal P2 (Sasol, Brunsbüttel, Germany) were calcined at 800 °C (heating rate 5 °C/min) for 90 min in air. Afterwards, the as-obtained Al$_2$O$_3$ was impregnated with a freshly prepared Rh(NO$_3$)$_3$ solution. Accordingly, 1.9 g RhCl$_3$·4.5H$_2$O (Alfa Aesar, Karlsruhe, Germany) were dissolved in 100 mL of water. The solution was heated and 20 mL of aqueous 20 M KOH (Fisher Chemical, Loughborough, UK) were added dropwise under reflux. The precipitated Rh(OH)$_3$ was centrifuged and washed four times with hot water. The resulting Rh(OH)$_3$ precipitate was then dissolved in 5 mL of concentrated HNO$_3$ (Fisher Chemical, Loughborough, UK) and subsequently diluted in 20 mL of deionized water to obtain a Rh(NO$_3$)$_3$ concentration of about 0.35 mol/L. As a last step, 2.85 mL of this solution was further diluted in 47.2 mL of deionized water and used for the wet impregnation of 10 g of Al$_2$O$_3$. This slurry was stirred for 1 h, water was subsequently removed using a rotary evaporator, and the resulting catalyst was dried over night at 110 °C. Finally, calcination of the material was performed at 700 °C in air for 60 min (heating rate 5 °C/min). The catalysts were abbreviated as follows: Rh1/Al$_2$O$_3$ for 1.0 mol% Rh on alumina. Several other loadings of Rh on alumina were abbreviated as Rh$_X$/Al$_2$O$_3$ with (x = 0.01, 0,1, and 0.5 in mol% Rh, determined via inductively coupled plasma - optical emission spectroscopy ICP-OES). For comparison, two commercial catalysts consisting of 0.5 % Rh (206172) on alumina and 5% Rh (C-301099-5) on alumina were purchased from Sigma-Aldrich (Steinheim, Germany) and Alfa Aesar (Karlsruhe, Germany), respectively, and tested under comparable conditions.

2.2. Catalyst Characterization

The BET (Brunauer–Emmett–Teller) specific surface area of the porous silica and the catalysts were measured by N$_2$ adsorption using a NOVA 4200e instrument from Quantachrome (Odelzhausen, Germany). As a pre-treatment, samples were outgassed and dried for 2 h at 200 °C at reduced pressure.

Powder X-ray diffraction (XRD) patterns of the calcined samples were measured in the angle range 5–80° 2Theta scale on a Theta/Theta diffractometer X'Pert Pro (Panalytical, Almelo, Netherlands) using a Ni-filtered Cu–K$_\alpha$ radiation (λ = 1.5418 Å, 40 kV, 40 mA). The data were recorded with the X'Celerator (RTMS) detector.

Transmission electron microscopy (TEM) measurements at 200 kV were performed on a JEM-ARM200F (JEOL) with aberration-correction by a CESCOR (CEOS) for the scanning transmission (STEM) applications (JEOL, Corrector: CEOS, Tokyo, Japan).

A Varian 715-ES ICP-OES (Inductively Coupled Plasma-Optical Emission Spectrometer) (Varian Palo Alto, CA, USA) was used for the determination of the elemental composition of the catalysts. Before analysis, the catalysts were completely dissolved in a solution containing 8 mL of aqua regia and 2 mL of hydrofluoric acid.

The H$_2$-TPR (temperature programmed reduction with H$_2$) experiments were done as described in the following. 160 mg of the respective sample was loaded in a u-shaped quartz reactor and heated from RT (room temperature) to 500 °C at 20 K/min in air, then cooled to RT and flushed with Ar flow (50 mL/min) for 30 min. H$_2$-TPR of Rh1/Al$_2$O$_3$ samples were carried out from 0 to 800 °C in a flow of 5% H$_2$/Ar (20 mL/min) with a heating rate of 5 K/min. The temperature was held at 800 °C for 1 h. The hydrogen consumption peaks were recorded simultaneously via a thermal conductivity detector (TCD, ChemiSorb 2920-Instrument, Mircomeritics, Norcross, GA, USA).

2.3. Membrane Separation

Pressure-stable and defect-free MFI membranes at the inner side of porous alumina tubes (l = 125 mm, d$_{outer}$ = 10 mm, d$_{inner}$ = 8 mm) were prepared via a secondary growth procedure previously reported by our group in Reference [31]. The tubes were sealed with glass at both ends and embedded in stainless steel permeation cells with Viton O-rings. LPG enrichment from natural gas using the MFI membranes was performed at $p_{permeate}$ = 0.17 bar, p_{feed} = 7 bar, and T = 75 °C. Compositions of permeate and retentate were analyzed continuously using an online coupled capillary GC HP 6890 from Hewlett Packard (Santa Clara, CA, USA).

2.4. Catalytic Steam Reforming Tests

Steam reforming of alkanes, alkane mixtures, and simulated natural gas (before and after membrane enrichment) were performed in a vertical fixed bed plug flow quartz reactor (l = 260 mm, d_{inner} of 8 mm) at 1 bar (detailed composition of the reaction mixtures are given in Table S1 and S2). If not otherwise stated, 150 mg of catalyst was fixed with quartz wool at the centre of the reactor tube and heated in a furnace at temperatures ranging from 200 to 850 °C. If not otherwise stated, the holding time spent for each temperature set point was about 35 min. Temperature was controlled by two thermocouples (ThermoExpert, Stapelfeld, Germany) at the outer reactor wall and in the middle of the catalyst bed. The gas flow was controlled by mass flow controllers (MKS, Andover, MA, USA). The total gas flow under ambient conditions of 100 cm^3 min^{-1} consisted of 25 cm^3 min^{-1} of reactant gas (steam and alkanes) diluted in 75 cm^3 min^{-1} of nitrogen to reach near isothermal conditions. If not otherwise stated, GHSV (gas hourly space velocity) was about 8000 h^{-1}. The required water was dosed to a vaporizer using a syringe pump at the front inlet of the reactor. All the transfer lines were heated at 130 °C. At the reactor outlet, water was condensed from the product gas stream in a cold trap at 0 °C. Gas phase products were analyzed by an online-GC 7890A (Agilent, Santa Clara, CA, USA). H_2, N_2, and CO were separated with a molsieve column 5Å from Agilent (CP-1306) (Santa Clara, CA, USA) and detected with a TCD. The alkanes and CO_2 were separated with a GS-Q column from Agilent (113-3432) (Santa Clara, CA, USA) and detected with a flame ionization detector (FID, Agilent, Santa Clara, CA, USA) with a methanation unit. For quantification of H_2, N_2, CO, CO_2, and C_1–C_5 alkanes, an external calibration was done with several test gas mixtures (supplied from Linde Group, Pullach, Germany) and their dilutions in N_2, as well as by measuring the pure gases. For discussion, H_2O and N_2 were excluded in the given volumetric gas compositions.

3. Results

3.1. Rh1/Al$_2$O$_3$ Catalyst

Rh single-atoms were deposited on a high surface alumina support of 181 m^2/g. The BET surface areas and pore volumes of the obtained Rh1/Al$_2$O$_3$ catalyst, given in Table 1, are only less affected after calcination, indicating the integrity of the support material after thermal treatment. The molar fraction of Rh in Rh1/Al$_2$O$_3$ was 1.0 mol%, which is in the range of the targeted value. X-ray diffraction patterns of the catalyst (Figure S1) shows only typical reflections of different alumina phases, namely, γ-, δ-, and θ-Al$_2$O$_3$ (JCPDS 29-0063 [46], PDF 46-1215 [47], and ICSD#082504 [48]).

Table 1. BET (Brunauer–Emmett–Teller) specific surface area (S$_{BET}$), total pore volume (V$_t$), and average pore diameter (D), as well as ICP-OES (inductively coupled plasma - optical emission spectroscopy) data of support and catalysts.

Samples	BET			ICP
	S$_{BET}$ (m^2/g)	V$_t$ (mL/g)	D (nm)	Rh (mol%)
Al$_2$O$_3$	181	0.413	9.1	-
Rh1/Al$_2$O$_3$	169	0.406	9.6	1.0

High-angle annular dark-field—scanning transmission electron microscopy (HAADF-STEM) images of fresh and spent catalysts are shown in Figure 1a,b. The spent catalysts were isolated after the light-off test with permeate gas at S/C = 4, running the reaction from 200 to 850 °C. In fresh Rh1/Al$_2$O$_3$, many visible Rh atoms well-dispersed over the γ-Al$_2$O$_3$ support and islands of single-sites can be detected (bright spots in yellow circles). Rh nanoparticles in the size of 1–3 nm are visible in the spent $Rh1/Al_2O_3$ catalyst, indicating particle formation from Rh single-sites during the steam reforming process.

H_2-TPR was performed with fresh and spent catalysts. The spent catalysts were isolated again after the light-off tests with permeate gas at 850 °C (S/C = 4). As can be concluded from Figure 1c, the absence of typical reduction peaks for particulate RhO_x species in the range from 80 to 250 °C [49,50] indicate the nature of the sites in fresh $Rh1/Al_2O_3$. Most Rh species in the fresh catalysts are present as single atoms, because the adsorption and dissociation of H_2 to H atoms cannot occur over single atoms [51]. H_2-TPR over spent $Rh1/Al_2O_3$ catalysts is in accordance with the TEM results. It is obvious that the single-sites in the fresh catalysts were transformed to nanoparticle species during an initial activation phase. The formed Rh particles in spent $Rh1/Al_2O_3$ show an extraordinary reducibility below 50 °C and were the active catalyst species.

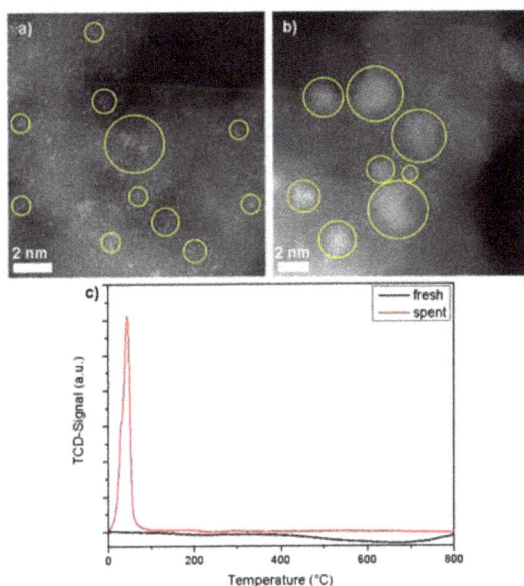

Figure 1. Catalyst characterization. High-angle annular dark-field—scanning electron microscopy (HAADF-STEM) images: (**a**) fresh $Rh1/Al_2O_3$, (**b**) spent $Rh1/Al_2O_3$, and (**c**) H_2-TPR (temperature programmed reduction with H_2) profiles of fresh and spent $Rh1/Al_2O_3$. Spent catalysts were isolated after hydrogen production from liquefied petroleum gas (LPG)-enriched permeate natural gas with a steam to carbon ratio (S/C) = 4, heating from 200 to 850 °C.

3.2. Steam Reforming of pure C_{1-4} Alkanes and Mixtures of C_{2-4} with Methane over $Rh1/Al_2O_3$

Steam reforming of the single alkanes methane, ethane, propane, *n*-butane, and *i*-butane were performed over $Rh1/Al_2O_3$ using S/C ratio and temperature as a measure for catalyst activity (Figure 2) and selectivity (Figure 3). At a GHSVs of 8000 h^{-1}, 1 mol% Rh appeared as sufficient loading to reach nearly equilibrium methane conversion [22] (Figure S2). At lower Rh loading, GHSVs have to be reduced to fulfil this task (e.g., Figures S3 and S4 for $Rh_{0.5}/Al_2O_3$). In the temperature range 250–400 °C, increasing S/C ratios lead to higher H_2 fractions in the product gas and higher alkanes yield higher H_2 fractions than methane reforming. Above 500 °C, increasing methane conversion leads to higher H_2 fractions due to higher hydrogen/carbon-ratios in such mixtures. Between 550 and 850 °C, the respective H_2 fractions remain nearly constant between 70 vol% and 80 vol% at S/C = 4.

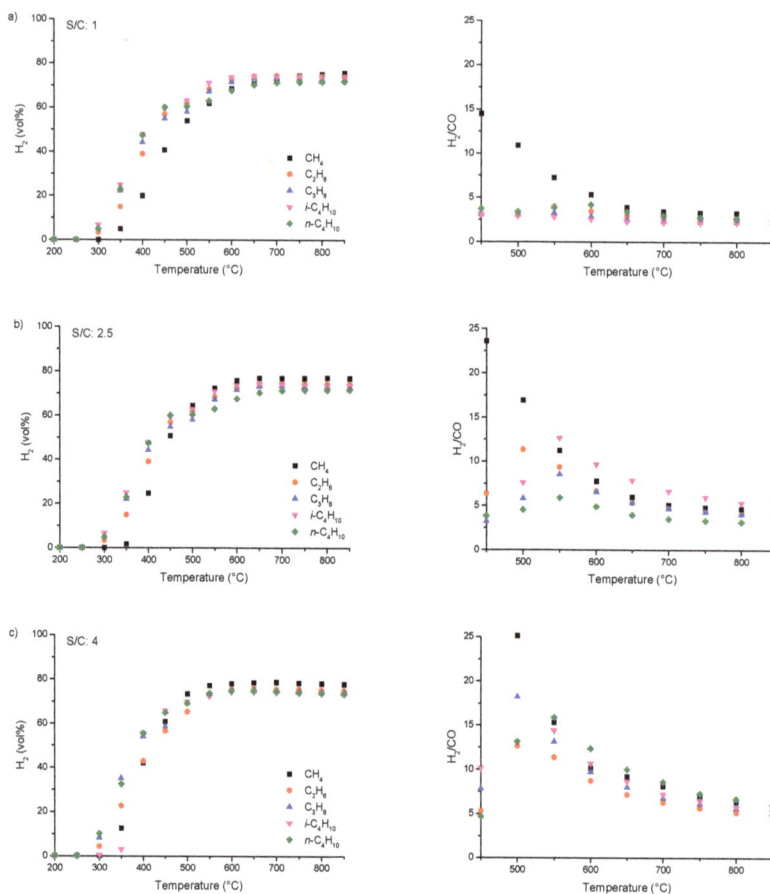

Figure 2. H_2 volumetric gas content (water and inert gas excluded) and H_2/CO ratio in the steam reforming of single alkanes over $Rh1/Al_2O_3$ at GHSV (gas hourly space velocity) of 8000 h^{-1} and different values of temperature and S/C ratios of (**a**) 1, (**b**) 2.5, and (**c**) 4.

The obtained H_2/CO ratios (Figure 2) also depend strongly on the S/C ratio and temperature. The H_2/CO ratios from methane steam reforming decrease with increasing temperature and decreasing S/C ratios. A more difficult behavior was found for C_{2-4} alkanes. For many of those, a maximum H_2/CO ratio was observed around 500–600 °C, whereby the values increase with increasing S/C ratios. This region relates to the aforementioned side reactions, namely the water–gas shift reaction as well as the formation of CH_4. Concerning the latter, the amount of formed methane from C_2–C_4 alkanes dependent on the temperature and S/C ratio is depicted in Figure 3. At S/C = 1, the potential to form methane goes along the following order of the alkanes: Ethane~propane > i-butane > n-butane, reaching values up to ~30 vol% methane, and is most pronounced between temperatures of 500 and 600 °C. For higher S/C ratios another trend is observed according to ethane~propane > i-butane~n-butane. The CH_4 formation can be suppressed by increasing the amount of steam. Different mechanisms can be responsible for CH_4 formation. Schädel et al. [20] observed CH_4 formation during the steam reforming of different higher alkanes over Rh catalysts and concluded that CH_4 is formed faster from higher hydrocarbons than decomposed, due to its lower reactivity at low temperatures. In addition, CO (and CO_2) hydrogenation by formed H_2 or the so-called methanation, also contributes to CH_4

production. Although the hydrogenolysis reaction is discussed as mostly being responsible for CH_4 production [52], methanation can also contribute to a significant loss of CO and consequently very high H_2/CO ratios. This feature could be suppressed to a certain extend by LPG enrichment as discussed later. As a direct consequence of CH_4 formation from the higher alkanes, two temperature regions can be defined for our approach: (i) before CH_4 formation in a low-temperature region, e.g., up to 350 °C and (ii) a high-temperature region when the formed CH_4 is completely reformed starting at 650 °C.

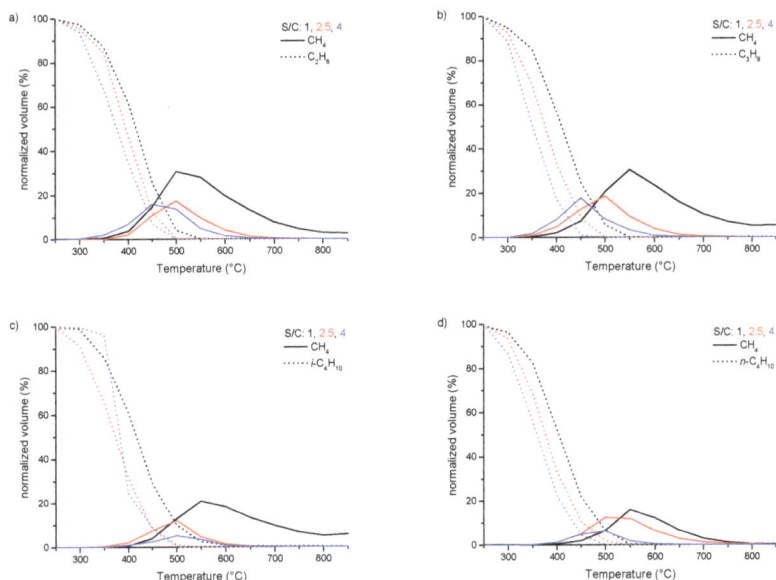

Figure 3. Formation of methane during steam reforming of (**a**) ethane, (**b**) propane, (**c**) *i*-butane and (**d**) *n*-butane over Rh1/Al_2O_3 at various S/C ratios and temperatures at a constant GHSV of ~8000 h^{-1}.

Finally, steam reforming tests of binary alkane mixtures with methane, namely, ethane/methane, propane/methane, and *n*-butane/methane, were performed at 350 °C over Rh1/Al_2O_3 to demonstrate the impact of higher alkanes on the achievable H_2 fractions and H_2/CO ratios (Figures S4 and S5). At this temperature, no methane is being formed from the higher alkanes. In general, an increasing content of higher alkanes causes a rise in the respective H_2 fractions. In the case of ethane/methane mixtures, for instance, the H_2 fractions increase from 5.0 vol% to 8.7 vol% (S/C = 1) when the volume fraction of ethane is increased from 0 to 100%. At the S/C ratio of 2.5, the H_2 fraction can be increased from 8.4 vol% to 15.0 vol% and even from 12.6 vol% to 22.6 vol% when the S/C ratio is further increased to 4. However, in the same way, the H_2/CO ratios decrease from 3.3 to 2.4 (S/C = 1), 4.6 to 3.9 (S/C = 2.5), and 5.9 to 4.8 (S/C = 4). Lower H_2/CO ratios at low methane content are attributed to (i) the reduced hydrogen/carbon-ratios and (ii) high CO selectivity of Rh1/Al_2O_3. The latter has already been demonstrated for the *n*-butane steam reforming in low-temperature range over Rh catalysts [53].

3.3. Enrichment of LPG from Natural Gas Using MFI-Membranes

Compared to presently used methods of recovering heavier hydrocarbons from methane [54], a membrane process offers a low energy consuming alternative. In this work, a wet natural gas (raw gas) was enriched with propane and *n*-butane (permeate gas; enriched with LPG) using a pressure stable MFI zeolite membrane. The compositions of the raw gas (sour gas and inert gas depleted, water-free) and permeate gas are depicted in Scheme 1. The subsequent steam reforming was performed in a separate reactor, since both operations require different working temperatures [55] (Scheme 1).

In the membrane process, pores are blocked for methane due to the preferred adsorption of LPG alkanes in the MFI structure and permeation is induced by a present gradient of the chemical potential, preferably by differences between permeate and feed pressure. At $p_{permeate} = 0.17$ bar and $p_{feed} = 7$ bar (T = 75 °C), propane and *n*-butane enrichment of 37.6 vol% and 22.6 vol% in permeate from initial 18.9 vol% and 4.9 vol%, respectively, was obtained. *i*-butane hardly passes the MFI membrane as it is trapped in the zick-zack-channels of the zeolite [56], which is reflected in a lower amount of this component in the permeate.

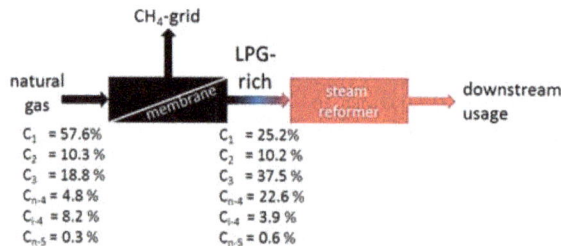

Scheme 1. Two-step process for the generation of hydrogen from wet natural gas, including membrane-based LPG-enrichment and further steam reforming.

3.4. Steam Reforming of Real Natural Gas Mixtures over Rh1/Al₂O₃

The product gas distributions obtained after steam reforming of the two gas mixtures over Rh1/Al$_2$O$_3$ (S/C = 1 and 4, GHSV~8000 h^{-1}) are displayed dependent on temperature in Figure 4. At S/C = 1 the reforming of raw gas starts at 350 °C and full conversion of higher alkanes is achieved at temperatures above 450 °C. With elevation of the reaction temperature, further methane conversion and rising H$_2$ and CO$_x$ concentrations are observed. Nearly full methane conversion is reached at a temperature of 850 °C. At S/C = 4, complete conversion of higher alkanes and methane occurs at lower temperatures compared to the case of S/C = 1, and the respective H$_2$ concentration exhibits 76 vol% at 550 °C and levels off at higher reaction temperatures.

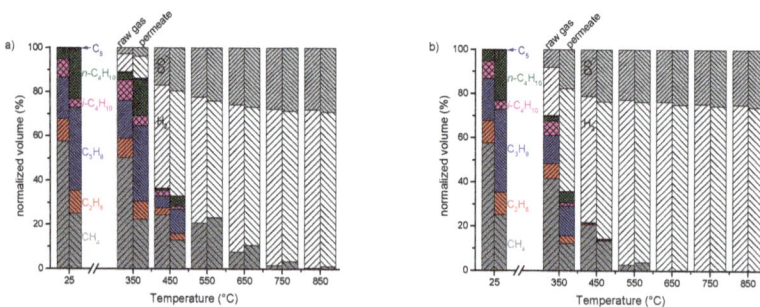

Figure 4. Volumetric gas contents of raw natural gas and LPG-enriched permeate natural gas from membrane pre-separation and after steam reforming over Rh1/Al$_2$O$_3$ at (**a**) S/C ratio of 1 and (**b**) S/C ratio of 4.

The product gas distribution resulting from the steam reforming of the permeate gas differs positively from that of the raw gas at low temperature, due to the higher LPG fraction in the mixture. Compared to the raw gas reforming, twice the concentration of H_2 is achieved from permeate gas reforming at 350 °C and S/C = 4 (H_2 concentration of 46.5%). Under such reaction conditions, $Rh1/Al_2O_3$ performs even better than a commercial catalyst consisting of 5% Rh on alumina (Figure S6). At 550 °C higher alkanes are nearly completely converted.

The related H_2 fractions, H_2/CO ratios, and CO/CO_2 volumetric contents are additionally depicted in Figures S7–S9. For both raw gas as well as permeate gas, the achievable H_2/CO ratios show a maximum between 500 and 550 °C, and gradually decrease with the elevation of temperatures above 550 °C. H_2/CO ratios obtained from permeate gas reforming are slightly lower than those from raw gas transformation. At S/C = 4 the CO/CO_x-ratio (Figure S9) presents the most significant minimum at 500 °C, leading to H_2/CO ratios far above 10, which point to the side reactions which were described earlier.

Time-on-stream performance of $Rh1/Al_2O_3$ in the steam reforming of LPG-enriched permeate natural gas at 350 °C and 850 °C (S/C ratio = 4) is displayed in Figure 5. The activity of $Rh1/Al_2O_3$ increases during the first hour on stream and is nearly stable in the subsequent reaction time.

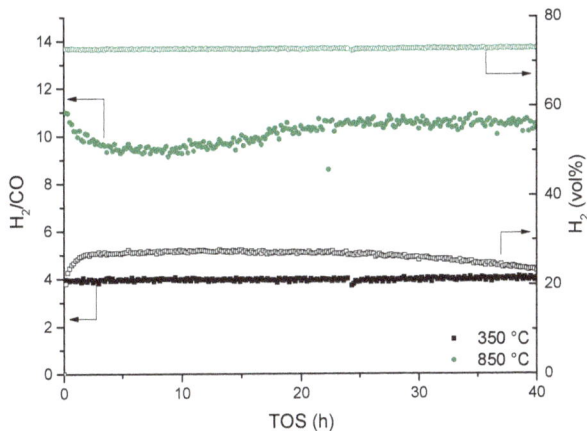

Figure 5. Steam reforming of permeate natural gas over $Rh1/Al_2O_3$ at S/C ratio of 4 and temperatures of 350 and 850 °C.

This behaviour is not in accordance with general considerations regarding catalyst deactivation [57]. Commonly, particle deactivation involves (i) a loss in surface area in the beginning of the reaction, (ii) slowed sintering on stream, and (iii) reaching a stable performance at a certain time. In the present case, the active phase, e.g., formed nanoparticles from Rh single-sites, is being created during the initial stage and seems to possess particularly suitable features for the reaction at low temperatures. At 850 °C, $Rh1/Al_2O_3$ shows no deactivation over 40 h on stream, the hydrogen yield remains at ~73 vol%, and the H_2/CO ratio was constant at 4.

4. Conclusions

A highly active Rh catalyst was formed in situ on Al_2O_3 during the reaction of C_1–C_5 alkanes with steam. The active phase in $Rh1/Al_2O_3$ were Rh nanoparticles in the range 1–3 nm formed from single Rh-atoms. The product spectrum of steam reformed alkanes over $Rh1/Al_2O_3$ strongly depends on S/C ratios and temperatures. With this catalyst, an alkane mixture of an LPG-rich natural gas containing 57.8 vol% CH_4 can be fully reformed at 650 °C at S/C = 4. Below 650 °C, the higher alkanes showed a positive impact on the hydrogen production from binary mixtures at low temperatures but contributed to methanation. Furthermore, maximum H_2/CO ratios (>10) were detected between 350 and 650 °C. Consequently, pre-reforming of LPG-rich natural gas should be performed under very mild conditions, no higher than 350 °C. Therefore, the application of a membrane-based pre-separation represents a feasible concept to obtain (i) easily activatable LPG-enriched natural gas for hydrogen production and (ii) a purified methane fraction for possible injection into the gas grid. Exemplarily, the enrichment of the C_2–C_5 alkane fraction (remaining 25.2 vol% CH_4) in the permeate gas leads to further enhancement of the steam reforming reaction. At low reaction temperatures as low as 350 °C ($S/C = 4$), the volumetric hydrogen content in the product was increased from 21.8% to 46.5 vol% at a reasonable H_2/CO ratio of 4.7.

Supplementary Materials: The following are available online at http://www.mdpi.com/2227-9717/6/12/263/s1: Table S1: Composition of the reaction gas mixtures comprising single alkanes or natural gas alkanes in 75% inert gas; Table S2: Composition of the reaction gas mixtures comprising binary in 75% inert gas; Figure S1. XRD patterns obtained from Rh1/Al_2O_3, the support precursor (γ–AlOOH), and the pure support (mixture of γ-, θ- and δ-alumina phases); Figure S2. (a) CH_4 and (b) H_2 volumetric gas contents (water and inert gas excluded), (c) CO/COx ratio, and d) H_2/CO ratio in the steam reforming of methane over Rhx/Al_2O_3 at various Rh loadings ($S/C = 4$); Figure S3. (a) CH_4 and (b) H_2 volumetric gas contents (water and inert gas excluded), (c) CO/COx ratio, and (d) H_2/CO ratio in the steam reforming of methane over $Rh_{0.5}$/Al_2O_3 at various Rh loadings at various GHSVs ($S/C = 4$); Figure S4. H_2 volumetric gas contents (water and inert gas excluded) in the steam reforming of C_2–C_4 alkanes in mixture with methane over Rh1/Al_2O_3 at 350 °C and S/C of (a) 1, (b) 2.5 and (c) 4; Figure S5. H_2/CO ratios in the steam reforming of C_2–C_4 alkanes in mixture with methane over Rh1/Al_2O_3 at 350 °C and S/C of (a) 1, (b) 2.5 and (c) 4; Figure S6. Volumetric gas contents of permeate gas from membrane pre-separation and after subsequent steam reforming at 350 °C and a S/C ratio of 4 over Rh1/Al_2O_3 and commercial catalysts; Figure S7. H_2 volumetric gas contents (water and inert gas excluded) in the steam reforming of methane over Rh1/Al_2O_3 at various temperatures and S/C ratios of (a) 1,(b) 2.5 and (c) 4; Figure S8. H_2/CO ratios in the product gases in the steam reforming of methane over Rh1/Al_2O_3 at various temperatures and S/C ratios of (a) 1, (b) 2.5 and (c) 4; Figure S9. Volumetric CO- and CO_2-contents in the product gases (water and inert gas excluded) in the steam reforming of methane over Rh1/Al_2O_3 at various temperatures and S/C ratios of (a) 1, (b) 2.5 and (c) 4.

Author Contributions: Conceptualization, D.S. and S.W.; methodology, D.S., R.D. and S.W.; investigation, D.S., R.D., H.A. (Hanan Atia), H.A. (Hadis Amani), M.-M.P., G.G., S.K.; writing—original draft preparation, D.L.; writing—review and editing, D.S., D.L. and S.W.; supervision, S.W.

Funding: This research was partially funded by Sino-German (CSC-DAAD) Postdoc Scholarship Program (57251553) funded by CSC (China Scholarship Council), DAAD (Deutscher Akademischer Austausch Dienst).

Conflicts of Interest: The authors declare no conflict of interest.

References

1. McFarland, E. Unconventional Chemistry for Unconventional Natural Gas. *Science* **2012**, *338*, 340–342. [CrossRef]
2. Tang, P.; Zhu, Q.; Wu, Z.; Ma, D. Methane activation: The past and future. *Energy Environ. Sci.* **2014**, *7*, 2580–2591. [CrossRef]
3. Horn, R.; Schlögl, R. Methane Activation by Heterogeneous Catalysis. *Catal. Lett.* **2015**, *145*, 23–39. [CrossRef]
4. Olivos-Suarez, A.I.; Szécsényi, À.; Hensen, E.J.M.; Ruiz-Martinez, J.; Pidko, E.A.; Gascon, J. Strategies for the Direct Catalytic Valorization of Methane Using Heterogeneous Catalysis: Challenges and Opportunities. *ACS Catal.* **2016**, *6*, 2965–2981. [CrossRef]
5. Schwach, P.; Pan, X.; Bao, X. Direct Conversion of Methane to Value-Added Chemicals over Heterogeneous Catalysts: Challenges and Prospects. *Chem. Rev.* **2017**, *117*, 8497–8520. [CrossRef]
6. Kondratenko, E.V.; Peppel, T.; Seeburg, D.; Kondratenko, V.A.; Kalevaru, N.; Martin, A.; Wohlrab, S. Methane conversion into different hydrocarbons or oxygenates: Current status and future perspectives in catalyst development and reactor operation. *Catal. Sci. Technol.* **2017**, *7*, 366–381. [CrossRef]
7. Ross, J.R.H.; van Keulen, A.N.J.; Hegarty, M.E.S.; Seshan, K. The catalytic conversion of natural gas to useful products. *Catal. Today* **1996**, *30*, 193–199. [CrossRef]
8. Lunsford, J.H. Catalytic conversion of methane to more useful chemicals and fuels: A challenge for the 21st century. *Catal. Today* **2000**, *63*, 165–174. [CrossRef]
9. Aasberg-Petersen, K.; Dybkjær, I.; Ovesen, C.V.; Schjødt, N.C.; Sehested, J.; Thomsen, S.G. Natural gas to synthesis gas—Catalysts and catalytic processes. *J. Nat. Gas Sci. Eng.* **2011**, *3*, 423–459. [CrossRef]

10. Baliban, R.C.; Elia, J.A.; Weekman, V.; Floudas, C.A. Process synthesis of hybrid coal, biomass, and natural gas to liquids via Fischer–Tropsch synthesis, ZSM-5 catalytic conversion, methanol synthesis, methanol-to-gasoline, and methanol-to-olefins/distillate technologies. *Comput. Chem. Eng.* **2012**, *47*, 29–56. [CrossRef]
11. Wender, I. Reactions of synthesis gas. *Fuel Process. Technol.* **1996**, *48*, 189–297. [CrossRef]
12. Song, X.P.; Guo, Z.C. Technologies for direct production of flexible H_2/CO synthesis gas. *Energy Convers. Manag.* **2006**, *47*, 560–569. [CrossRef]
13. Rostrup-Nielsen, J.R. New aspects of syngas production and use. *Catal. Today* **2000**, *63*, 159–164. [CrossRef]
14. Joensen, F.; Rostrup-Nielsen, J.R. Conversion of hydrocarbons and alcohols for fuel cells. *J. Power Sources* **2002**, *105*, 195–201. [CrossRef]
15. Avcı, A.K.; Trimmb, D.L.; Aksoylu, A.E.; Önsan, Z.I. Hydrogen production by steam reforming of *n*-butane over supported Ni and Pt-Ni catalysts. *Appl. Catal. A* **2004**, *258*, 235–240. [CrossRef]
16. Hotz, N.; Stutz, M.J.; Loher, S.; Stark, W.J.; Poulikakos, D. Syngas production from butane using a flame-made $Rh/Ce_{0.5}Zr_{0.5}O_2$ catalyst. *Appl. Catal. B* **2007**, *73*, 336–344. [CrossRef]
17. Von Rickenbach, J.; Nabavi, M.; Zinovik, I.; Hotz, N.; Poulikakos, D. A detailed surface reaction model for syngas production from butane over Rhodium catalyst. *Int. J. Hydrogen Energy* **2011**, *36*, 12238–12248. [CrossRef]
18. Pasel, J.; Wohlrab, S.; Kreft, S.; Rotov, M.; Löhken, K.; Peters, R.; Stolten, D. Routes for deactivation of different autothermal reforming catalysts. *J. Power Sources* **2016**, *325*, 51–63. [CrossRef]
19. Auprêtre, F.; Descorme, C.; Duprez, D. Bio-ethanol catalytic steam reforming over supported metal catalysts. *Catal. Commun.* **2002**, *3*, 263–267. [CrossRef]
20. Schädel, B.T.; Duisberg, M.; Deutschmann, O. Steam reforming of methane, ethane, propane, butane, and natural gas over a rhodium-based catalyst. *Catal. Today* **2009**, *142*, 42–51. [CrossRef]
21. Karakaya, C.; Maier, L.; Deutschmann, O. Surface Reaction Kinetics of the Oxidation and Reforming of CH_4 over Rh/Al_2O_3 Catalysts. *Int. J. Chem. Kinet.* **2016**, *48*, 144–160. [CrossRef]
22. Angeli, S.D.; Monteleone, G.; Giaconia, A.; Lemonidou, A.A. State-of-the-art catalysts for CH_4 steam reforming at low temperature. *Int. J. Hydrogen Energy* **2014**, *39*, 1979–1997. [CrossRef]
23. Supat, K.; Chavadej, S.; Lobban, L.L.; Mallinson, R.G. Combined Steam Reforming and Partial Oxidation of Methane to Synthesis Gas under Electrical Discharge. *Ind. Eng. Chem. Res.* **2003**, *42*, 1654–1661. [CrossRef]
24. Bakker, W.J.W.; Kapteijn, F.; Poppe, J.; Moulijn, J.A. Permeation characteristics of a metal-supported silicalite-1 zeolite membrane. *J. Membr. Sci.* **1996**, *117*, 57–78. [CrossRef]
25. Krishna, R.; Vandenbroeke, L.J.P. The Maxwell-Stefan Description of Mass-Transport across Zeolite Membranes. *Chem. Eng. J. Biochem. Eng. J.* **1995**, *57*, 155–162. [CrossRef]
26. Vroon, Z.A.E.P.; Keizer, K.; Gilde, M.J.; Verweij, H.; Burggraaf, A.J. Transport properties of alkanes through ceramic thin zeolite MFI membranes. *J. Membr. Sci.* **1996**, *113*, 293–300. [CrossRef]
27. Keizer, K.; Burggraaf, A.J.; Vroon, Z.A.E.P.; Verweij, H. Two component permeation through thin zeolite MFI membranes. *J. Membr. Sci.* **1998**, *147*, 159–172. [CrossRef]
28. Van de Graaf, J.M.; Kapteijn, F.; Moulijn, J.A. Methodological and operational aspects of permeation measurements on silicalite-1 membranes. *J. Membr. Sci.* **1998**, *144*, 87–104. [CrossRef]
29. Gump, C.J.; Lin, X.; Falconer, J.L.; Noble, R.D. Experimental configuration and adsorption effects on the permeation of C_4 isomers through ZSM-5 zeolite membranes. *J. Membr. Sci.* **2000**, *173*, 35–52. [CrossRef]
30. Arruebo, M.; Coronas, J.; Menendez, M.; Santamaria, J. Separation of hydrocarbons from natural gas using silicalite membranes. *Sep. Purif. Technol.* **2001**, *25*, 275–286. [CrossRef]
31. Wohlrab, S.; Meyer, T.; Stöhr, M.; Hecker, C.; Lubenau, U.; Oßmann, A. On the performance of customized MFI membranes for the separation of *n*-butane from methane. *J. Membr. Sci.* **2011**, *369*, 96–104. [CrossRef]
32. Dragomirova, R.; Stohr, M.; Hecker, C.; Lubenau, U.; Paschek, D.; Wohlrab, S. Desorption-controlled separation of natural gas alkanes by zeolite membranes. *RSC Adv.* **2014**, *4*, 59831–59834. [CrossRef]
33. Neubauer, K.; Dragomirova, R.; Stöhr, M.; Mothes, R.; Lubenau, U.; Paschek, D.; Wohlrab, S. Combination of membrane separation and gas condensation for advanced natural gas conditioning. *J. Membr. Sci.* **2014**, *453*, 100–107. [CrossRef]
34. Dragomirova, R.; Jorabchi, M.N.; Paschek, D.; Wohlrab, S. Operational Criteria for the Separation of Alkanes by Zeolite Membranes. *Chem. Ing. Tech.* **2017**, *89*, 926–934. [CrossRef]

35. Angeli, S.D.; Pilitsis, F.G.; Lemonidou, A.A. Methane steam reforming at low temperature: Effect of light alkanes' presence on coke formation. *Catal. Today* **2015**, *242*, 119–128. [CrossRef]
36. Guo, X.; Fang, G.; Li, G.; Ma, H.; Fan, H.; Yu, L.; Ma, C.; Wu, X.; Deng, D.; Wei, M.; et al. Direct, Nonoxidative Conversion of Methane to Ethylene, Aromatics, and Hydrogen. *Science* **2014**, *344*, 616–619. [CrossRef] [PubMed]
37. Huang, W.; Zhang, S.; Tang, Y.; Li, Y.; Nguyen, L.; Li, Y.; Shan, J.; Xiao, D.; Gagne, R.; Frenkel, A.I.; et al. Low-Temperature Transformation of Methane to Methanol on Pd_1O_4 Single Sites Anchored on the Internal Surface of Microporous Silicate. *Angew. Chem.* **2016**, *128*, 13639–13643. [CrossRef]
38. Vanelderen, P.; Vancauwenbergh, J.; Tsai, M.-L.; Hadt, R.G.; Solomon, E.I.; Schoonheydt, R.A.; Sels, B.F. Spectroscopy and Redox Chemistry of Copper in Mordenite. *Chem. Phys. Chem.* **2014**, *15*, 91–99. [CrossRef]
39. Grundner, S.; Markovits, M.A.C.; Li, G.; Tromp, M.; Pidko, E.A.; Hensen, E.J.M.; Jentys, A.; Sanchez-Sanchez, M.; Lercher, J.A. Single-site trinuclear copper oxygen clusters in mordenite for selective conversion of methane to methanol. *Nat. Commun.* **2015**, *6*, 7546. [CrossRef]
40. Grundner, S.; Luo, W.; Sanchez-Sanchez, M.; Lercher, J.A. Synthesis of single-site copper catalysts for methane partial oxidation. *Chem. Commun.* **2016**, *52*, 2553–2556. [CrossRef]
41. Kulkarni, A.R.; Zhao, Z.-J.; Siahrostami, S.; Nørskov, J.K.; Studt, F. Monocopper Active Site for Partial Methane Oxidation in Cu-Exchanged 8MR Zeolites. *ACS Catal.* **2016**, *6*, 6531–6536. [CrossRef]
42. Wallis, P.; Schonborn, E.; Kalevaru, V.N.; Martin, A.; Wohlrab, S. Enhanced formaldehyde selectivity in catalytic methane oxidation by vanadia on Ti-doped SBA-15. *RSC Adv.* **2015**, *5*, 69509–69513. [CrossRef]
43. Wallis, P.; Wohlrab, S.; Kalevaru, V.N.; Frank, M.; Martin, A. Impact of support pore structure and morphology on catalyst performance of VO_x/SBA-15 for selective methane oxidation. *Catal. Today* **2016**, *278*, 120–126. [CrossRef]
44. Dang, T.T.H.; Seeburg, D.; Radnik, J.; Kreyenschulte, C.; Atia, H.; Vu, T.T.H.; Wohlrab, S. Influence of V-sources on the catalytic performance of VMCM-41 in the selective oxidation of methane to formaldehyde. *Catal. Commun.* **2018**, *103*, 56–59. [CrossRef]
45. Duarte, R.B.; Krumeich, F.; van Bokhoven, J.A. Structure, Activity, and Stability of Atomically Dispersed Rh in Methane Steam Reforming. *ACS Catal.* **2014**, *4*, 1279–1286. [CrossRef]
46. Lian, J.; Ma, J.; Duan, X.; Kim, T.; Li, H.; Zheng, W. One-step ionothermal synthesis of g-Al_2O_3 mesoporous nanoflakes at low temperature. *Chem. Commun.* **2010**, *46*, 2650–2652. [CrossRef]
47. Fargeot, D.; Mercurio, D.; Dauger, A. Structural characterization of alumina metastable phases in plasma sprayed deposits. *Mater. Chem. Phys.* **1990**, *24*, 299–314. [CrossRef]
48. Husson, E.; Repelin, Y. Structural studies of transition aluminas. Theta alumina. *Eur. J. Solid State Inorg. Chem.* **1996**, *33*, 1223–1231.
49. Weng, W.Z.; Pei, X.Q.; Li, H.M.; Luo, C.R.; Liu, Y.; Lin, H.Q.; Huang, C.J.; Wan, H.L. Effects of calcination temperatures on the catalytic performance of Rh/Al_2O_3 for methane partial oxidation to synthesis gas. *Catal. Today* **2006**, *117*, 53–61. [CrossRef]
50. Yao, H.C.; Japar, S.; Shelef, M. Surface interactions in the system $RhAl_2O_3$. *J. Catal.* **1977**, *50*, 407–418. [CrossRef]
51. Guan, H.L.; Lin, J.; Qiao, B.T.; Miao, S.; Wang, A.Q.; Wang, X.D.; Zhang, T. Enhanced performance of Rh_1/TiO_2 catalyst without methanation in water-gas shift reaction. *AIChE J.* **2017**, *63*, 2081–2088. [CrossRef]
52. Graf, P.O.; Mojet, B.L.; van Ommen, J.G.; Lefferts, L. Comparative study of steam reforming of methane, ethane and ethylene on Pt, Rh and Pd supported on yttrium-stabilized zirconia. *Appl. Catal. A* **2007**, *332*, 310–317. [CrossRef]
53. Igarashi, A.; Ohtaka, T.; Motoki, S. Low-temperature steam reforming of *n*-butane over Rh and Ru catalysts supported on ZrO_2. *Catal. Lett.* **1992**, *13*, 189–194. [CrossRef]
54. Alcheikhhamdon, Y.; Hoorfar, M. Natural gas quality enhancement: A review of the conventional treatment processes, and the industrial challenges facing emerging technologies. *J. Nat. Gas Sci. Eng.* **2016**, *34*, 689–701. [CrossRef]
55. Dragomirova, R.; Wohlrab, S. Zeolite Membranes in Catalysis—From Separate Units to Particle Coatings. *Catalysts* **2015**, *5*, 2161–2222. [CrossRef]

56. Krishna, R.; Paschek, D. Separation of hydrocarbon mixtures using zeolite membranes: A modelling approach combining molecular simulations with the Maxwell–Stefan theory. *Sep. Purif. Technol.* **2000**, *21*, 111–136. [CrossRef]

57. Hansen, T.W.; DeLaRiva, A.T.; Challa, S.R.; Datye, A.K. Sintering of Catalytic Nanoparticles: Particle Migration or Ostwald Ripening? *Acc. Chem. Res.* **2013**, *46*, 1720–1730. [CrossRef] [PubMed]

processes

MDPI

Review

Approaches to Suppress CO$_2$-Induced Plasticization of Polyimide Membranes in Gas Separation Applications

Moli Zhang [1], Liming Deng [1], Dongxiao Xiang [1], Bing Cao [1], Seyed Saeid Hosseini [2,*] and Pei Li [1,*]

[1] College of Materials Science and Engineering, Beijing University of Chemical Technology, Beijing 100029, China; mollyzhang6@163.com (M.Z.); dlm0517@163.com (L.D.); xdx_xiang@163.com (D.X.); bcao@mail.buct.edu.cn (B.C.)

[2] Membrane Science and Technology Research Group, Department of Chemical Engineering, Tarbiat Modares University, Tehran 14115, Iran

* Correspondence: saeid.hosseini@modares.ac.ir (S.S.H.); lipei@mail.buct.edu.cn (P.L.); Tel.: 86-10-6441-3857 (S.S.H. & P.L.)

Received: 9 November 2018; Accepted: 15 January 2019; Published: 21 January 2019

Abstract: Polyimides with excellent physicochemical properties have aroused a great deal of interest as gas separation membranes; however, the severe performance decay due to CO$_2$-induced plasticization remains a challenge. Fortunately, in recent years, advanced plasticization-resistant membranes of great commercial and environmental relevance have been developed. In this review, we investigate the mechanism of plasticization due to CO$_2$ permeation, introduce effective methods to suppress CO$_2$-induced plasticization, propose evaluation criteria to assess the reduced plasticization performance, and clarify typical methods used for designing anti-plasticization membranes.

Keywords: polyimide; plasticization; membrane; gas separation; carbon dioxide

1. Introduction

1.1. Natural Gas Process

Narrowly speaking, natural gas—a term in the energy field—is referred to as a mixture of hydrocarbons and non-hydrocarbon gases that naturally exist underground. The typical composition of raw natural gas is shown in Table 1, based on samples from eight different locations worldwide [1,2]. Methane is the main component of raw natural gas—one of the raw materials. It is consumed during combustion for heat release or chemical production. When producing the same amount of energy, CO$_2$ emissions by natural gas are ~26% less than oil and coal [3]. Therefore, natural gas is regarded as a source for the cleanest and lowest carbon emitting fossil fuel in the world [4].

Table 1. Typical composition of raw natural gas [1,2].

Component	Composition Range (mol%)
CH$_4$	29.98–90.12
C$_2$H$_6$	0.55–14.22
C$_3$H$_8$	0.23–12.54
C$_4$H$_{10}$	0.14–8.12
C$_{5+}$	0.037–3.0
N$_2$	0.21–26.10
H$_2$S	0.0–3.3
CO$_2$	0.06–42.66
He	0.0–1.8

According to the BP (British Petroleum) Statistical Review of World Energy (2018), 2017 witnessed the fastest rates of rapid increases in natural gas consumption (3.0%; 96 billion cubic meters) and production (4.0%; 131 billion cubic meters) [5]. Furthermore, Figure 1 clearly illustrates that in 2017, the share of natural gas in global primary energy consumption by fuel consistently increased to 23.4% [6]. Undoubtedly, strong natural gas growth can be expected in the near future. This is supported by an increase in the levels of industrialization and power demand, continued coal-to-gas switching, and the growing availability of low-cost supplies [7]. In conclusion, the rapid increase in natural gas consumption has driven the necessity for natural gas processing to meet pipeline requirements.

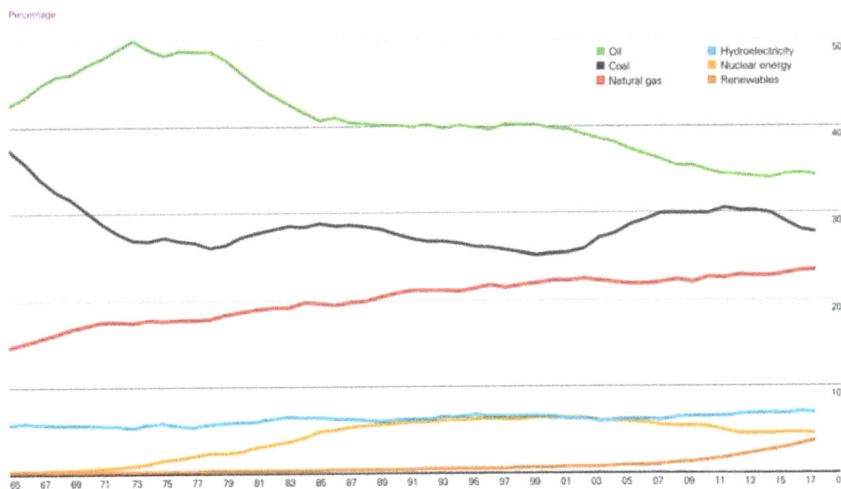

Figure 1. Shares of global primary energy consumption by fuel [6].

It has been widely recognized that variations in raw natural gas consumption are considerably region-dependent [8], which is supported by the worldwide composition data of raw natural gas (shown in Table 2). Nevertheless, nations tightly control the composition of natural gas prior to entry into the industrial pipeline grids. Take the U.S. for example, where Table 3 shows a whole set of criteria defining the upper limits of common impurities. Therefore, to remove the undesired components and ultimately meet the requirements of target composition specifications, preconditioning is required for all raw natural gases. Furthermore, the greenhouse effect generated by CH_4 must also be taken into consideration. Generally speaking, if CH_4 is released during the purification and shipment of natural gas, it may cause a severe greenhouse effect.

Table 2. Composition of natural gas reservoirs (volume percent) in some regions of the world [9].

Component	Groningen (Netherlands)	Laeq (France)	Uch (Pakistan)	Uthmaniyah (Saudi Arabia)	Ardjuna (Indonesia)
CH_4	81.3	69	27.3	55.5	65.7
C_2H_6	2.9	3	0.7	18	8.5
C_3H_8	0.4	0.9	0.3	9.8	14.5
C_4H_{10}	0.1	0.5	0.3	4.5	5.1
C_{5+}	0.1	0.5	-	1.6	0.8
N_2	14.3	1.5	25.2	0.2	1.3
H_2S	-	15.3	-	1.5	-
CO_2	0.9	9.3	46.2	8.9	4.1

Table 3. Composition specifications for natural gas delivery to the U.S. national pipeline grid [10].

Component	Specification
CO_2	<2%
H_2O	<120 ppm
H_2S	<4 ppm
C_3+ content	950–1050 Btu/scf; dew point: <−20 °C
total inert gases (N_2, He)	<4%

Notes: ppm, parts per million; and scf, standard cubic feet.

CO_2 can reduce the calorific value of natural gas and result in global warming. CO_2 and H_2S contribute to the corrosion of natural gas pipelines; therefore, numerous studies have been conducted to remove such acidic gases [11,12]. The most extensively used methods include cryogenic distillation [13,14], amine absorption [15], pressure swing adsorption (PSA) [16,17], and membrane separation technologies. Of these, cryogenic distillation requires a large amount of energy to condense permanent gases such as CO_2 [16]. Therefore, a hybrid system combining amine absorption with membrane separation is preferred due to its high energy efficiency [18].

The amine absorption process has displayed favorable performance in extracting CO_2 from the CO_2/CH_4 gas mixtures. However, the high capital and operating costs remain a problem. Membrane technology is a good alternative for gas separation, ascribed to its low capital investment and ease of operation. It also exhibits the efficiency of the membrane units, and advantages of design flexibility and compactness [19]. This technology is expected to be the superior gas separation method when CO_2/CH_4 selectivity is increased to 40 [20]. However, the presence of CO_2 and/or other highly sorbing components can result in membrane plasticization and swelling.

1.2. Membrane Materials Used for Gas Separation

Progress in membrane technology has been largely dependent on high-performance membrane materials. Inorganic membranes are not discussed in this paper due to their expensive module fabrication and limited industrial applications, even though considerable progress has been achieved in this field. This review focuses on polymer membranes. To date, a variety of promising polymer membranes, including cellulose acetate (CA) [21,22], polycarbonate (PC) [23–27], polysulfone (PSF) [28–30], polymethyl methacrylate (PMMA) [31–33], polyimide (PI), and polymer of intrinsic microporosity (PIMs) [34–37], have been developed to meet the demands of the gas separation industry.

There are many promising polymer membrane materials, yet the gas separation market is only dominated by a few. This can be explained by the different operating conditions between research laboratories and industrial applications. Current commercial membranes are used under high pressure, and high concentrations of plasticizing impurities such as water, BTEX aromatics (The term "BTEX aromatics" refers to benzene, toluene, ethylbenzene, and xylene), and other heavy hydrocarbons [10]. Performance stability, chemical resistance, and low capital cost should also be incorporated into the criteria of producing industrially available membranes [20].

In particular, cellulose acetate, polyimides and perfluoropolymers are commercially available polymers used for fabricating CO_2 removal membranes [10]. The performance of CA-based membranes, and a range of polyimide membranes, have been summarized in detail by Scholes et al. [38] and will not be discussed in this review.

Throughout this paper, the term polyimide represents aromatic polyimide. Aromatic polyimides contain the imide group—a constitutional unit in the polymer backbone (illustrated in Figure 2). It is a linear and heterocyclic aromatic polymer, widely used to synthesize films, fibers, molding powders, coatings, and composites. Its outstanding heat resistance, relatively high resistance to chemical solvents, strong mechanical strength, and high selectivity in major gas pairs (such as CO_2/CH_4 and O_2/N_2), make polyimide an attractive candidate amid many polymeric membranes. Since the manufacture of commercial polyimides membranes, many have been developed for gas separation, and used in the separation process by the DuPont (USA) and Ube (Japan) industries [39].

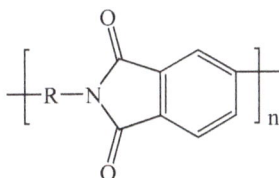

Figure 2. Chemical structure of polyimides.

Strong plasticization resistance is required in several gas separation applications such as CO_2/CH_4, propylene/propane, and butadiene/butane separations, which can be attributed to the presence of highly sorbing components. The CO_2/CH_4 separation containing CO_2 sorbates, is more frequently encountered in industrial applications such as high-pressure natural gas sweetening and enhanced oil recovery (EOR). Hence, there is a growing concern over the development of polyimide membranes with a reduced plasticization effect and minimal loss of selectivity. Numerous studies over the past decade have been carried out to investigate the mechanisms and suppression methods of plasticization [40].

1.3. Plasticization in Polyimides

Plasticization is usually defined as the increased segmentation mobility of polymer chains [41]. CO_2 permeability will rise with increased feed pressure after a critical point, which is generally regarded as the mark of being plasticized. Simultaneously, an undesired loss of gas pair selectivity is often observed. Under the high pressure of CO_2, the polymer matrix swells, subsequently resulting in possible free volume changes and inter-chain spacing. Consequently, the mobility of the polymer segments increases, thus weakening the size-sieving ability of polyimide membranes. The plasticization phenomena can be demonstrated by the obvious loss of selectivity. Therefore, developing membrane materials that can maintain gas selectivity in the presence of aggressive feed streams is of utmost importance.

Some intriguing methods for prominently improving the plasticization resistance of polyimide membranes are illustrated in this article. Among them, crosslinking, such as thermal crosslink, diamine crosslink, diol crosslink, semi-interpenetrating networks, ultraviolet (UV) crosslink, and hydrogen bonding, is a practical and widely used method described in the literature. Furthermore, mixed matrix membranes (MMMs) or polymer blending have also been proposed to resist plasticization.

This review mainly focuses on approaches to suppress the notorious CO_2-induced plasticization of polyimide membranes. Furthermore, the mechanism, attractive advantages, and unavoidable weakness of each method are specified in the corresponding sections and illustrated in some cited literature. Some recent achievements and representative work are discussed in detail. It is fundamental to understand plasticization mechanisms and principles in order to design polyimide membranes with less plasticization effect. In the end, a promising prospect is provided to instruct further research and propose some constructive suggestions.

2. Plasticization Mechanism

Plasticization often takes place in the presence of highly condensable gases (such as CO_2), especially under aggressive conditions (like high pressure and low temperature). There have been numerous studies on plasticization during the separation process of mixed gases such as CO_2/CH_4, propylene/propane [42–46], and ethylene/ethane [47,48]. This review mainly focuses on the CO_2-induced plasticization phenomenon, which usually happens during the natural gas sweetening process where CO_2 acts as a plasticizer.

Plasticization can be observed through multiple measures. The most commonly used approach is monitoring CO_2 permeability while increasing CO_2 feed pressure to observe the changes in polymer chain segmental mobility. During the gas permeation process, CO_2 permeability coefficients of most glassy polyimides will descend first due to the saturation of Langmuir sites. The polymer matrix will swell when CO_2 concentrations reach a critical point. Then, an obvious acceleration in gas permeability is observed with increasing CO_2 partial pressure [49] (Figure 3 [41]). Feed pressure at the minimum point of a CO_2 permeability curve is commonly defined as the plasticization pressure, which is an important indicator of plasticization. Simultaneously, a loss in gas pair selectivity occurs.

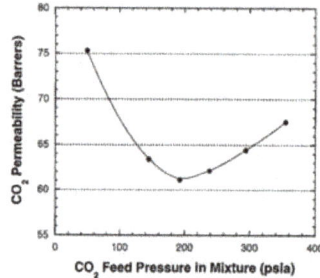

Figure 3. Mixed gas separation characteristics for 6FDA-6FpDA using 50:50 CO_2/CH_4 mixtures at 35 °C [41]. Notes: 6FDA, 4,4'-(hexafluoroisopropylidene)diphthalic anhydride; and 6FpDA, 4,4'-(9-fluorenylidene) bis (2-methyl-6-isopropylaniline).

Changes in the physical properties of membranes such as glass transition temperature (T_g), membrane thickness, and refractive index, can also indicate plasticization [50]. However, the occurrence of one sign is not necessarily accompanied by changes in other chemical or physical properties. In the case of polyethersulfone (PES), for example, although the decreased CO_2/CH_4 selectivity indicates that the polymer is plasticized, CO_2 permeability at 25 °C decreases by 54% as the CO_2 feed pressure increases from 0 to 27 atm [51]. This indicates that polymer matrix plasticization will not necessarily lead to an increase in permeability of all gases.

The polymer matrix can absorb a certain amount of CO_2 at modest pressure, leading to a significant reduction in the glass transition temperature [52]. In the range of low pressure, decrease in the permeability coefficient with increasing partial pressure is caused by the diminution of the solubility coefficient. In the range of high pressure, the positive correlation between the permeability coefficient and partial pressure is typical plasticizing behavior, which cannot be described by the dual mode mobility model [53].

After a systematic study of the plasticizing phenomena caused by CO_2 in 11 different glassy polymers, Bos et al. believed that the plasticization of a glass polymer could be defined as a function of the increase in the CO_2 permeability coefficient to the pressure of the feed gas, and the minimum pressure required to increase the permeability coefficient was defined as the plasticizing pressure [54].

It is generally believed that the diffusion coefficient increases when CO_2 concentrations exceed the critical value required by plasticization (namely, when the feed pressure exceeds the plasticizing pressure), subsequently leading to the increment of the permeability coefficient [55]. For glass polymers, the plasticization phenomenon occurs when polymer CO_2 concentrations reach critical levels. However, the critical pressure is polymer-dependent. Plasticization pressures of different polyimide membranes reported in the literature are listed in Table 4.

Plasticization induces chain relaxation and results in increased diffusivity of all gases in the feed mixture. Therefore, the reduction in membrane selectivity can be observed. A plasticization problem is that it can be inhibited by limiting the flexibility of polymer segments. There are many approaches that are currently being used to inhibit polymer segment flexibility, which are critically discussed in the subsequent sections.

Table 4. Different plasticization pressures reported in the literature.

CL Method	SU-1 or BC-1	SU-2 or BC-2	SU-3 or BC-3	Crosslinker or Fillers	Plasticization Pressure (atm)	Ref.
Decarboxylation CL	6FDA	DAM	DABA	-	~34.02	[56]
Decarboxylation CL	6FDA	DAM	DABA	-	-	[57]
Decarboxylation CL	6FDA	CADA1/CADA2	BTDA/DSDA	-	30	[58]
Decarboxylation CL	6FDA	MPP	PP	-	30.62	[59]
Decarboxylation CL	6FDA	DAT	DATCA	-	30	[60]
Heat at 350 °C	Matrimid	-	-	-	>39.48	[61]
TR	-	-	-	-	20	[62]
TR	6FDA	HAB	-	-	(partial pressure) >25 (fugacity)	[63]
Diamino	6FDA	durene	-	EDA	48.99	[64]
Diamino	6FDA	durene	-	CHBA	48.99	[65]
Diamino	6FDA	durene	-	PAMAM/DAB	30	[66]
Diamino	6FDA	Matrimid	-	p-xylenediamine	>32	[67]
Diol	6FDA	mPD	DABA	EG	35	[19]
Diol	6FDA	6FpDA	DABA	EG	-	[41]
Diol	6FDA	DAM	DABA	CHDM	-	[68]
Diol	6FDA	DAM	DABA	EG/CHDM	40	[69]
Semi-IPN	Matrimid	-	-	Thermid FA-700	>49.35	[70]
Semi-IPN	Matrimid	-	-	DCFT	>30	[71]
Ionic	6FDA	NDA	TMPDA	azide	30	[72]
MMMs	6FDA	durene	DABA	N,N'-dimethylpiperazine	25	[73]
MMMs	w-PS	durene	-	ZIF-8	30	[74]
MMMs	6FDA	durene	-	ZIF-8	-	[75]
MMMs	6FDA	durene	-	zeolite T MIL-53(Al)	19.74	[76]
MMMs	Matrimid	-	-	ZIF-8 Cu_3BTC_2	-	[77]
MMMs	Matrimid 6FDA	DAT durene	DAM	Ni_2(dobdc)	-	[78]
Blending	Matrimid	PSF	-	-	35	[79]
Blending	Matrimid	P84	-	-	14.80	[80]
Blending	Matrimid	cPIM-1	-	-	20	[81]
Blending	Torlon	cPIM-1	-	-	30	[82]

In this table: CL, crosslink; SU, structure unit; BC, blending component; TR, thermal rearrangement; DAM, 2,4,6-trimethyl-1,3-phenylenediamine; DABA, 3,5-diaminobenzoic acid; CADA1, carboxylic acid-containing diamine with -CF$_3$ group; CADA2, carboxylic acid-containing diamine with -H group; BTDA, 3,3',4,4'-benzophenonetetracarboxylic dianhydride; DSDA, 3,3',4,4'-diphenylsulfonetetracarboxylic dianhydride; MPP, 3,3-bis[4-(4-amino-3-methylphenoxy)phenyl]phthalide; PP, 3,3-bis[4-(4-aminophenoxy)phenyl]phthalide; DAT, 2,6-diaminotriptycene; DATCA, 2,6-diaminotriptycene-14-carboxylic acid; HAB, 3,3'-dihydroxy-4,4'-diamino-biphenyl; EDA, 1,2-ethylene diamine; CHBA, 1,3-cyclohexanebis(methylamine); PAMAM, polyamidoamine; DAB, diaminobutane; mPD, m-phenylene diamine; EG, ethylene glycol; CHDM, cyclohexane-1,4-diyldimethanol; DCFT, phenolphthalein dicyanate; NDA, 1,5-naphthalenediamine; TMPDA, 2,3,5,6-tetramethyl-1,4-phenylenediamine; ZIF-8, zeolite imidazole framework-8; w-PS, waste polystyrene; MIL-53(Al), aluminum terephthalate; Cu_3BTC_2, copper benzene-1,3,5-tricarboxylate; PSF, polysulfone; and cPIM-1, carboxylated polymer of intrinsic microporosity-1.

3. Methods to Reduce CO_2-Induced Plasticization

Extensive studies have been carried out and have successfully achieved suppression of polyimide membrane plasticization. Many criteria are available to classify these approaches, which are considerably involved in the formation of crosslink networks. Some crosslink networks are formed firmly by physical bonds such as hydrogen bonds, while others are formed by chemical bonds. A wide range of methods can achieve chemical bonds of polymer chains such as thermal treatment, chemical crosslinking using diamine or diol, formation of semi-interpenetrating networks, and UV crosslinking. Crosslinking is a practical and widely used approach to inhibit the plasticization of membrane materials. Some typical technologies of crosslinking are introduced in the following sections.

3.1. Thermal-Induced Crosslink

Rapidly quenching polyimide polymers can significantly suppress undesirable plasticization and introduce excess free volume into the polyimide matrix, thereby leading to greater stabilization in typical solvents. To illustrate, aromatic carboxylic acids along backbones can be dehydrated to form anhydride, which is subsequently decomposed into aryl radicals. A hydrogen atom will be abstracted from the backbone to bond covalently with the aryl radical, which is likely to occur in specific reactive sites.

Steric hindrance of both functional and pendant groups must be taken into consideration while designing the crosslink points, since some structures (such as –CF_3) are too large for steric reactivity. Fine-tuning crosslink networks is not possible using this method, because the sites and extent of the crosslink are not measurable or quantifiable as a result of the non-definitive and insufficiently understood mechanism.

Matrimid 5218 has been utilized as the model polyimide to investigate the plasticization phenomenon [61]. After thermal treatment at 350 °C, the 15 and 30 min heat-treated Matrimid films did not plasticize in the whole range of CO_2 feed pressure (up to about 40 bar, i.e., 39.48 atm). Noticeably, CO_2 permeability gradually approached a limiting constant value with increasing feed pressure. All thermally-treated films became insoluble. Upon pressure elevation, the 15 and 30 min heat-treated films showed up to 11% and 30% higher selectivity, respectively, compared with the untreated membranes. This can be attributed to the densification of the polymer matrix.

3.1.1. Decarboxylation-Induced Crosslinking

A novel polyimide 6FDA-DAM-DABA (2:1) has been successfully synthesized [56]. The sample quenched from above T_g exhibited enhanced plasticization resistance, where plasticization pressure is about 34.02 atm (500 psi). Quenching polyimide brings a large free volume into the polyimide matrix and endows it with enhanced plasticization resistance. Through decarboxylation at high temperature (slightly below the T_g), thermally crosslinked polyimide linked by covalent bonds has more free volume compared to diol or diamine crosslinking agents. Thermal gravimetric analyzer (TGA) and C^{13}-NMR (nuclear magnetic resonance) measurements prove that the free acid polymer undergoes decarboxylation under such conditions. Then, the residual aryl radicals react through some generated active sites to form the crosslinked structure. Figure 4 shows possible crosslinking sites through diamines in the free acid polymer.

The 6FDA-DAM:DABA(3:2) polyimide was prepared by tuning the proportion of monomers in order to study the effect of thermal treatment on plasticization resistance [57]. The results revealed that thermally crosslinked membranes showed no signs of plasticization up to 47.63 atm (700 psi) for pure CO_2 gas, or 1000 psi for 50/50 CO_2/CH_4 mixed-gas separation.

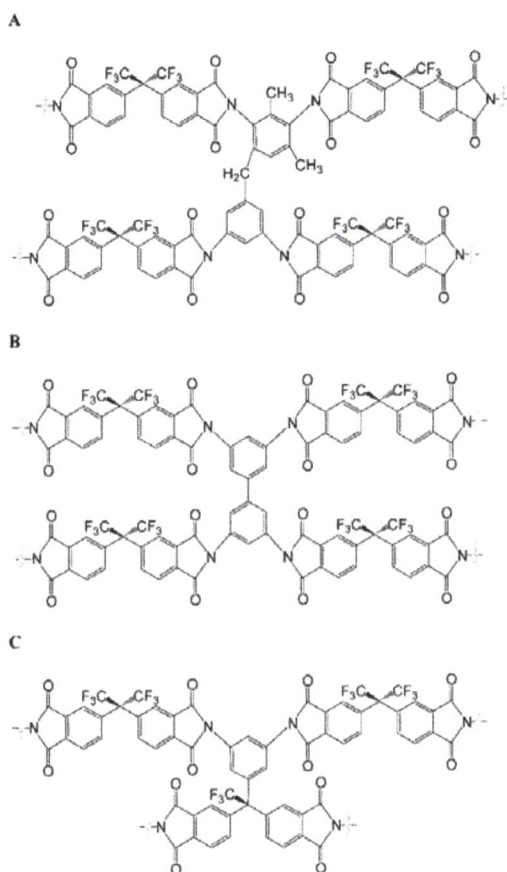

Figure 4. Possible crosslinking sites through diamines in the free acid polymer: (**A**) through the DAM methyl; (**B**) biphenyl crosslink; and (**C**) at the cleaved CF$_3$ site [56].

Triptycene—a rigid three-dimensional (3D) symmetric structure with a 120° included angle of two benzene rings—can dramatically disturb polymer chain packing and enlarge polymer free volumes. Therefore, introducing triptycene building units into polymer chains should enhance permeability performance. Three triptycene-based polymers, including 6FDA-DAT, 6FDA-DAT/DATCA (9:1), and 6FDA-DAT/DATCA (8:2), were synthesized by our group [60] (DAT, 2,6-diaminotriptycene; and DATCA, 2,6-diaminotriptycene-14-carboxylic acid). Figure 5 shows chemical structures of 6FDA-DAT and 6FDA-DAT/DATCA. No plasticization was observed for crosslinked 6FDA-DAT/DATCA (9:1) and 6FDA-DAT/DATCA (8:2) at CO$_2$ pressure up to 30 atm for pure gas, and a partial CO$_2$ pressure up to 20 atm for (CO$_2$:CH$_4$ 1:1) mixed gases. In addition, 6FDA-DAT/DATCA (9:1)-450 had the highest CO$_2$ pure gas permeability of 305.8 Barrer with a CO$_2$/CH$_4$ ideal selectivity of 27.8, while 6FDA-DAT/DATCA (8:2)-350 had the highest CO$_2$/CH$_4$ ideal selectivity of 43.7 with a CO$_2$ pure gas permeability of 58.5 Barrer.

Two carboxylic acid-containing diamines, CADA1 and CADA2, were successfully synthesized to polymerize with 6FDA, BTDA, and DSDA, respectively [58]. The synthesis route is illustrated in Figure 6. The CO$_2$-induced plasticization phenomenon was not observed at a CO$_2$ pressure up to 30 atm for 6FDA-CADA1-425, which had a CO$_2$ permeability of 917.4 Barrer with a CO$_2$/CH$_4$ ideal selectivity of 28.11.

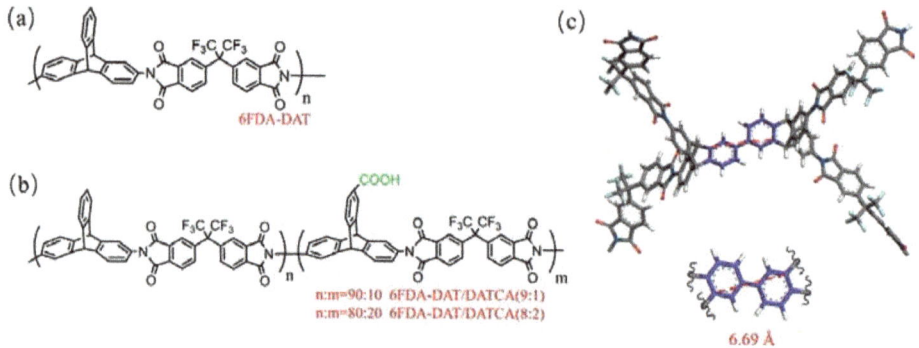

Figure 5. Chemical structures of (**a**) 6FDA-DAT; (**b**) 6FDA-DAT/DATCA; and (**c**) estimated distance (6.69 Å) between the crosslinked chain segments of 6FDA-DAT/DATCA (energy of chain conformations was minimized using Material Studio 7.0) [60].

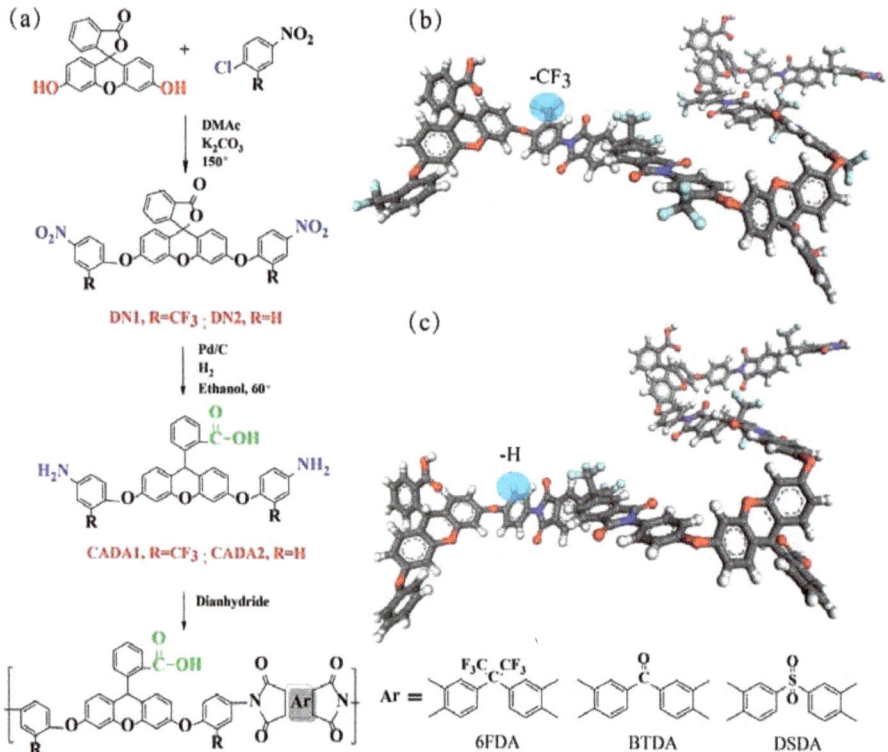

Figure 6. (**a**) Synthesis routes of dinitro monomers, diamines as well as and 6FDA-, BTDA-, and DSDA-based carboxylic acid-containing polyimides, (**b**) and (**c**) show the energy-minimized chain conformations of 6FDA-CADA1 and 6FDA-CADA2 polymers, respectively (using Material Studio 7.0) [58].

In the above work, the temperature sufficient to complete the crosslink reaction was 425 °C, which is higher than the T_g of most polymers. Therefore, the follow-up objective was to lower the crosslinking temperature, by adopting two phenolphthalein-based polyimides, 6FDA-MPP and 6FDA-PP [59]. Figure 7 shows chemical structures of 6FDA-CADA2, 6FDA-MPP, and 6FDA-PP. No plasticization was observed in the 6FDA-MPP-400 and 6FDA-PP-400 polyimides at a CO_2 pressure up to 30.62 atm (450 psi). The results found that 6FDA-MPP-400 obtained the best performance with a CO_2/CH_4 ideal selectivity of 39.2 and a CO_2 permeability of 193.8 Barrer.

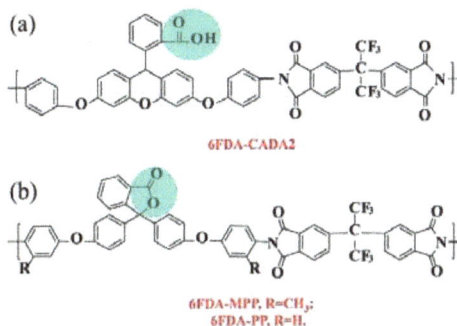

Figure 7. Chemical structures of (a) 6FDA-CADA2 [26], and (b) 6FDA-MPP, 6FDA-PP [59].

Figure 8 illustrates two different mechanisms, revealing how polymer chains crosslink under thermal treatment. The free carboxylic acid groups exist along the polymer chains, allowing the polymers to go through a series of changes (Figure 8a). Polymers containing lactone rings are supposed to cleave their lactone ring, which directly decomposes causing polymer chain breaks. Such types of crosslink reactions occur in the presence of oxygen (heated in air atmosphere). This thermal oxidative crosslinking method is widely applied during the preparation of carbon molecular sieve (CMS) membranes. In our current work, the crosslinking temperature of the lactone ring containing polyimides significantly decreased to 350 °C, which is lower than the pre-polymer's glass transition temperature. If the lactone ring-based polyimide is adopted to spin hollow fiber membranes, the porous substrate of the membrane will be preserved during thermal oxidative crosslinking.

3.1.2. Thermally Rearranged (TR) Polymers

Thermally rearranged (TR) polymers are highly permeable polymers that, at high temperatures, convert polyimide or polyamide into heterocyclic polybenzothiazole (PBT) or polybenzoxazole (PBO). It was originally proposed that polyimides containing ortho-positioned functional groups (for example –OH) can be used to prepare rearranged polymers with altered chain pain packing, with confined spatial location due to the thermal rearrangement [62]. Figure 9 shows changes in conformation–polymer chains and spatial relocation due to chain rearrangement in confinement. As a result, TR polymers show excellent resistance to plasticization at CO_2 partial pressure as high as 20 atm. In contrast to glassy polymers, TR polymer membranes do not exhibit substantially reduced selectivity, even at elevated CO_2 fugacity (~15 atm).

Some TR polymers show strong resistance to plasticization. For the polyimide synthesized from 3,3'-dihydroxy-4,4'-diamino-biphenyl (HAB) and 6FDA (hexafluoroisopropylidene-diphtalic anhydride), pure CO_2 permeation curves pass through a minimum pressure at about 20 atm, which then begins to increase [63]. However, when the HAB–6FDA polyimide precursor were converted to polybenzoxazole (PBO) by heating under flowing N_2, plasticization pressure points were not observed for CO_2 up to 25 atm. In addition, more TR conversion would result in a higher permeability of CO_2 and CH_4, as well as higher plasticization fugacity than those of the polyimide.

(a)

Two adjacent carboxylic acid groups

Anhydride

Two free phenyl radicals

A biphenyl linkage between two polymer chains

(b)

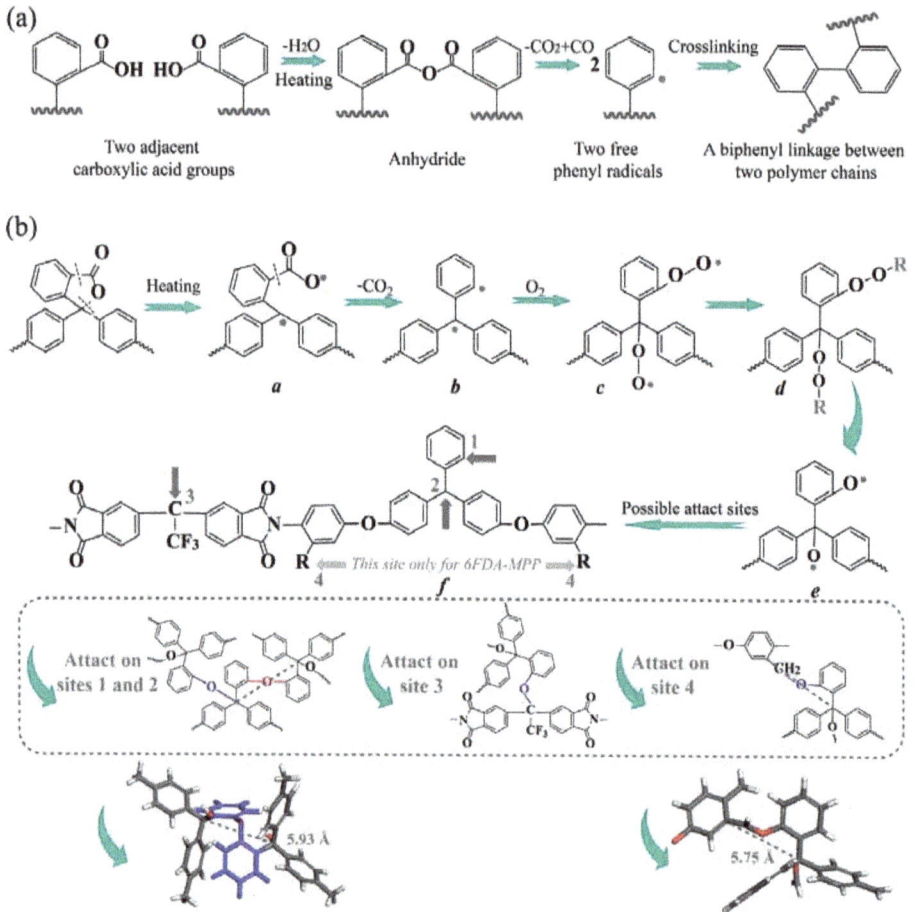

Figure 8. (**a**) The thermal crosslinking mechanism of carboxylic acid-containing polyimides, and (**b**) the thermal oxidative crosslinking mechanism of phenolphthalein-based polyimides as well as the energy-minimized distances between the possible crosslinking sites estimated using Material Studio 7.0 [59].

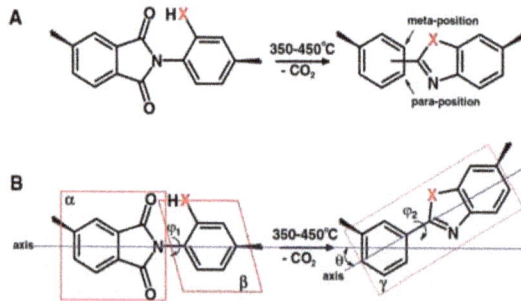

Figure 9. (**A**) Changes in conformation–polymer chains consisting of meta- and/or para-linked chain conformations can be created via rearrangement. (**B**) Spatial relocation due to chain rearrangement in confinement, which may lead to the generation of free-volume elements [62].

3.2. Chemical Crosslinking

When polymer chains are chemically crosslinked by covalent bonds, the crosslink networks are quite stable in harsh environments and can sufficiently inhibit the rotation of polymer chains. Polyimides contain many activated sites on polymer chains that are able to establish covalent bonds.

Typical reactive sites are imide bonds. In theory, all polyimides can be crosslinked on their imide groups and show improved plasticization resistance. Crosslink networks can also be tailored according to the characteristic functional group of specific polymers. This method with an encouraging prospect is shown in Section 3.2.1.

Beyond using imide bonds as crosslinking sites, the crosslinking reaction can also take place through reactive functional groups and other crosslinking agents as introduced in Sections 3.2.2–3.2.4.

3.2.1. Polyamine Crosslinking

Imide bonds in polyimide membranes can react with specific diamine monomers to cleave the imide ring, thereby crosslinking the polymer chains. This is a promising approach to crosslink polyimide membranes, which possess the imide bond. Introducing amine bonds further introduces improved properties such as hydrophilicity and stability in solvents.

To collectively optimize the permeability and selectivity, the molecular structure and steric constraints must be considered as key factors. In addition, different kinds of diamines react distinctly when crosslinked with polyimide chains. Some factors that are less related to diamines themselves, may also play critical roles in the performance of polyimide membranes such as side-reactions and degree of crosslinking.

Previous work has reported on the crosslinking of monomeric and polymeric diamines with polyimide membranes. Figure 10 illustrates some diamines acting as the crosslinkers reported in the literature. Monomeric diamines fall into two classes: aliphatic and aromatic diamines.

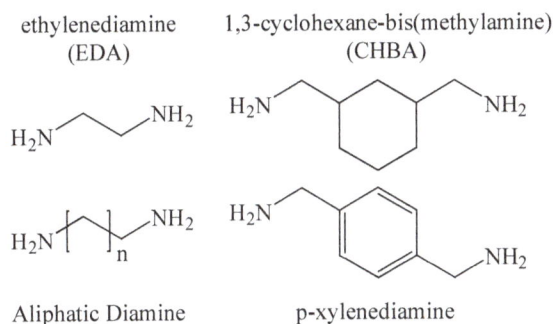

Figure 10. Diamine monomers used in the literature.

This crosslinking method is often followed by thermal treatment to enhance CO_2 plasticization resistance. However, thermal treatment may lead to the release of diamine monomers and decomposition of the crosslinked polymer networks.

Diamine-Monomer Crosslinking

Shao et al. chose 1,2-ethylene diamine (EDA)—a linear aliphatic diamine—to prepare crosslinked 6FDA-durene, and then treated the membrane material with thermal annealing at 250 °C [64]. The modified membrane material exhibited plasticization pressure at 48.99 atm (720 psi), while the original 6FDA-durene membrane was plasticized by CO_2 at 300 psi. The imide bonds reacted with EDA, resulting in imide ring cleavage. EDA monomers then connected two polymer chains through their amide bonds. The insertion of crosslinking agents into polymer chains significantly decreased the d-space, bringing about the structure-tightening effect. Figure 11 shows possible reaction mechanisms during 1,2-ethylene diamine (EDA)-induced cross-linking and thermal annealing.

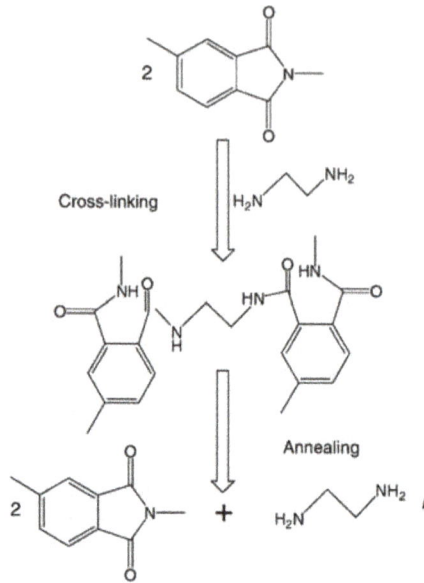

Figure 11. Possible reaction mechanisms during 1,2-ethylene diamine (EDA)-induced cross-linking and thermal annealing [64].

The subsequent thermal annealing triggers the imidization reaction of crosslinking networks, accompanied by EDA release at specific temperatures. Higher thermal annealing temperatures achieve stronger plasticization resistance.

At the specific high temperature stage, coupling effects of EDA crosslinking and thermal annealing accelerate the formation of charge transfer complexes (CTCs). As a consequence, polymer chain mobility decreases, the membrane matrix becomes denser, and the membranes' color changes from transparent to yellow. Shao et al. ultimately found two approaches to enhance the anti-plasticization properties: (1) A shorter EDA crosslinking time followed by a higher annealing temperature, or (2) a longer EDA crosslinking time followed by a lower annealing temperature.

It is a remarkable fact that a single treatment, polyamine crosslinking or heating, cannot strongly suppress the plasticization. Plasticization resistance will be enhanced only if the polymers are crosslinked by polyamine molecules first, followed by thermal treatment. This is proven by several experimental phenomena. The original 6FDA-durene was still transparent after heating at 250 °C in vacuum, which indicates that 250 °C thermal annealing on the original 6FDA-durene could not induce CTC formation. Compared to the original, easier membrane plasticization was achieved after 1-min EDA cross-linking. In summary, only the coupling effects of EDA cross-linking and thermal annealing may facilitate the formation of CTCs in bulk 6FDA-durene, to achieve the desired anti-plasticization effect.

6FDA-durene membranes can also be crosslinked with 1,3-cyclohexanebis (methylamine) (CHBA); the modified membranes showed no obvious plasticization up to 48.99 atm (720 psi) [65]. The reaction mechanism during the chemical crosslinking and thermal annealing is similar to membranes modified by EDA. Although the thermal treatment regenerates imide groups from amide groups in the crosslinked networks, polyimide membranes micro-structures exhibit irreversible changes. Unlike EDA-modified membranes, crosslinking in the CHBA/methanol solution facilitates the diffusion of crosslinking agents through the membranes to react with the bulky body, because methanol swells the membranes. Thermal annealing almost homogeneously changes the chemical composition and structure of the crosslinked polyimide matrix. Therefore, in this case, membranes are not only modified on the surface, but also in the body. Diamine monomers open the main chains in the membranes' matrix, increasing polymer chain flexibility. This relative flexibility tends to achieve configurational rearrangement and forms CTCs, especially at higher temperatures (200 °C in this case). Figure 12 shows charge transfer complex model of 6FDA-durene polyimides. Thermal annealing induces the regeneration of chemical composition from amide groups to imide groups. However, the formation of CTCs restricts the mobility of polymer chains. These two factors contribute to enhancing anti-plasticization characteristics.

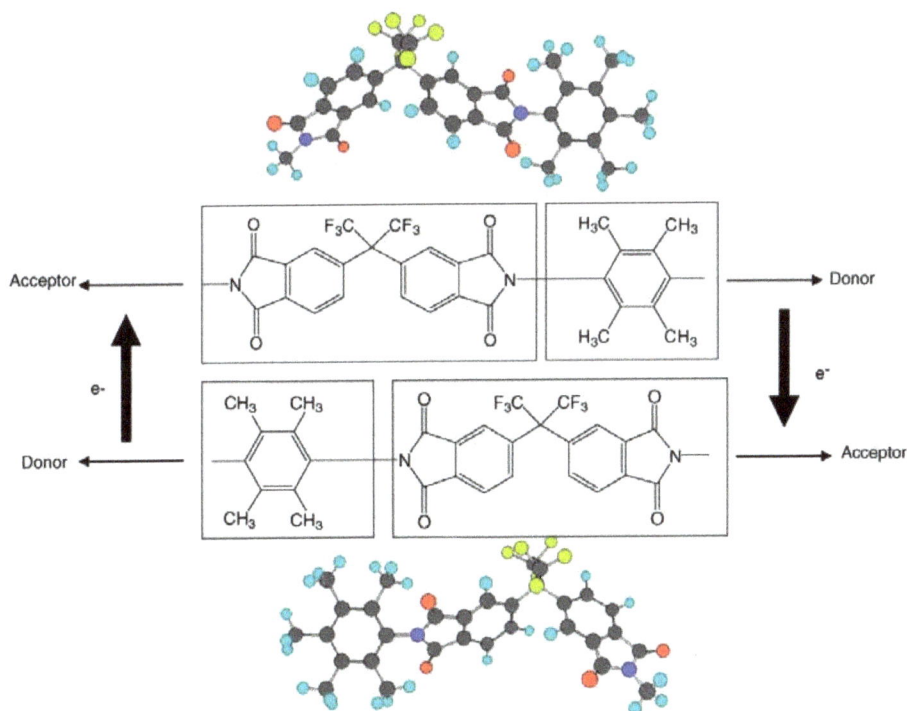

Figure 12. Charge transfer complex model of 6FDA-durene polyimides [65].

The commercially available polyimide, Matrimid, can be crosslinked with *p*-xylenediamine to restrict plasticization [67]. The permeability of 30-day crosslinked membranes remains constant in the whole range of CO_2 feed pressure (from 3.55 to 32.4 atm). The crosslink reaction can be detected by measuring the gel content. Compared with fluoropolyimides such as 6FDA-durene, Matrimid has less free volume and swelling degree in methanol [83]. Therefore, it takes the Matrimid membrane matrix longer to swell in methanol. Furthermore, membrane swelling in methanol is the rate-determining step, thus the crosslinking rate of Matrimid is much lower than those of fluoropolyimides.

Amine-Tetramer, Dendrimer Crosslinking

Diamines and tetramines—dendrimers containing four free amine groups—can be used as crosslinkers such as polyamidoamine (PAMAM) and diaminobutane (DAB). Their chemical structures are shown in Figure 13.

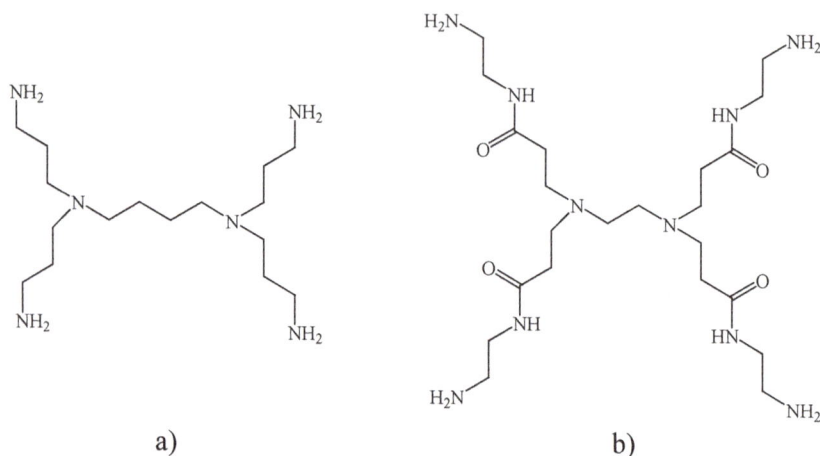

Figure 13. Chemical structures of (**a**) diaminobutane (DAB), and (**b**) polyamidoamine (PAMAM).

Xiao et al. combined PAMAM crosslinking with thermal treatment methods, but did not observe the plasticization phenomenon up to 30 atm [66]. The start temperature of PAMAM decomposition was previous to the thermal degradation of PAMAM-modified polyimide membranes. In addition, thermal stability decrease of the PAMAM-modified polyimide from 593 to 579 °C was unexpected. The X-ray photoelectron spectroscopy (XPS) data proved that thermal treatment of PAMAM-modified polyimide films at 250 °C brought about imidization of poly(amic amide), and the thermal decomposition of the PAMAM dendrimer. ^{1}H nuclear magnetic resonance (NMR) indicated that PAMAM dendrimers are cleaved into small molecules at 250 °C. When heating membrane materials, PAMAM presence in polyimide degrades polyimide chains to low molecular weight fragments with lower thermal stability. Heat treatment increases inter-segmental mobility, thus increasing the likelihood of PAMAM dendrimer free primary amine groups meeting the imide groups in the polyimide backbone. However, high temperature treatments induce significant dendrimer decomposition at the amide linkage, leading to crosslinked structure breakdown. They also result in the degradation of polyimide, which may produce a heterodispersed population of compounds with lower molecular weights. A high temperature treatment induces a higher degree of crosslinking, which tightens the polymer chains and limits intersegmental chain mobility. Therefore, thermal treatments at 120 °C offer better plasticization resistance than simple immersion modifications. Interestingly, modified films treated at 250 °C exhibited stronger plasticization compared with the original polyimide films. The most likely explanation is that the 250 °C treatment rebuilds the imide ring in the main polymer backbone, and enhances chain rigidity. At the same time, CTCs formation between neighboring polyimide chains helps decrease CO_2 sorption, reduce chain mobility, and stabilizes the structure.

Although many polyamine-crosslinked polyimide membranes have been successfully synthesized, the plasticization response has not been reported, indicating a need for further research in this area. Interestingly, investigating the effect of crosslinking reagents on different polyimides has been made [84].

Immersing the membrane in a dendrimer-methanol solution can induce polymer chains crosslink. Chung et al. used PAMAM to crosslink 6FDA-durene membranes [85]. Compared with diamines, with two functional end groups, the bulk teramines exert a space-filling effect on the polymer chains to reduce d-spacing, which leads to free volume decrease. Furthermore, due to the steric hindrance of the large dendrimer molecules, it takes more time for them to diffuse into the polymer matrix. This may result in asymmetric crosslinking throughout the bulky membrane matrix. In addition, the large molecular size makes the decreasing rate of the diffusion coefficients of PAMAM crosslinked polymers significantly slower than using other crosslinking agents containing amino groups. The initial increase in gas solubility with immersion time may be attributed to two factors: one is that PAMAM contains all kinds of amine groups such as –NH$_2$, –NH–, or –N<. These amines groups have strong interactions with CO$_2$, resulting in the increase of CO$_2$ solubility of polyimides; the other possible reason is that the big molecular size of dendrimers may slightly increase the interstitial space, which may result in greater adsorption. However, the plasticization response was not reported in this work.

The effect of modification time and the generations of the PAMAM dendrimer on the properties of modified polyimide films has been reported [86]. When it is only dried at room temperature, some free primary amine groups for the loaded PAMAM dendrimers remain on the surface of polyimide films.

DAB dendrimers, a crosslinking agent for polyimides, can perform crosslinking reactions at room temperature and improve gas separation performance of polyimide membranes, especially for H$_2$/N$_2$, He/N$_2$, H$_2$/CO$_2$ [87].

Polyamine Crosslinking

Compared with monomer polyamines, there have been fewer investigations on crosslinking polyimide membranes by polymeric polyamine. For instance, poly(propylene glycol) block poly(ethylene glycol) block poly(propylene glycol) diamine (PPG/PEG/PPGDA) was successfully employed to chemically crosslink Matrimid 5218 at room temperature [88].

Brief Summary

Based on the above discussion, some insights can be given into polyamine crosslinking. Only the integration of thermal annealing with polyamine crosslinking can significantly improve plasticization resistance. As polyamine crosslinking reduces the interchain spacing, the polymer chains become closer to each other. Thermal annealing enhances chain mobility, so more CTCs will be formed between the polymer chains. In conclusion, the essence of this method is to facilitate the CTCs' formation.

Finding a suitable polyamine and probing proper reaction protocols will be the future trend in polyamine crosslinking aimed improving plasticization resistance. Therefore, the molecular size, number of free primary amine groups, molecular rigidity and polarization, etc. must be taken into consideration.

To illustrate, consider the molecular size first: If the polyamine molecule is too small, it will decrease the d-space of the polymer chains, generating the decline of permeability; conversely, it takes a large polyamine molecule more time to achieve the desired crosslinking effect because its larger molecular size makes it harder to penetrate into the polyimide matrix. Therefore, a polyamine with a moderate molecular size is assumed to space apart the chains while stabilizing the membranes. They should be able to penetrate deeper into the polymer matrix and thus construct a thicker crosslinking layer on the surface of polyimide films.

3.2.2. Diol Crosslinking of Carboxylated and Sulfonated Polyimide Membranes

If the polyimides have carboxylate or sulfonate groups, they can be crosslinked by diols or polyols through esterification. Unlike amine crosslinking, some diols or polyols can suppress the plasticization, while maintaining good permeability and selectivity of the original polymers. Common diol crosslinkers are shown in Figure 14.

Similar to crosslinking by polyamines, crosslinking by diols is sometimes followed by thermal treatment. Furthermore, hydrogen bond formation cannot be negligible due to the presence of carboxyl groups along the polymer backbones. The contribution made by hydrogen bonding to plasticization suppression will be discussed in Section 3.3.

Ethylene Glycol (EG)	Buthylene Glycol (BG)	butane-1,4-diol

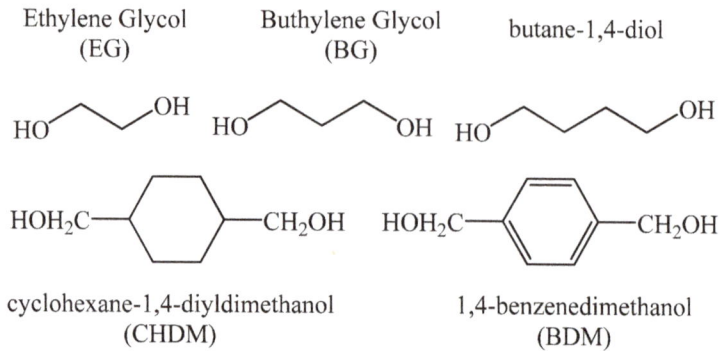

cyclohexane-1,4-diyldimethanol (CHDM)	1,4-benzenedimethanol (BDM)

Figure 14. Common diol crosslinkers.

Diol Crosslinking through Carboxylic Acid Groups

Staudt-Bickel C et al. introduced –COOH groups to the polyimide backbones and crosslinked the polymer with ethylene glycol [19]. The highlight of this crosslinked polymer with exquisite structure was the little plasticization up to 35 atm pure CO_2 pressure accompanied by increased CO_2/CH_4 selectivity. The strong crosslinked 6FDA-DABA exhibited much better CO_2/CH_4 separation properties with an ideal selectivity of 87 and a CO_2 permeability of 10.4 Barrer at 3.74 atm and 308 K. 6FDA-6FpDA/DABA crosslinked with ethylene glycol (EG) can stabilize the membrane much more effectively, leading to plasticization pressure increase [41].

The cyclohexane-1,4-diyldimethanol (CHDM) monoester can also be used in esterification crosslinking. The plasticization pressure correlated with a sorbed CO_2 partial molar volume of $29 \pm 2 \ cm^3/mol$ in the polymer [68]. However, the exact plasticization pressure was not reported.

To find the initial principle of choosing a diol crosslinker, the effects of length and flexibility of various diol crosslinking agents has been investigated [69]. In view of high permeability with reasonable CO_2/CH_4 selectivity, 6FDA-DAM:DABA 2:1 was used as a model polymer. The esterification reaction was shown to be effective in stabilizing membranes against CO_2 plasticization up to 40 atm feed pressure. Compared with untreated 6FDA-DAM:DABA membranes, the incorporation of DABA increased selectivity. 6FDA-DAM:DABA crosslinked with buthylene glycol (BG) obtained the highest selectivity of 34 among the crosslinked samples.

6FDA-DAM:DABA and 6FDA-6FpDA:DABA copolymers have been synthesized for crosslinking with BG and CHDM [89]. CO_2 permeabilities increased by factors of 4.1 and 2.4, respectively, at 20 atm feed pressure, without loss in selectivity. Furthermore, selectivity increased with higher crosslinking density for feed pressure below 20 atm. The 6FDA-DAM:DABA 2:1 membranes crosslinked with CHDM showed a CO_2/CH_4 selectivity of 32-33, while the membranes crosslinked with BG exhibited similar selectivity performance with a CO_2/CH_4 selectivity of 33.

Diol Crosslinking through Sulfonic Acid Groups

Sulfonic acid groups along the polymer chains can react with diol molecules to form ester bonds linking the two polymer chains. The reaction mechanism of the sulfonic acid group resembles the mechanism of the carboxylic acid group discussed above. Although this crosslinking method has been successfully applied in various separation processes [90,91], no report has been found in the field of polyimide membranes in gas separation applications.

3.2.3. Semi-Interpenetrating Network

The construction of crosslink networks can be obtained from multiple types of polymer chains, and by crosslinking polymer chains with another polymer network. Some reactive monomers containing functional groups can chemically crosslink or physically interlock the linear polyimide chains, while solely reacting with its species to form a polymer network. The introduction of these reactive monomers can tune the size and quantity of the cavities. Furthermore, the free volume distribution of the pseudo-interpenetrating networks (pseudo-IPNs) may be beneficial for the separation of small molecules by polymeric membranes, due to both the mean cavity size and the quantity of free volume influencing polymeric membranes permeability.

In situ polymerization of azido monomers tends to form oligomer networks. Simultaneously, functional groups within the network react with specific sites on polymer chains to form semi-interpenetrating networks, resulting in a membrane with no signs of plasticization up to 30 atm [72]. The chemical structure of the azide used in this paper is shown in Figure 15. Azide introduces the azido group, a linear 1,3-dipolar structure, into the membrane matrix. Upon heating or under ultraviolet radiation, azido groups are able to create nitrene—a highly reactive, short-lived intermediate. Nitrenes insert into C–H bonds and add to unsaturated C–C bonds such as alkenes, alkynes, and arenes due to their electrophilicity. These data indicate that every kind of polyimide has the potential to react with azide. In general, the reactive priority of the C–H insertion follows this sequence: tertiary > secondary > primary C–H bonds. Functional groups cable of reacting with nitrene are: (1) Phenyl groups of azide or host polyimide; (2) –CH$_3$ substituent groups of azide or host polyimide; (3) the –CH$_2$– group in the cyclohexanone structure of azide; and (4) α,β-unsaturated ketone of azide. First, C–H bonds of phenyl groups are not easily attacked by nitrene radicals, owing to the benzene ring stability, which comprises the highly conjugated π electrons. Although the unfavorable addition of nitrene to C=C bonds has been reported, the absence of the characteristic peak of tertiary amine indicates that the cycloaddition of nitrenes to the phenyl ring and the C=C bonds has not occurred. Second, due to steric hindrance, there is no reaction with alkenes. Azide can, therefore, react with its own species to form polymer networks, and with the functional groups of the host polyimide to form an interconnected pseudo-IPN. Therefore, for those polyimide membranes containing no functional groups to react with nitrenes, such as 6FDA-NDA, the combination of polyimide and poly(azide) only results in a pseudo-IPN without any interconnection (NDA is short for "1,5-naphthalenediamine").

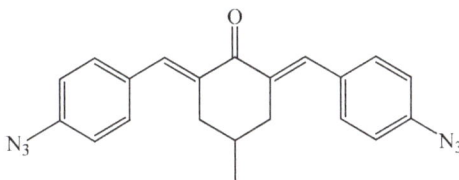

Figure 15. Chemical structure of 2,6-bis(4-azidobenzylidene)-4-methylcyclohexanone.

As the insoluble gel in the film reflects the formation of crosslinked networks, it can be inferred that nitrene radicals react with the methyl C–H bonds of 6FDA-TMPDA to form chemical crosslinks (TMPDA is short for "2,3,5,6-tetramethyl-1,4-phenylenediamine"). The formation of denser interconnected pseudo-IPNs restricts chain movement and enhances the anti-plasticization property of the material.

Another crosslinker, phenolphthalein dicyanate (DCFT), has been shown to improve plasticization resistance. Chemical structures of DCFT and DCFT-resin are shown in Figure 16. DCFT-modified membranes did not show any signs of plasticization in CO$_2$ feed pressures ranging from 1 to 30 atm [71]. Dicyanate cyclotrimerized into a uniform three-dimensional network of oxygen-linked triazine rings (cyanurate ester resin). The absence of by-products was indicated by the ^{1}H NMR, elemental analysis, and mass spectrometry data. It is worth noting that copper naphthenate and nonylphenol serving as a catalyst must be added to mitigate curing temperatures from 280 °C to 180 °C. The results of Fourier transform infrared spectroscopy (FTIR) and differential scanning calorimetry (DSC) showed that after heat treatment at 180 °C for 60 min, dicyanate seemed to be entirely converted into the corresponding cyanurate ester resin. DCFT introduces cyanurate ester resin into the Matrimid matrix, which causes a compacting effect. Hence, all semi-IPNs showed a higher density compared to Matrimid films, indicating the densification of semi-IPNs. Consequently, this densification effect successfully restricts plasticization by hindering polymer chain mobility.

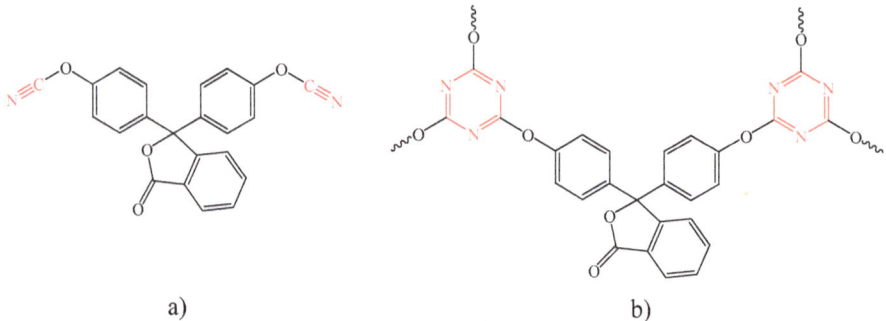

a)   b)

Figure 16. Chemical structures of (**a**) phenolphthalein dicyanate (DCFT), and (**b**) cyanurate ester resin (DCFT-Resin).

Except for highly reactive monomers, oligomer-containing functional groups can also be used to form semi-IPNs together with host polyimides. It is well-known that the use of longer oligomer chains results in improved toughness. Bos et al. used Thermid FA-700 as a crosslinker to resist plasticization effectively [36]. Figure 17 illustrates chemical structure of Thermid FA-700. Modified films showed a constant permeability at elevated pressure in the pressure range (up to 50 bar, i.e., 49.35 atm) of the CO_2 permeability test. It has been already proven that the acetylenic end group can be crosslinked by heating at 250–275 °C without the evolution of volatile products, while the oligomer polymerizes [92]. Two kinds of crosslinking reactions are likely to happen: (1) the benzenoid cyclization (crosslinking) reaction, i.e., trimerization reaction of the acetylene groups; and (2) the naphthalene formation (polymerization) reaction. These reactions form semi-IPNs. Compared with the Matrimid/Thermid film without heat treatment, thermal annealing triggers a crosslinking reaction between the oligomer and Matrimid, which leads to the obvious suppression of plasticization.

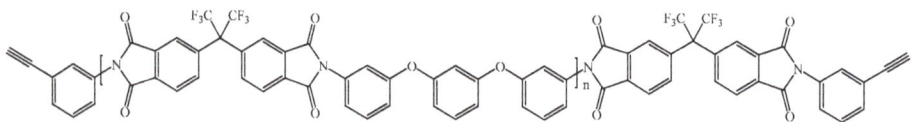

Figure 17. Chemical structure of Thermid FA-700.

3.2.4. Ionic Bonding

Apart from inserting covalent bonds into polymer chains, an alternative strategy entails connecting polymer chains with ionic bonds. A prerequisite for the formation of ionic bonds is the presence of an electron donor or acceptor along the polymer backbone, which can be carried out by the substitution reaction.

It has been reported that the introduction of bromine into polymer constitutional units by bromination of 6FDA-durene, allows crosslinking of the original membrane to N,N'-dimethylpiperazine [73]. This electrostatic chemical bond effectively restricts polymer chain mobility and somewhat stabilizes the CO_2/N_2 selectivity. As a result, membranes did not show an upturn of permeability in the isotherm within the experimental scale of 25 atm of CO_2. It is remarkable that the pendant piperazinium-mediated crosslinked polyimide membranes displayed a high CO_2

permeability of 475.5 Barrer, together with high CO_2/CH_4 and CO_2/N_2 permselectivities of 34.5 and 18, respectively.

3.3. Physical Crosslinking

The introduction of polar functional groups, such as hydroxyl and carboxyl groups, that can form hydrogen bonds, leads to CTCs and pseudo-crosslink networks formation. To present a larger picture of the physical crosslinking method, mixed matrix membranes are also discussed in this section.

3.3.1. Hydroxyl Group

Alaslai et al. prepared three polyimides with different numbers of hydroxyl groups (shown in Figure 18), and demonstrated that polyimides with strong polar –OH groups could mitigate plasticization when tested under high-pressure binary CO_2/CH_4 mixed-gas conditions due to strong chain interactions by inter-chain hydrogen bonding and CTC formation [93]. Compared with 6FDA-mPDA, hydroxyl-containing polyimide membranes maintained very high CO_2/CH_4 selectivity (~75 at CO_2 partial pressure of 10 atm) due to high CO_2 plasticization resistance when tested under high-pressure mixed-gas conditions.

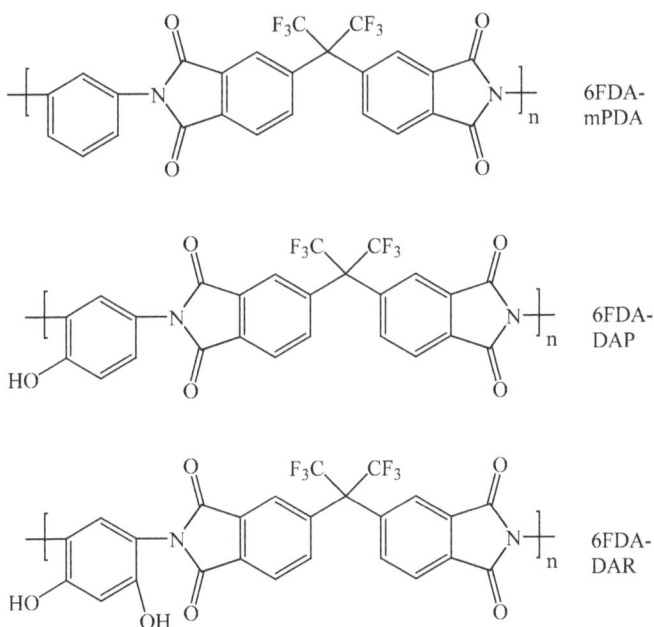

Figure 18. Chemical structures of 6FDA-mPDA, 6FDA-DAP, and 6FDA-DAR.

3.3.2. Carboxyl Group

It is well-known that the majority of carboxylic acids often exist as dimeric pairs in non-polar media. Structure of associated carboxylic acids is depicted in Figure 19. These strong polar-associating functional groups, without the formation of any covalent bonds, participate in a reversible process and are subject to media polarity and system temperature.

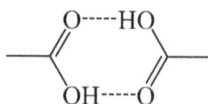

Figure 19. Structure of associated carboxylic acids.

Staudt-Bickel et al. found that polymer chains of polyimides were crosslinked by hydrogen bonding between the free carboxylic acid groups, which resulted in a strong reduction in permeability [19]. The effect of such "virtual" hydrogen bonded crosslinks on plasticization has not been reported. Additionally, an increased CO_2/CH_4 selectivity corresponds to an increased degree of crosslinking, attributed to reduced swelling and polymer chain mobility. Even with a 10% degree of crosslinking, a 20% increased selectivity can be seen in comparison to the reference polyimide. However, hydrogen bonding accounting for the formation of CTCs in the blends does well to inhibit plasticization. This will be elaborated in Section 3.5.

3.3.3. Mixed Matrix Membranes (MMMs)

Polymer membranes suffer not only from plasticization, but also permeability–selectivity trade-off limitations. These undesired phenomena can be eliminated by preparing mixed matrix membranes [94]. MMMs reported in the literature can be classified as: solid/polymer; liquid/polymer, and solid/liquid–polymer MMMs [95]. For solid/polymer MMMs, the integration of organic and/or inorganic fillers such as carbon nanotubes, carbon molecular sieves [96,97], activated carbon [98–100], zeolites [101,102], silica [103–105], and metal organic frameworks (MOFs) [106–108] can improve plasticization resistance.

Zeolitic imidazolate frameworks (ZIFs) are a sub-class of metal organic frameworks (MOFs). ZIFs show superior thermal and chemical stability, and are inherently more compatible with polymer matrix [109]. Some studies have found that the introduction of ZIF-8 (Zn (2-methylimidazole)$_2$) can improve the anti-plasticization performance of pure polymer membranes [74,75].

6FDA-durene/DABA co-polyimides were selected as polymer matrix due to their impressive performance during CO_2/CH_4 and C_3H_6/C_3H_8 separation [74]. MMMs synthesized from cross-linkable co-polyimides (6FDA-Durene/DABA (9/1) and 6FDA-Durene/DABA (7/3)), showed significant enhancements in plasticization suppression characteristics up to a CO_2 pressure of 30 atm after annealing at 400 °C due to the cross-linking reaction of the carboxyl in the DABA moiety. After being thermally annealed at 400 °C, the MMM made of 6FDA-Durene/DABA (9/1) with 20 wt% ZIF-8, showed a CO_2/CH_4 selectivity of 19.61 and an impressive CO_2 permeability 728 Barrer in the mixed gas tests.

Instead of a sharp ZIF-8 crystal, blending 10 wt% blunt ZIF-8 in the waste polystyrene (*w*-PS) container-derived membrane matrix showed good aging resistance and improved anti-plasticization gas separation performance [75]. CO_2 permeability of pure PS membranes increased as CO_2 pressure increased to 3 bar (i.e., 2.96 atm). In contrast, MMMs displayed slightly increased permeability at increased operating pressure.

Except for ZIF-8, zeolite T was used as a filler embedded in the membrane matrix. A quantity of 1 wt% loaded zeolite T/6FDA-durene MMM showed improvement at a CO_2 plasticization pressure to 20 bar (i.e., 19.74 atm) when compared to the pristine 6FDA-durene membrane, which was plasticized at a CO_2 pressure of 5 bar [76]. CO_2 permeability and CO_2/CH_4 ideal selectivity of the MMM also increased to 843.6 Barrer and 19.1, respectively.

Investigations show that the addition of three distinctively different MOFs (MIL-53(Al) (breathing MOF), ZIF-8 (flexible MOF), and Cu_3BTC_2 (rigid MOF)) dispersed in a Matrimid polyimide can somewhat alleviate plasticization [77]. At pressures higher than the plasticization pressure of neat Matrimid (10–12 bar), the permeabilities of all MOF–MMMs slightly increased when compared with the native polymer. Additionally, the plasticization pressure increased with the MOF loading.

Incorporating Ni$_2$(dobdc) (dobdc^{4-} = dioxidobenzenedicarboxylate) metal organic framework nanocrystals into various polyimides (Matrimid, 6FDA-DAT:DAM (1:1), and 6FDA-DAM) improves the performance of their anti-plasticization properties [78]. Compared with the pristine polyimides, plasticized at around 10–15 bar, Ni$_2$(dobdc)/polyimide composites showed no sign of plasticization at a CO$_2$ partial pressure <30 bar. The separation performance of composite membranes was closer to the Robeson upper bound of the CO$_2$/CH$_4$ gas pair when compared with the pristine polymer, especially at high CO$_2$ pressure, thus indicating better anti-plasticization properties. These, along with the solution processability of the mixed-matrix format, make Ni$_2$(dobdc)/polyimides intriguing materials for commercial membrane applications.

3.4. Ultraviolet (UV) Radiation Crosslinking

UV radiation is a post-treatment, which is used after membrane fabrication [110,111]. When the main chain of polyimide contains a specific structure like benzophenone, it can be crosslinked by UV radiation. Take 3,3′,4,4′-benzophenonetetracarboxylicdianhydride (BTDA) for example, the carbonyl bridge of the dianhydride is a photoactive group and serves as a UV crosslinking site. Figure 20 shows chemical structure of BTDA. The reaction mechanism of UV crosslinking of BTDA is shown in Figure 21. Radiation time, intensity, and the distance between the membrane and light source will all affect the degree of crosslinking. Thus, the degree of crosslinking of these UV-crosslinked membrane materials strongly depends on the experimental conditions.

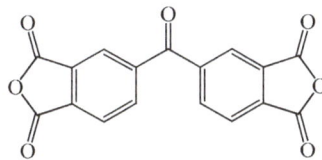

3,3′,4,4′-Benzophenonetetracarboxylicdianhydride (BTDA)

Figure 20. Chemical structure of 3,3′,4,4′-Benzophenonetetracarboxylicdianhydride (BTDA).

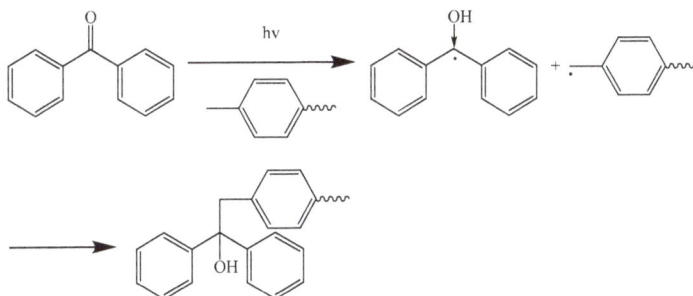

Figure 21. The reaction mechanism of UV crosslinking-modified polyimide containing BTDA.

Hays et al. prepared a series of semi-flexible aromatic polyimides by polycondensation of dianhydrides with phenylene diamines having alkyl substituents on all ortho positions to the amine functions incorporating at least in part 3,3′,4,4′-benzophenone tetracarboxylic dianhydride [112]. The photochemical crosslinked polyimides membranes reveal high permeation to gases while still being able to effectively separate several combinations of gases. For instance, the oxygen permeability of this kind of polyimide does not decline too much, while the selectivity of O$_2$/N$_2$ significantly increased.

In contrast to the findings of Hays et al., Kita et al. observed that the permeability coefficients decreased with increasing UV irradiation time as a result of crosslinking [113]. This decrease can be explained by the change in diffusivity, while solubility is not greatly affected. In addition, the largest change in selectivity was obtained for the H_2/CH_4 gas pair, which had a large difference in the molecular sizes of the constituents. It is worth noticing that the selectivity for the H_2/CH_4 separation increased by a factor of 50 after 30 min of UV irradiation with a decrease in the H_2 permeability by a factor of 5.

Liu et al. investigated a series of UV irradiation modified copolyimides prepared from 2,4,6-trimethyl-1,3-phenylenediamines (3MPDA), BTDA, and pyromellitic dianhydride (PMDA) [114]. Photochemically crosslinking modification resulted in the increasing of gas selectivity of most gas pairs and declination of gas permeability. However, for the H_2/N_2 gas pair, the crosslinked copolymer exhibited a higher H_2 permeability and H_2/N_2 selectivity.

Liu et al. also synthesized another polyimide prepared from BTDA and 2,3,5,6-tetramethyl-1,4-diphenylenediamine (4MPDA), and compared two crosslinking modification methods: putting the polyimide in an ambient environment for four months, and under UV irradiation for two or eight hours [115]. The results found that the crosslinking reaction by UV irradiation only occurred in the surface layer, but crosslinking at an ambient environment for a long time took place in the whole film. This led to a much higher crosslinking density on the membrane surface and higher gas selectivity. Therefore, the resulting crosslinked polymer had an asymmetric structure.

Although plenty of meaningful and interesting work has been done in the UV radiation crosslinking of polyimide, few researchers have focused on the anti-plasticization performance after UV radiation crosslinking. Therefore, the effect of UV crosslinking on plasticization resistance needs further investigation.

3.5. Blending

Given the obvious drawbacks of low permeability and plasticization resistance, it is natural to conceive the idea of blending polyimides with other polymers exhibiting high permeability and anti-plasticization to overcome the corresponding disadvantages.

Through blending, materials may show new characteristics different from the pristine materials. Blending easily plasticized membrane materials with anti-plasticized membrane materials may improve plasticization resistance. Furthermore, blending one relatively cheaper polymer component with another more expensive component can cut down on the capital cost. Based on the above advantages, blending can be used to tailor the performance of the original single polyimide membranes.

However, not all polymers can be blended homogeneously on a molecular level. The miscibility is the most challenging problem when blending different polymers. Some coexistence modes of different polymers are well-established. For instance, one polymer may act as a dispersed or co-continuous phase in another bulky polymer matrix. Another possibility is that a single polymer may exist as an individual block mass due to the heterogeneous dispersion.

In order to prepare a blend material with superior properties, many well-designed experiments have been conducted. Polyimides, such as Matrimid, Torlon, Ultem, etc. have the ability to be blended homogeneously with a small loading of other polymers. The chemical structure of polyimides showing miscibility are depicted in Figure 22. Homogeneous blend membranes are often transparent, exhibit a single glass transition temperature, and present a clear and single phase in the polarized light microscopy (PLM) image.

To begin with, Matrimid, a commercially available polyimide, is often used in blends with other polymers to modify its low permeability. Matrimid/PSF (20/80 wt%) shows plasticization pressures up to above 30 atm [79]. The blends have been characterized by optical, morphological, and thermodynamic measurements to test the compatibility of two candidate polymer materials. No phase separation was observed with optical microscopy. DSC scans of PSF/PI blends with different compositions indicated one single glass transition temperature. It is already known that frequency shifts of miscible blends often indicate specific interactions between the characteristic groups of pure polymers. FTIR spectra shifts and intensity changes suggest PSF and PI interactions, and mixing at the molecular level. These results show that PSF and PI blending is complete and homogeneous.

Furthermore, the phase state of PSF/PI blends has been studied further by differential scanning calorimetry, rheology, and x-ray scattering [116].

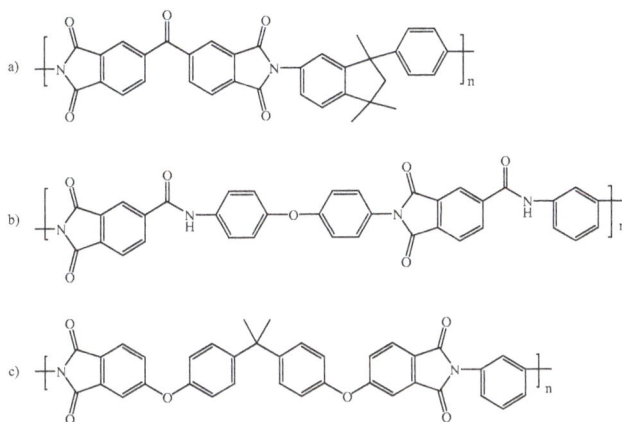

Figure 22. Chemical structures of polyimides: (**a**) Matirmid® 5218, (**b**) Torlon, and (**c**) Ultem.

Bos et al. blended PSF and P84 (copolyimide of 3,3′,4,4′-benzophenone tetracarboxylic dianhydride and 80% methylphenylene-diamine + 20% methylene diamine) with Matrimid, respectively [80]. The proven homogenous blends, Matrimid/PSF (50/50 wt%) and Matrimid/P84 (60/40 wt%) were cast to form transparent membranes. Matrimid showed relatively high permeability when compared with P84, and P84 did not display an apparent signal of plasticization up to 30 atm, much higher than the critical plasticization pressure for Matrimid. Therefore, the latter blend showed a plasticization pressure of 14.80 atm (15 bar). The incomplete suppression of CO_2-plasticization at room temperature was attributed to the concentration of P84 in the blend. Moreover, the CO_2/CH_4 mixed-gas experiment of the two blends revealed a strong resistance to plasticization. The two blends showed no signs of CO_2-induced plasticization with the increasing partial CO_2 pressure in the mix gas separation test. CH_4 permeability showed a slight increase, indicating a small tendency to plasticize.

Matrimid can also be blended with carboxylated polymers of intrinsic microporosity (cPIM-1) at the molecular level [81]. The addition of cPIM-1 in Matrimid significantly enhanced the plasticization pressure for all blended membranes. A small loading of 5–10 wt% of cPIM-1 in Matrimid improved the plasticization pressure from less than 10 atm to 15 atm, while a higher loading of cPIM-1 shifted the plasticization pressure to 20 atm. The interstitial space measured by X-ray diffraction (XRD) indicated an interaction among the polymer chains between cPIM-1 and Matrimid with a reduction in interstitial space, and an enhancement in chain packing. The presence of hydrogen bonding promoted compatibility between these two polymers. The newly formed hydrogen bonds between cPIM-1 and Matrimid, integrated with the rigid polymer backbone of cPIM-1 to suppress the plasticization.

Except for Matrimid, Torlon, a rigid polyimide with superior gas-pair selectivity and intrinsic superior anti-plasticization property, can be blended with PIM-1 showing outstanding gas permeability [82]. The Torlon/cPIM-1 blends displayed plasticization pressures up to 30 atm, when compared with the pristine cPIM-1 membrane, it exhibited a plasticization pressure at around 20 atm. Hydrogen bonding between Torlon and cPIM-1 promotes polymer chain packing and reduces the fractional free volume (FFV), which accounts for this enhancement of anti-plasticization performance. The introduction of Torlon formed partial miscible blends with cPIM-1, and reduced the inter-segment mobility in the polymer matrix. It is noteworthy that cPIM-1/Torlon blend membranes, with a small amount of cPIM-1 or Torlon (5–10 wt%), can hold a homogeneous morphology. The principle of polymer blend miscibility can be testified by PLM, FTIR, DSC, TGA analyses and UV absorbance tests. A homogenous blend may present a clear and single phase in PLM images. C=O band shifts of cPIM-1 from 1700 cm^{-1} to a lower band, with an increase in Torlon loading, indicates hydrogen bonding interactions between cPIM-1 and Torlon. Moreover, the CTCs formation between cPIM-1/Torlon is demonstrated by the fact that the wavelength of the UV absorbance band of cPIM-1/Torlon blends exceeds the predicted data. The strong hydrogen bonding couples with CTC interactions to promote better compatibility between the two polymers.

Although Ultem/PIM-1 polymer blends have been reported, the plasticization resistance has not been investigated yet [117].

4. Conclusions and Prospects

In the present literature, various ways have been thoroughly discussed to restrain the typical CO_2-induced plasticization of polyimide membranes, which contribute to providing a big picture of the research findings in this field. To improve the currently existing methods and develop novel methods, the mechanisms and processes of plasticization were explored.

Taken together, there are rarely any common methods that can be applied to the majority of polyimide membranes to suppress plasticization. The most effective ways vary from membrane to membrane. Therefore, researchers should conceive of specific methods according to the specific chemical molecular structure.

Many novel and effective crosslinking methods for polyimide membranes, such as hyperbranched polyimides have yielded good performance in gas separation. Nevertheless, the plasticization response has not been reported. Therefore, this unexploited area may be one of the directions that will contribute to synthesizing the next potential polyimide membranes with good plasticization resistance.

Author Contributions: M.Z. wrote the whole paper and collected most of the literature except for Section 3.4. L.D. wrote Section 3.4. D.X. collected papers on Section 2 and set the title of this review. P.L. was responsible for major revision. B.C. and S.S.H. provided many significant comments and suggestions.

Acknowledgments: This work was funded by the National Natural Science Foundation of China (grand No: 51773011).

Conflicts of Interest: The authors declare no conflict of interest.

References

1. Dung, E.J.; Bombom, L.S.; Agusomu, T.D. The effects of gas flaring on crops in the niger delta, nigeria. *GeoJournal* **2008**, *73*, 297–305. [CrossRef]
2. Kidnay, A.; Kidnay, A.; Parrish, W.; McCartney, D. *Fundamentals of Natural Gas Processing*, 2nd ed.; CRC Press: Boca Raton, FL, USA, 2011.
3. Balzani, V.; Armaroli, N. *Energy for A Sustainable World: From the Oil Age to a Sun-Powered Future*; John Wiley & Sons: Hoboken, NJ, USA, 2010; p. 390.
4. Adewole, J.K.; Ahmad, A.L.; Ismail, S.; Leo, C.P. Current challenges in membrane separation of CO_2 from natural gas: A review. *Int. J. Greenh. Gas Control* **2013**, *17*, 46–65. [CrossRef]
5. Analysis–Spencer Dale, Group Chief Economist. Available online: https://www.bp.com/en/global/corporate/energy-economics/statistical-review-of-world-energy/natural-gas.html (accessed on 1 August 2018).
6. Bp Statistical Review of World Energy. Available online: https://www.bp.com/content/dam/bp/en/corporate/pdf/energy-economics/statistical-review/bp-stats-review-2018-full-report.pdf (accessed on 1 August 2018).
7. Natural Gas. Available online: https://www.bp.com/en/global/corporate/energy-economics/energy-outlook/demand-by-fuel/natural-gas.html (accessed on 1 August 2018).
8. Faramawy, S.; Zaki, T.; Sakr, A.A.E. Natural gas origin, composition, and processing: A review. *J. Nat. Gas Sci. Eng.* **2016**, *34*, 34–54. [CrossRef]
9. Shimekit, B.; Mukhtar, H. *Natural Gas Purification Technologies–Major Advances for CO_2 Separation and Future Directions*; InTech: London, UK, 2012.
10. Baker, R.W.; Lokhandwala, K. Natural gas processing with membranes: An overview. *Ind. Eng. Chem. Res.* **2008**, *47*, 2109–2121. [CrossRef]

11. Bhide, B.D.; Voskericyan, A.; Stern, S.A. Hybrid processes for the removal of acid gases from natural gas. *J. Membr. Sci.* **1998**, *140*, 27–49. [CrossRef]
12. George, G.; Bhoria, N.; AlHallaq, S.; Abdala, A.; Mittal, V. Polymer membranes for acid gas removal from natural gas. *Sep. Purif. Technol.* **2016**, *158*, 333–356. [CrossRef]
13. Xu, G.; Liang, F.; Yang, Y.; Hu, Y.; Zhang, K.; Liu, W. An improved CO_2 separation and purification system based on cryogenic separation and distillation theory. *Energies* **2014**, *7*, 3484–3502. [CrossRef]
14. Maqsood, K.; Pal, J.; Turunawarasu, D.; Pal, A.J.; Ganguly, S. Performance enhancement and energy reduction using hybrid cryogenic distillation networks for purification of natural gas with high CO_2 content. *Korean J. Chem. Eng.* **2014**, *31*, 1120–1135. [CrossRef]
15. Mandal, B.P.; Guha, M.; Biswas, A.K.; Bandyopadhyay, S.S. Removal of carbon dioxide by absorption in mixed amines: Modelling of absorption in aqueous mdea/mea and amp/mea solutions. *Chem. Eng. Sci.* **2001**, *56*, 6217–6224. [CrossRef]
16. Siriwardane, R.V.; Shen, M.-S.; Fisher, E.P.; Poston, J.A. Adsorption of CO_2 on molecular sieves and activated carbon. *Energy Fuels* **2001**, *15*, 279–284. [CrossRef]
17. Cavenati, S.; Grande, C.A.; Rodrigues, A.E. Removal of carbon dioxide from natural gas by vacuum pressure swing adsorption. *Energy Fuels* **2006**, *20*, 2648–2659. [CrossRef]
18. Peters, L.; Hussain, A.; Follmann, M.; Melin, T.; Hägg, M.B. CO_2 removal from natural gas by employing amine absorption and membrane technology—A technical and economical analysis. *Chem. Eng. J.* **2011**, *172*, 952–960. [CrossRef]
19. Staudt-Bickel, C.; Koros, W.J. Improvement of CO_2/CH_4 separation characteristics of polyimides by chemical crosslinking. *J. Membr. Sci.* **1999**, *155*, 145–154. [CrossRef]
20. Baker, R.W. Future directions of membrane gas separation technology. *Ind. Eng. Chem. Res.* **2002**, *41*, 1393–1411. [CrossRef]
21. Bikson, B.; Nelson, J.K.; Muruganandam, N. Composite cellulose acetate/poly(methyl methacrylate) blend gas separation membranes. *J. Membr. Sci.* **1994**, *94*, 313–328. [CrossRef]
22. Gantzel, P.K.; Merten, U. Gas separations with high-flux cellulose acetate membranes. *Ind. Eng. Chem. Process Des. Dev.* **1970**, *9*, 331–332. [CrossRef]
23. Kim, M.H.; Kim, J.H.; Kim, C.K.; Kang, Y.S.; Park, H.C.; Won, J.O. Control of phase separation behavior of pc/pmma blends and their application to the gas separation membranes. *J. Polym. Sci. Part B Polym. Phys.* **1999**, *37*, 2950–2959. [CrossRef]
24. Şen, D.; Kalıpçılar, H.; Yilmaz, L. Development of polycarbonate based zeolite 4a filled mixed matrix gas separation membranes. *J. Membr. Sci.* **2007**, *303*, 194–203. [CrossRef]
25. Hellums, M.W.; Koros, W.J.; Husk, G.R.; Paul, D.R. Fluorinated polycarbonates for gas separation applications. *J. Membr. Sci.* **1989**, *46*, 93–112. [CrossRef]
26. Lai, J.-Y.; Liu, M.-J.; Lee, K.-R. Polycarbonate membrane prepared via a wet phase inversion method for oxygen enrichment from air. *J. Membr. Sci.* **1994**, *86*, 103–118. [CrossRef]
27. Ward, W.J.; Browall, W.R.; Salemme, R.M. Ultrathin silicone/polycarbonate membranes for gas separation processes. *J. Membr. Sci.* **1976**, *1*, 99–108. [CrossRef]
28. Kesting, R.E.; Fritzsche, A.K.; Murphy, M.K.; Cruse, C.A.; Handermann, A.C.; Malon, R.F.; Moore, M.D. The second-generation polysulfone gas-separation membrane. I. The use of lewis acid: Base complexes as transient templates to increase free volume. *J. Appl. Polym. Sci.* **1990**, *40*, 1557–1574. [CrossRef]
29. Gür, T.M. Permselectivity of zeolite filled polysulfone gas separation membranes. *J. Membr. Sci.* **1994**, *93*, 283–289. [CrossRef]
30. Ahn, J.; Chung, W.-J.; Pinnau, I.; Guiver, M.D. Polysulfone/silica nanoparticle mixed-matrix membranes for gas separation. *J. Membr. Sci.* **2008**, *314*, 123–133. [CrossRef]
31. Hu, C.-C.; Liu, T.-C.; Lee, K.-R.; Ruaan, R.-C.; Lai, J.-Y. Zeolite-filled pmma composite membranes: Influence of coupling agent addition on gas separation properties. *Desalination* **2006**, *193*, 14–24. [CrossRef]
32. Jao-Ming, C.; Da-Ming, W.; Fung-Ching, L.; Juin-Yih, L. Formation and gas flux of asymmetric pmma membranes. *J. Membr. Sci.* **1996**, *109*, 93–107. [CrossRef]
33. Lai, J.-Y.; Huang, S.-J.; Chen, S.-H. Poly(methyl methacrylate)/(dmf/metal salt) complex membrane for gas separation. *J. Membr. Sci.* **1992**, *74*, 71–82. [CrossRef]

34. Budd, P.M.; Elabas, E.S.; Ghanem, B.S.; Makhseed, S.; McKeown, N.B.; Msayib, K.J.; Tattershall, C.E.; Wang, D. Solution-processed, organophilic membrane derived from a polymer of intrinsic microporosity. *Adv. Mater.* **2004**, *16*, 456–459. [CrossRef]

35. Budd, P.M.; Msayib, K.J.; Tattershall, C.E.; Ghanem, B.S.; Reynolds, K.J.; McKeown, N.B.; Fritsch, D. Gas separation membranes from polymers of intrinsic microporosity. *J. Membr. Sci.* **2005**, *251*, 263–269. [CrossRef]

36. Zhang, C.; Yan, J.; Tian, Z.; Liu, X.; Cao, B.; Li, P. Molecular design of tröger's base-based polymers containing spirobichroman structure for gas separation. *Ind. Eng. Chem. Res.* **2017**, *56*, 12783–12788. [CrossRef]

37. Zhang, C.; Fu, L.; Tian, Z.; Cao, B.; Li, P. Post-crosslinking of triptycene-based tröger's base polymers with enhanced natural gas separation performance. *J. Membr. Sci.* **2018**, *556*, 277–284. [CrossRef]

38. Scholes, C.A.; Stevens, G.W.; Kentish, S.E. Membrane gas separation applications in natural gas processing. *Fuel* **2012**, *96*, 15–28. [CrossRef]

39. Liaw, D.-J.; Wang, K.-L.; Huang, Y.-C.; Lee, K.-R.; Lai, J.-Y.; Ha, C.-S. Advanced polyimide materials: Syntheses, physical properties and applications. *Progr. Polym. Sci.* **2012**, *37*, 907–974. [CrossRef]

40. Wang, S.; Li, X.; Wu, H.; Tian, Z.; Xin, Q.; He, G.; Peng, D.; Chen, S.; Yin, Y.; Jiang, Z.; Guiver, M.D. Advances in high permeability polymer-based membrane materials for CO_2 separations. *Energy Environ. Sci.* **2016**, *9*, 1863–1890. [CrossRef]

41. Wind, J.D.; Staudt-Bickel, C.; Paul, D.R.; Koros, W.J. The effects of crosslinking chemistry on CO_2 plasticization of polyimide gas separation membranes. *Ind. Eng. Chem. Res.* **2002**, *41*, 6139–6148. [CrossRef]

42. Swaidan, R.J.; Ma, X.; Litwiller, E.; Pinnau, I. Enhanced propylene/propane separation by thermal annealing of an intrinsically microporous hydroxyl-functionalized polyimide membrane. *J. Membr. Sci.* **2015**, *495*, 235–241. [CrossRef]

43. Staudt-Bickel, C.; Koros, W.J. Olefin/paraffin gas separations with 6fda-based polyimide membranes. *J. Membr. Sci.* **2000**, *170*, 205–214. [CrossRef]

44. Das, M.; Koros, W.J. Performance of 6fda–6fpda polyimide for propylene/propane separations. *J. Membr. Sci.* **2010**, *365*, 399–408. [CrossRef]

45. Krol, J.J.; Boerrigter, M.; Koops, G.H. Polyimide hollow fiber gas separation membranes: Preparation and the suppression of plasticization in propane/propylene environments. *J. Membr. Sci.* **2001**, *184*, 275–286. [CrossRef]

46. Velioğlu, S.; Ahunbay, M.G.; Tantekin-Ersolmaz, S.B. Propylene/propane plasticization in polyimide membranes. *J. Membr. Sci.* **2016**, *501*, 179–190. [CrossRef]

47. Salinas, O.; Ma, X.; Litwiller, E.; Pinnau, I. Ethylene/ethane permeation, diffusion and gas sorption properties of carbon molecular sieve membranes derived from the prototype ladder polymer of intrinsic microporosity (pim-1). *J. Membr. Sci.* **2016**, *504*, 133–140. [CrossRef]

48. Xu, L.; Rungta, M.; Koros, W.J. Matrimid®derived carbon molecular sieve hollow fiber membranes for ethylene/ethane separation. *J. Membr. Sci.* **2011**, *380*, 138–147. [CrossRef]

49. Zhou, S.; Stern, S.A. The effect of plasticization on the transport of gases in and through glassy polymers. *J. Polym. Sci. Part B Polym. Phys.* **1989**, *27*, 205–222. [CrossRef]

50. Tiwari, R.R.; Smith, Z.P.; Lin, H.; Freeman, B.D.; Paul, D.R. Gas permeation in thin films of "high free-volume" glassy perfluoropolymers: Part ii. CO_2 plasticization and sorption. *Polymer* **2015**, *61*, 1–14. [CrossRef]

51. Sanders, E.S. Penetrant-induced plasticization and gas permeation in glassy polymers. *J. Membr. Sci.* **1988**, *37*, 63–80. [CrossRef]

52. Chiou, J.S.; Barlow, J.W.; Paul, D.R. Plasticization of glassy polymers by CO_2. *J. Appl. Polym. Sci.* **1985**, *30*, 2633–2642. [CrossRef]

53. Li, P.; Chung, T.S.; Paul, D.R. Gas sorption and permeation in pim-1. *J. Membr. Sci.* **2013**, *432*, 50–57. [CrossRef]

54. Bos, A.; Pünt, I.G.M.; Wessling, M.; Strathmann, H. CO_2-induced plasticization phenomena in glassy polymers. *J. Membr. Sci.* **1999**, *155*, 67–78. [CrossRef]

55. Petropoulos, J.H. Plasticization effects on the gas permeability and permselectivity of polymer membranes. *J. Membr. Sci.* **1992**, *75*, 47–59. [CrossRef]

56. Kratochvil, A.M.; Koros, W.J. Decarboxylation-induced cross-linking of a polyimide for enhanced CO_2 plasticization resistance. *Macromolecules* **2008**, *41*, 7920–7927. [CrossRef]

57. Qiu, W.; Chen, C.-C.; Xu, L.; Cui, L.; Paul, D.R.; Koros, W.J. Sub-tg cross-linking of a polyimide membrane for enhanced CO_2 plasticization resistance for natural gas separation. *Macromolecules* **2011**, *44*, 6046–6056. [CrossRef]

58. Zhang, C.; Li, P.; Cao, B. Decarboxylation crosslinking of polyimides with high CO_2/CH_4 separation performance and plasticization resistance. *J. Membr. Sci.* **2017**, *528*, 206–216. [CrossRef]

59. Zhang, C.; Cao, B.; Li, P. Thermal oxidative crosslinking of phenolphthalein-based cardo polyimides with enhanced gas permeability and selectivity. *J. Membr. Sci.* **2018**, *546*, 90–99. [CrossRef]

60. Tian, Z.; Cao, B.; Li, P. Effects of sub-tg cross-linking of triptycene-based polyimides on gas permeation, plasticization resistance and physical aging properties. *J. Membr. Sci.* **2018**, *560*, 87–96. [CrossRef]

61. Bos, A.; Pünt, I.G.M.; Wessling, M.; Strathmann, H. Plasticization-resistant glassy polyimide membranes for $CO_2/co4$ separations. *Sep. Purif. Technol.* **1998**, *14*, 27–39. [CrossRef]

62. Park, H.B.; Jung, C.H.; Lee, Y.M.; Hill, A.J.; Pas, S.J.; Mudie, S.T.; Van Wagner, E.; Freeman, B.D.; Cookson, D.J. Polymers with cavities tuned for fast selective transport of small molecules and ions. *Science* **2007**, *318*, 254–258. [CrossRef]

63. Gleason, K.L.; Smith, Z.P.; Liu, Q.; Paul, D.R.; Freeman, B.D. Pure- and mixed-gas permeation of CO_2 and CH_4 in thermally rearranged polymers based on 3,3′-dihydroxy-4,4′-diamino-biphenyl (hab) and 2,2′-bis-(3,4-dicarboxyphenyl) hexafluoropropane dianhydride (6fda). *J. Membr. Sci.* **2015**, *475*, 204–214. [CrossRef]

64. Shao, L.; Chung, T.; Goh, S.; Pramoda, K. Polyimide modification by a linear aliphatic diamine to enhance transport performance and plasticization resistance. *J. Membr. Sci.* **2005**. [CrossRef]

65. Shao, L.; Chung, T.-S.; Goh, S.H.; Pramoda, K.P. The effects of 1,3-cyclohexanebis(methylamine) modification on gas transport and plasticization resistance of polyimide membranes. *J. Membr. Sci.* **2005**, *267*, 78–89. [CrossRef]

66. Xiao, Y.; Shao, L.; Chung, T.-S.; Schiraldi, D.A. Effects of thermal treatments and dendrimers chemical structures on the properties of highly surface cross-linked polyimide films. *Ind. Eng. Chem. Res.* **2005**, *44*, 3059–3067. [CrossRef]

67. Tin, P.S.; Chung, T.S.; Liu, Y.; Wang, R.; Liu, S.L.; Pramoda, K.P. Effects of cross-linking modification on gas separation performance of matrimid membranes. *J. Membr. Sci.* **2003**, *225*, 77–90. [CrossRef]

68. Wind, J.D.; Sirard, S.M.; Paul, D.R.; Green, P.F.; Johnston, K.P.; Koros, W.J. Carbon dioxide-induced plasticization of polyimide membranes: Pseudo-equilibrium relationships of diffusion, sorption, and swelling. *Macromolecules* **2003**, *36*, 6433–6441. [CrossRef]

69. Wind, J.D.; Staudt-Bickel, C.; Paul, D.R.; Koros, W.J. Solid-state covalent cross-linking of polyimide membranes for carbon dioxide plasticization reduction. *Macromolecules* **2003**, *36*, 1882–1888. [CrossRef]

70. Bos, A.; Pünt, I.G.M.; Wessling, M.; Strathmann, H. Suppression of CO_2-plasticization by semiinterpenetrating polymer network formation. *J. Polym. Sci. Part B Polym. Phys.* **1998**, *36*, 1547–1556. [CrossRef]

71. Maya, E.M.; Sanchez, M.L.; Marcos, A.; de la Campa, J.G.; de Abajo, J. Preparation and properties of catalyzed polyimide/dicyanate semi-interpenetrating networks for polymer gas membrane with suppressed CO_2-plasticization. *J. Appl. Polym. Sci.* **2011**, *124*, 713–722. [CrossRef]

72. Low, B.T.; Chung, T.S.; Chen, H.; Jean, Y.-C.; Pramoda, K.P. Tuning the free volume cavities of polyimide membranes via the construction of pseudo-interpenetrating networks for enhanced gas separation performance. *Macromolecules* **2009**, *42*, 7042–7054. [CrossRef]

73. Kammakakam, I.; Wook Yoon, H.; Nam, S.; Bum Park, H.; Kim, T.-H. Novel piperazinium-mediated crosslinked polyimide membranes for high performance CO_2 separation. *J. Membr. Sci.* **2015**, *487*, 90–98. [CrossRef]

74. Askari, M.; Chung, T.-S. Natural gas purification and olefin/paraffin separation using thermal cross-linkable co-polyimide/zif-8 mixed matrix membranes. *J. Membr. Sci.* **2013**, *444*, 173–183. [CrossRef]

75. Zhuang, G.-L.; Tseng, H.-H.; Uchytil, P.; Wey, M.-Y. Enhancing the CO_2 plasticization resistance of ps mixed-matrix membrane by blunt zeolitic imidazolate framework. *J. CO_2 Utilization* **2018**, *25*, 79–88. [CrossRef]

76. Jusoh, N.; Yeong, Y.F.; Lau, K.K.; Shariff, A.M. Enhanced gas separation performance using mixed matrix membranes containing zeolite t and 6fda-durene polyimide. *J. Membr. Sci.* **2017**, *525*, 175–186. [CrossRef]

77. Shahid, S.; Nijmeijer, K. Performance and plasticization behavior of polymer–mof membranes for gas separation at elevated pressures. *J. Membr. Sci.* **2014**, *470*, 166–177. [CrossRef]

78. Bachman, J.E.; Long, J.R. Plasticization-resistant ni2(dobdc)/polyimide composite membranes for the removal of CO_2 from natural gas. *Energy Environ. Sci.* **2016**, *9*, 2031–2036. [CrossRef]
79. Kapantaidakis, G.C.; Kaldis, S.P.; Dabou, X.S.; Sakellaropoulos, G.P. Gas permeation through psf-pi miscible blend membranes. *J. Membr. Sci.* **1996**, *110*, 239–247. [CrossRef]
80. Bos, A.; Pünt, I.; Strathmann, H.; Wessling, M. Suppression of gas separation membrane plasticization by homogeneous polymer blending. *AIChE J.* **2004**, *47*, 1088–1093. [CrossRef]
81. Yong, W.F.; Chung, T.-S. Miscible blends of carboxylated polymers of intrinsic microporosity (cpim-1) and matrimid. *Polymer* **2015**, *59*, 290–297. [CrossRef]
82. Yong, W.F.; Li, F.Y.; Chung, T.S.; Tong, Y.W. Molecular interaction, gas transport properties and plasticization behavior of cpim-1/torlon blend membranes. *J. Membr. Sci.* **2014**, *462*, 119–130. [CrossRef]
83. Liu, Y.; Wang, R.; Chung, T.-S. Chemical cross-linking modification of polyimide membranes for gas separation. *J. Membr. Sci.* **2001**, *189*, 231–239. [CrossRef]
84. Shao, L.; Liu, L.; Cheng, S.-X.; Huang, Y.-D.; Ma, J. Comparison of diamino cross-linking in different polyimide solutions and membranes by precipitation observation and gas transport. *J. Membr. Sci.* **2008**, *312*, 174–185. [CrossRef]
85. Chung, T.-S.; Chng, M.L.; Pramoda, K.P.; Xiao, Y. Pamam dendrimer-induced cross-linking modification of polyimide membranes. *Langmuir* **2004**, *20*, 2966–2969. [CrossRef]
86. Xiao, Y.; Chung, T.-S.; Chng, M.L. Surface characterization, modification chemistry, and separation performance of polyimide and polyamidoamine dendrimer composite films. *Langmuir* **2004**, *20*, 8230–8238. [CrossRef]
87. Shao, L.; Chung, T.-S.; Goh, S.H.; Pramoda, K.P. Transport properties of cross-linked polyimide membranes induced by different generations of diaminobutane (dab) dendrimers. *J. Membr. Sci.* **2004**, *238*, 153–163. [CrossRef]
88. Zhao, H.-Y.; Cao, Y.-M.; Ding, X.-L.; Zhou, M.-Q.; Liu, J.-H.; Yuan, Q. Poly(ethylene oxide) induced cross-linking modification of matrimid membranes for selective separation of CO_2. *J. Membr. Sci.* **2008**, *320*, 179–184. [CrossRef]
89. Wind, J.D.; Paul, D.R.; Koros, W.J. Natural gas permeation in polyimide membranes. *J. Membr. Sci.* **2004**, *228*, 227–236. [CrossRef]
90. Park, H.B.; Lee, C.H.; Sohn, J.Y.; Lee, Y.M.; Freeman, B.D.; Kim, H.J. Effect of crosslinked chain length in sulfonated polyimide membranes on water sorption, proton conduction, and methanol permeation properties. *J. Membr. Sci.* **2006**, *285*, 432–443. [CrossRef]
91. Seo, J.; Jang, W.; Lee, S.; Han, H. The stability of semi-interpenetrating polymer networks based on sulfonated polyimide and poly(ethylene glycol) diacrylate for fuel cell applications. *Polym. Degrad. Stab.* **2008**, *93*, 298–304. [CrossRef]
92. Alam, S.; Kandpal, L.D.; Varma, I.K. Ethynyl-terminated imide oligomers. *J. Macromol. Sci. Part C* **1993**, *33*, 291–320. [CrossRef]
93. Alaslai, N.; Ghanem, B.; Alghunaimi, F.; Litwiller, E.; Pinnau, I. Pure- and mixed-gas permeation properties of highly selective and plasticization resistant hydroxyl-diamine-based 6fda polyimides for CO_2/CH_4 separation. *J. Membr. Sci.* **2016**, *505*, 100–107. [CrossRef]
94. Vinoba, M.; Bhagiyalakshmi, M.; Alqaheem, Y.; Alomair, A.A.; Pérez, A.; Rana, M.S. Recent progress of fillers in mixed matrix membranes for CO_2 separation: A review. *Sep. Purif. Technol.* **2017**, *188*, 431–450. [CrossRef]
95. Rezakazemi, M.; Ebadi Amooghin, A.; Montazer-Rahmati, M.M.; Ismail, A.F.; Matsuura, T. State-of-the-art membrane based CO_2 separation using mixed matrix membranes (mmms): An overview on current status and future directions. *Progr. Polym. Sci.* **2014**, *39*, 817–861. [CrossRef]
96. Vu, D.Q.; Koros, W.J.; Miller, S.J. Mixed matrix membranes using carbon molecular sieves: I. Preparation and experimental results. *J. Membr. Sci.* **2003**, *211*, 311–334. [CrossRef]
97. Singh, A.; Koros, W.J. Significance of entropic selectivity for advanced gas separation membranes. *Ind. Eng. Chem. Res.* **1996**, *35*, 1231–1234. [CrossRef]
98. Anson, M.; Marchese, J.; Garis, E.; Ochoa, N.; Pagliero, C. Abs copolymer-activated carbon mixed matrix membranes for CO_2/CH_4 separation. *J. Membr. Sci.* **2004**, *243*, 19–28. [CrossRef]
99. García, M.G.; Marchese, J.; Ochoa, N.A. High activated carbon loading mixed matrix membranes for gas separations. *J. Mater. Sci.* **2012**, *47*, 3064–3075. [CrossRef]

100. Marchese, J.; Anson, M.; Ochoa, N.A.; Prádanos, P.; Palacio, L.; Hernández, A. Morphology and structure of abs membranes filled with two different activated carbons. *Chem. Eng. Sci.* **2006**, *61*, 5448–5454. [CrossRef]

101. Rostamizadeh, M.; Rezakazemi, M.; Shahidi, K.; Mohammadi, T. Gas permeation through h2-selective mixed matrix membranes: Experimental and neural network modeling. *Int. J. Hydrog. Energy* **2013**, *38*, 1128–1135. [CrossRef]

102. Adams, R.T.; Lee, J.S.; Bae, T.-H.; Ward, J.K.; Johnson, J.R.; Jones, C.W.; Nair, S.; Koros, W.J. CO_2–CH_4 permeation in high zeolite 4a loading mixed matrix membranes. *J. Membr. Sci.* **2011**, *367*, 197–203. [CrossRef]

103. Merkel, T.C.; He, Z.; Pinnau, I.; Freeman, B.D.; Meakin, P.; Hill, A.J. Effect of nanoparticles on gas sorption and transport in poly(1-trimethylsilyl-1-propyne). *Macromolecules* **2003**, *36*, 6844–6855. [CrossRef]

104. Ge, L.; Zhou, W.; Rudolph, V.; Zhu, Z. Mixed matrix membranes incorporated with size-reduced cu-btc for improved gas separation. *J. Mater. Chem. A* **2013**, *1*, 6350–6358. [CrossRef]

105. He, Z.; Pinnau, I.; Morisato, A. Nanostructured poly(4-methyl-2-pentyne)/silica hybrid membranes for gas separation. *Desalination* **2002**, *146*, 11–15. [CrossRef]

106. Yang, T.; Chung, T.-S. High performance zif-8/pbi nano-composite membranes for high temperature hydrogen separation consisting of carbon monoxide and water vapor. *Int. J. Hydrog. Energy* **2013**, *38*, 229–239. [CrossRef]

107. Japip, S.; Wang, H.; Xiao, Y.; Shung Chung, T. Highly permeable zeolitic imidazolate framework (zif)-71 nano-particles enhanced polyimide membranes for gas separation. *J. Membr. Sci.* **2014**, *467*, 162–174. [CrossRef]

108. Shahid, S.; Nijmeijer, K. High pressure gas separation performance of mixed-matrix polymer membranes containing mesoporous fe(btc). *J. Membr. Sci.* **2014**, *459*, 33–44. [CrossRef]

109. Caro, J. Are mof membranes better in gas separation than those made of zeolites? *Curr. Opin. Chem. Eng.* **2011**, *1*, 77–83. [CrossRef]

110. Matsui, S.; Ishiguro, T.; Higuchi, A.; Nakagawa, T. Effect of ultraviolet light irradiation on gas permeability in polyimide membranes. 1. Irradiation with low pressure mercury lamp on photosensitive and nonphotosensitive membranes. *J. Polym. Sci. Part B Polym. Phys.* **1998**, *35*, 2259–2269. [CrossRef]

111. Matsui, S.; Nakagawa, T. Effect of ultraviolet light irradiation on gas permeability in polyimide membranes. Ii. Irradiation of membranes with high-pressure mercury lamp. *J. Appl. Polym. Sci.* **1998**, *67*, 49–60. [CrossRef]

112. Hayes, R.A. EI du Pont de Nemours and Co. Polyimide Gas Separation Membranes. U.S. Patent 4,717,393, 1988.

113. Kita, H.; Inada, T.; Tanaka, K.; Okamoto, K.-I. Effect of photocrosslinking on permeability and permselectivity of gases through benzophenone- containing polyimide. *J. Membr. Sci.* **1994**, *87*, 139–147. [CrossRef]

114. Liu, Y.; Pan, C.; Ding, M.; Xu, J. Gas permeability and permselectivity of photochemically crosslinked copolyimides. *J. Appl. Polym. Sci.* **1999**, *73*, 521–526. [CrossRef]

115. Liu, Y.; Pan, C.; Ding, M.; Xu, J. Effect of crosslinking distribution on gas permeability and permselectivity of crosslinked polyimides. *Eur. Polym. J.* **1999**, *35*, 1739–1741. [CrossRef]

116. Kapantaidakis, G.C.; Kaldis, S.P.; Sakellaropoulos, G.P.; Chira, E.; Loppinet, B.; Floudas, G. Interrelation between phase state and gas permeation in polysulfone/polyimide blend membranes. *J. Polym. Sci. Part B Polym. Phys.* **1999**, *37*, 2788–2798. [CrossRef]

117. Hao, L.; Li, P.; Chung, T.-S. Pim-1 as an organic filler to enhance the gas separation performance of ultem polyetherimide. *J. Membr. Sci.* **2014**, *453*, 614–623. [CrossRef]

processes

MDPI

Article

Steam Explosion and Vibrating Membrane Filtration to Improve the Processing Cost of Microalgae Cell Disruption and Fractionation

Esther Lorente [1], Monika Haponska [2], Ester Clavero [1] , Carles Torras [2,*] and Joan Salvadó [2]

[1]   Catalonia Institute for Energy Research, IREC, Marcel·lí Domingo 2, 43007 Tarragona, Catalonia, Spain; elorenteroyo@gmail.com (E.L.); ester.clavero@gmail.com (E.C.)
[2]   Chemical Engineering Department, Universitat Rovira i Virgili, Av. Països Catalans 26, 43007 Tarragona, Catalonia, Spain; monika.haponska@gmail.com (M.H.); joan.salvado@urv.cat (J.S.)
*   Correspondence: carlestorrasfont@gmail.com; Tel.: +34-977-202-444

Received: 8 February 2018; Accepted: 20 March 2018; Published: 22 March 2018

Abstract: The aim of this study is to explore an innovative downstream route for microalgae processing to reduce cost production. Experiments have been carried out on cell disruption and fractionation stages to recover lipids, sugars, and proteins. Steam explosion and dynamic membrane filtration were used as unit operations. The species tested were *Nannochloropsis gaditana*, *Chlorella sorokiniana*, and *Dunaliella tertiolecta* with different cell wall characteristics. Acid-catalysed steam explosion permitted cell disruption, as well as the hydrolysis of carbohydrates and partial hydrolysis of proteins. This permitted a better access to non-polar solvents for lipid extraction. Dynamic filtration was used to moderate the impact of fouling. Filtration enabled two streams: A permeate containing water and monosaccharides and a low-volume retentate containing the lipids and proteins. The necessary volume of solvent to extract the lipids is thus much lower. An estimation of operational costs of both steam explosion and membrane filtration was performed. The results show that the steam explosion operation cost varies between 0.005 \$/kg and 0.014 \$/kg of microalgae dry sample, depending on the cost of fuel. Membrane filtration cost in fractionation was estimated at 0.12 \$/kg of microalgae dry sample.

Keywords: cell disruption; costs; dynamic membrane filtration; fractionation; microalgae; steam explosion

---

1. Introduction

Around 10 years ago, the idea of using microalgae as a very efficient photosynthetic crop to provide energy was re-adopted [1], following the results obtained in earlier studies [2]. Microalgae appeared as a good alternative to produce transportation fuels in the context of energy crisis and climate change.

Cost barriers in the several stages of mass production of energy vectors appeared. This resulted in having to re-address improvements in culture, harvesting, cell disruption, lipid extraction, and final production.

The production of biofuels from microalgae results in a variety of returns. These include a high lipid content, no competition for arable lands, and the use of a variety of water qualities, including wastewaters during the cultivation period [3]. However, it has become clear that the option to produce only fuel from microalgae is not economically viable [4].

Researchers have learned from the preliminary results that, apart from reducing the costs of microalgae production, benefits have to be obtained from all fractions while also looking for other side paybacks in order to have an economically feasible production [5].

To achieve a positive economic balance, several matters should be taken into account such as CO_2 capture [6,7], water quality improvement [8,9], procurement of commodities [10,11], and high-added value products [12,13].

The process unit operations needed in order to proceed to microalgae biorefining depend very much on the strain and products to be sought (i.e., commodity or high-added value products), but a typical sequence is: Culture in open ponds or photobioreactors [14,15], dewatering [16–21], cell disruption [22–27], and fractionation [13,20,28].

In the present study, the focus is on the operation of cell disruption by using steam explosion, and secondly, in the process of fractionation.

Although steam explosion has been in use from the beginning of the 20th century, it has only been used in a few cases for microalgae biorefining, and therefore a comparison of the results can hardly be performed [22,29]. As regards the results of previous work [26], steam explosion is used at relatively mild conditions to break the cell walls and produce the hydrolysis of carbohydrates.

Depending on the strain cell wall characteristics, cell disruption can be a cost-intensive operation and several procedures have been reported at the laboratory level, including the use of ultrasounds, microwave, or high pressure [25,30]. Steam explosion is proposed in this work as an innovative technique for this application and is easy to scale-up with pilot plant results because of the nature of the equipment and because it is widely used industrially [22,24]. Steam explosion has given the best results when compared with other methods for cell disruption such as ultrasonication, microwave, and autoclave [29]. Beyond breaking the cell wall, if a low concentration of acid is used, steam explosion can hydrolysate the poly-saccharides in the cell and produce sugars in a first stage of fractionation [29]. The main energy input for the steam explosion process is heat, thus reducing the cell disruption costs considering other techniques like sonication and that residual heat can be used. It should be stated that steam explosion is a commercial high-throughput available technology. A pilot plant with a capacity of 2 Tm/h has already been operated successfully with lignocellulosic materials from 1991 [31].

Membrane filtration and solvent extraction are methods to be used for fractionation [24,28]. In a first unit operation, membrane filtration can be used to obtain two streams: a retentate containing lipids and proteins and a permeate containing water with the hydrolyzed monosaccharides [24]. As in microalgae dewatering, fouling is a main drawback. To overcome this problem, dynamic filtration provides an adequate solution [32]. Also, the use of ultrafiltration membranes (instead of microfiltration) increases permeability [20]. In a second unit operation, sugars could be concentrated using nanofiltration membranes [33]. To recover non-polar lipids from the retentate stream of the first operation, a hexane extraction is used. In our previous work [24], the microalga *Nannochloropsis gaditana* was selected to investigate the fractionation strategy for lipids and carbohydrates recovery. In this study, we intend to validate the selected fractionation path when different common microalgae species were used: *Chlorella sorokiniana* [34], *Nannochloropsis gaditana* [35], and *Dunaliella tertiolecta* [36]. They are representative of different types of species of freshwater and marine strains. They have also been chosen because they represent different levels of strength in their cell walls. *N. gaditana* and *C. sorokiniana* are two species with recalcitrant cell walls, whereas *D. tertiolecta* lacks a cell wall. The cell wall of *N. gaditana* is primarily cellulose (75%) [37]. This inner cellulose layer is protected by an algaenan layer which is assumed to be primarily responsible for the wall's recalcitrance to breakage [37]. Besides, the *C. sorokiniana* cell wall contains little glucose [38] and therefore its cell wall may lack cellulose. On the other hand, the presence of algaenan in the *C. sorokiniana* cell wall may depend on the physiological state of the culture [39].

The study about the use of steam explosion will provide a basis of cost comparison with those technologies that use electrical power to operate.

2. Materials and Methods

2.1. Microalgae Samples

A semi-closed photobioreactor, with a 3050 L capacity and placed outdoors, was used for growing *Nannochloropsis gaditana* Lubián (strain CCMP1775, Provasoli—Guillard National Center for Marine Algae and Microbiota). A more detailed description of the photobioreactor is given in Nurra et al. [26]. Cultures were performed between May and July, when the mean temperature ranged from 27 °C to 33 °C. The medium for the *N. gaditana* culture consisted of seawater enriched with 0.3 mL/L of Codafol 14.6.5 (Coda Sustainable Agro Solution S.A.). This plant fertilizer contains, in w/w, 14% nitrogen, 6% P_2O_5, 5% K_2O, 0.1% Fe, 0.05% Zn, 0.05% Mn, 0.05% Cu, and 0.001% Mo.

Chlorella sorokiniana (strain CCAP 211/8k) and *Dunaliella tertiolecta* (strain CCAP19/6B) were grown indoors in column photobioreactors (300 L, 50 cm diam.) aerated with air and illuminated with Philips MASTER TLD 58 W/865 fluorescents giving an irradiance at the photobioreactor surface of 300 µmol photon/m^2/s. *C. sorokiniana* was cultured at 22 ± 3 °C in tap water enriched with the following nutrients: $NaNO_3$ (5.8 mM), $K_2HPO_4 \cdot 3H_2O$ (0.092 mM), KH_2PO_4 (0.28 mM), Na_2EDTA (0.045 mM), $FeCl_3 \cdot 6H_2O$ (17.9 µM), $ZnSO_4 \cdot 7H_2O$ (1.69 µM), $MnCl_2 \cdot 4H_2O$ (4.48 µM), $Na_2MoO_4 \cdot 2H_2O$ (0.10 µM), $CuSO_4 \cdot 5H_2O$ (0.17 µM), and $CoCl_2 \cdot 6H_2O$ (0.06 µM). *D. tertiolecta* was cultured at 20 ± 3 °C in artificial seawater prepared with tap water and 37 $g \cdot L^{-1}$ of Aquaforest Reef Salt® enriched with $NaNO_3$ (4.4 mM), $Na_2HPO_4 \cdot 2H_2O$ (0.04 mM), and the same micronutrient concentrations as in *C. sorokiniana*. Phosphate was fed-batch to increase the concentration of the culture to 3.2 µM to avoid precipitation, presumably with magnesium and calcium ions.

All the cultures were harvested some days after the stationary phase of growing was reached, except for the cultures of *D. tertiolecta* used in the steam explosion treatment without acid, which were harvested at the end of the log phase.

A continuous centrifuge (Clara 20 High Flow, Alfa-Laval, Lund, Sweden) was used to concentrate the microalgal biomass samples. The centrifuge was operated at 9060 rpm, using a counter pressure of 4 bar. A Seepex progressive cavity pump (BN series) was used to feed the sample with 1000 L/h of a nominal flow rate. After concentration, the samples *N. gaditana* and *C. sorokiniana* were frozen at −80 °C. For defrosting the samples, they were placed at 4 °C for two days, prior to the steam explosion procedure. *D. tertiolecta* was harvested and concentrated just before the biorefinery process to avoid extra actions that might break its naked cells.

2.2. Steam Explosion

The equipment for the steam explosion of microalgae consisted of a 16 L reactor, operated in batch, and a collection vessel. The generation of steam was achieved with an electric boiler (Boreal, 380 V/82 kW) and thermally isolated high-pressure pipes were used to conduct the steam to the reactor. This was regulated by two valves placed in series, which were used to control the entrance of steam into the reactor. In the upper part of the reactor, there was a valve (2″ diameter) for feeding the sample. In the bottom of the reactor, a flash valve allowed a fast decompression to the collecting tank at atmospheric pressure. The tank consisted of a cylinder with a capacity of 100 L and a diameter of 50 cm. It had two valves, one for steam release and another for the collection of sample in liquid phase.

In each experiment, 4 kg of microalgae was introduced into the reactor, which had been preheated. Some samples were previously impregnated with sulphuric acid at a concentration of 5% (w/w, wet sample basis) by mixing for 2 h at room temperature. The steam explosion pre-treatments were conducted at 150 °C (which corresponds to a saturated steam pressure of 4.7 bar) with a retention time of 5 min. The selection of the experimental conditions, which includes temperature, time, and acid concentration, was performed in a previous study [29]. After reaction and before the fractionation experiments, the exploded samples were collected and neutralized (to pH 5).

2.3. Filtration

A Vibratory Shear Enhanced Processing (VSEP, serie L, New Logic Research, Inc., Emeryville, CA, USA) system was used to perform dynamic membrane filtration experiments. A detailed description of this filtering system can be found elsewhere [19]. Approximately 6.0 kg of microalgal sample was used for each experiment, with a transmembrane pressure of 5 bar and a vibrational frequency of 55.4 ± 0.1 Hz.

Experiments were performed with PE5, a commercial polymeric membrane (Nanostone, Eden Prairie, MN, USA), manufactured from polyether-sulfone and with a molecular weight cut-off (MWCO) of 5000 Da. The filtration area was $0.0446 \, m^2$.

Water flux measurements were performed in order to determine the permeability of virgin membranes. After that, the steam exploded microalgae biomass was filtered and measurements of permeability vs. time were conducted during the experiment. The permeability with pre-treated algae was determined at the fixed time of 60 min. Finally, after cleaning, the system water permeability was measured again. The last step allowed for the determination of the irreversible fouling resistance of membranes. Also, two factors could be calculated, i.e., the irreversible fouling factor (IF), which is determined as the ratio of water permeabilities before and after the experiment, and total fouling factor (TF), consisting of the ratio between virgin membrane permeability with water and microalgae sludge permeability. In all cases, permeability was calculated from measurements of permeate mass weight progress with time. Permeate output was driven to a vessel placed on a scale, which was connected to a computer. An own-made software was recording and calculating permeability in real time to assess experimentation. Permeability was determined as follows. For water, measurements were performed at three different transmembrane pressures between the recommended range given by the manufacturer to ensure that a linear correlation between both parameters was achieved. For microalgae sludge, flow rate measurements were being performed with an interval of 10 s.

2.4. Lipid Extraction

The lipids from microalgal samples were extracted by contacting the same volume of sample and of *n*-hexane (20 mL). The extraction conditions were 60 °C and agitation at 800 rpm, for 2 h. After the contact time, separation was achieved by centrifugation at 4000 rpm for 10 min. The mixture partitioned into three fractions: organic phase, aqueous phase, and residual solid. To extract and quantify lipids, the top hexane phase was recovered and was then heated to complete dryness in the oven (at 70 °C).

2.5. Analytical Techniques

2.5.1. Light Microscope

A Zeiss Axio Scope A1 (Carl Zeiss Light Microscopy, Jena, Germany) microscope, equipped with Nomarski interference contrast optics, was used to check the effects of the steam explosion technique on cell morphology. A digital camera JENOPTIK ProgRes Speed Xtcore 3 was used to obtain the light micrographs. Objective magnifications from 10 to 100 were used.

2.5.2. Dry Matter and Ash Content (TGA)

Thermogravimetric analyses (TGA), with a LECO instrument (TGA701), were performed in order to determine the dry ash free (DAF) weight of the samples which allows us to verify the mass balance during the steam explosion and membrane filtration processes. The samples were dried in a nitrogen atmosphere at 105 °C to constant mass, for the dry matter content determination. After that, the atmosphere was changed to oxygen and the temperature was increased up to 550 °C, in order to determine the ash content.

2.5.3. Total Lipid Extraction with Bligh and Dyer Method

The Bligh and Dyer method was used to extract the lipids from the fresh and steam exploded microalgal biomass. This method is the most commonly used at the analytical level for the quantitative extraction of lipids from microalgae [40].

2.5.4. Analytical Acid Hydrolysis

In order to determine the total extractable sugars, analytical acid hydrolysis experiments were conducted with the fresh microalgal samples, following a standard procedure (ASTM D1106-84). Although this method was originally used with lignocellulosic materials, microalgal biomass has also been previously analyzed [41]. The process consists of sulphuric acid hydrolysis in two stages. In the first stage, the freeze dried algal biomass sample (300 mg) is placed in contact with 72% (w/w, wet basis) sulphuric acid in a water bath at 30 °C, for 1 h. In the second stage, the sample is diluted to a concentration of 4% (w/w, wet basis) sulphuric acid and placed in an autoclave at 120 °C, for 45 min. After hydrolysis, filtration is performed using glass fiber filters in order to separate the acid insoluble residues from the hydrolysate. Finally, HPLC (high performance liquid chromatography) analyses were performed to quantitatively determine the sugar contents.

2.5.5. Monosaccharides Analysis

HPLC analyses were conducted in order to identify and quantify the monosaccharides present in the microalgal samples in solution. A Biorad Aminex HPX-87H column (300 mm × 7.8 mm) at 50 °C was used, with a refraction index detector. Additionally, the mobile phase was a 5 mM solution of sulphuric acid with a flow rate of 0.5 mL/min. The identification of monomeric sugars was achieved by a comparison of retention times with those of the standards. The integration of peaks in the chromatograms allowed the quantification, using a calibration curve, which was previously prepared with the standards.

2.5.6. Protein Analysis

Two different methods were used for protein analysis, namely solubilization and hot NaOH. To quantify the proteins released by the steam explosion treatment, the solubilization method was used. In this method, proteins were suspended by mixing 0.2 mL of sample in 1 mL 0.1 N NaOH. After 1 h of incubation at room temperature, samples were centrifuged at 4000 rpm for 10 min. Protein in the supernatant was precipitated with trichloroacetic acid (TCA) to avoid interfering substances. Following Barbarino and Lourenço [42], proteins were precipitated with 25% TCA at the ratio of 2.5:1 (TCA:homogenate) and centrifuged at 4000 rpm. Pellets were consecutively re-suspended in 10% and 5% TCA and finally solubilized in 0.1 N NaOH for the Bicinchoninic acid protein assay (BCA kit, Sigma-Aldrich, St. Louis, MO, USA). Color development was measured as absorbance at 562 nm using a microplate reader (INFINITE M200 PRO, Tecan, Männedorf, Switzerland). Absorbance values were read against a standard curve generated with a protein standard (bovine serum albumin), and percentage protein was calculated on a dry weight basis.

Since cell disruption was not expected using the solubilization method, a stronger method (hot NaOH) that allowed cell wall disruption was also applied to the concentrated culture and the steam exploded sample to evaluate the effects of steam explosion. In this procedure, 0.5 mL samples were extracted with 0.5 mL 2 N NaOH with 0.5% β-mercaptoethanol (v/v) at 90 °C for 10 min and centrifuged at 4000 rpm. Proteins were precipitated with TCA and solubilized in 0.1 N NaOH for the Bicinchoninic acid assay, as explained previously. Both extraction methods were performed in triplicate.

2.5.7. Particle Size Distribution

A Malvern Mastersizer 2000 piece of equipment with the Hydro 2000 MU module for liquid samples was used for particle size distribution measurements. A blue laser light was used. The medium consisted of 500 mL of demineralized water and sludge sample drops were added without further treatment until obtaining an appropriate obscuration level (as stated by the equipment instructions).

Two different levels of sonication: 6 kHz and 24 kHz, were used in the measurements, in order to check if aggregation had occurred with particles present in the sludge.

2.5.8. Optical Density

Absorbance measurements at 750 nm were performed to estimate the turbidity of the permeate, which can confirm total particle or oil rejection after membrane filtration. Absorbance was measured using a microplate reader (INFINITE M200 PRO, Tecan), and 96 well plates were used for the absorbance determinations. The optical density (OD750 nm) values were obtained by dividing the raw values over the path-length, and using as a reference the OD750 nm of filtered (0.45 µm) seawater.

3. Results and Discussion

3.1. Steam Explosion Treatment of Studied Strains

A steam explosion experiment was performed for each microalgae sample, at 150 °C, for 5 min and using 5% sulphuric acid to impregnate the samples. An additional experiment was performed with *D. tertiolecta*, to analyze the effect of steam explosion without acid impregnation, since this microalga has no cell wall. By comparing the dry ash free weight values of the samples before and after the steam explosion treatment, good balance closures (>97%) were obtained for all the experiments.

3.1.1. Cell Morphology

The examination of cell morphology by light microscopy showed that *C. sorokiniana*, *N. gaditana*, and *D. tertiolecta* had experienced high levels of cell disruption after the steam explosion pretreatment (Figure 1). Original samples consisted of isolated cells, except for *C. sorokiniana*, which contained both single cells and cell aggregates, hence the bimodal distribution in Figure 2B. Sonication dispersed cells and most aggregates were disintegrated. Accordingly, after sonication, the peak centered in ca. 3 µm, matching the *C. sorokiniana* cell size, was much higher, and the peak centered at ca. 20 µm which corresponds to aggregates almost disappeared.

Although *C. sorokiniana* appeared slightly damaged after thawing, with the cytoplasm slightly shrunken and retracted from the smooth cell wall, it was the less injured of the three species after steam explosion. *C. sorokiniana* cells showed three different patterns of disruption. Cells could be totally disrupted, algal material appearing as granulated aggregates. Cells could also maintain their unity but have granular cytoplasm and wrinkled margins. In this condition, cells had a low contrast appearance, which reveals that shapes may be flatter, probably due to a thinner and softer cell wall. More often, *C. sorokiniana* cells maintained their unity and high contrast appearance with smooth margins, but the cellular content was homogeneous except for a central depression, and no intracellular organelles (like chloroplast or pyrenoid) could be detected. After thawing, *N. gaditana* cells had the same morphology as live cells. However, after steam explosion treatment, algal material was mostly unevenly distributed in aggregates. They correspond to particles of different sizes. Some of them presented a yellow-brown color and could correspond to chloroplast remains. In a few cases, cells were detected, but then they appeared with granular cytoplasm and wrinkled margins as the intermediate disruption pattern of *C. sorokiniana*. It should be noted that the cell disruption effect of steam explosion was not apparently enhanced by freezing because *N. gaditana*, the cell walled species whose morphology appeared more altered after thawing, was less affected by steam explosion. Naked cells of *D. tertiolecta* were strongly sensitive, even to the centrifugation process. After centrifugation, cells lost their internal structure or were totally disrupted. The steam explosion treatment further disintegrated the algal material and formed granulated aggregates. The same kind of cell debris was observed in the treatments with and without acid.

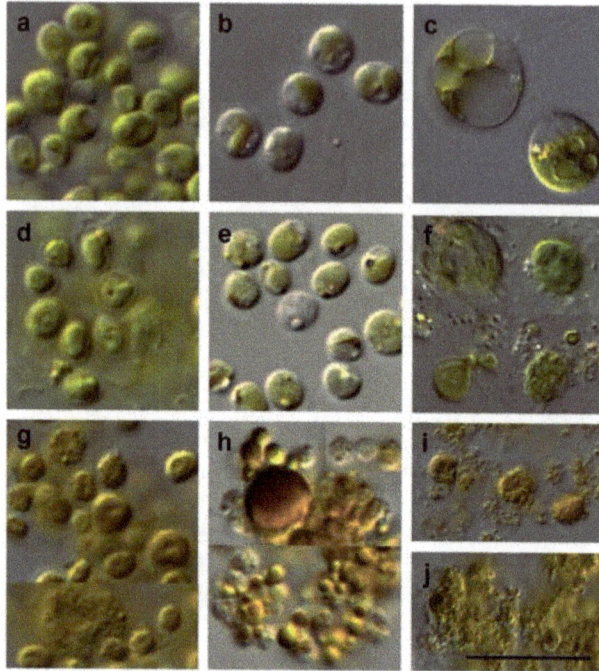

Figure 1. Light micrographs of *Chlorella sorokiniana* (**a,d,g**), *Nannochloropsis gaditana* (**b,e,h**), and *Dunaliella tertiolecta* (**c,f,i,j**) before and after steam explosion. (**a,b,c**) Live cells; (**d,e**) Thawed material; (**f**) *D. tertiolecta* after centrifugation; (**g,h,i**) Algal material after steam explosion with acid; (**j**) *D. tertiolecta* after steam explosion without acid. Scale bar corresponds to 10 μm in (**a–f**) and to 20 μm in (**g–j**).

3.1.2. Particle Size Distribution

Morphological characterization by means of microscopy was confirmed by the results obtained from particle size distribution (Figure 2).

Steam explosion produces aggregates when used with *N. gaditana* and *C. sorokiniana*. These aggregates disappear after filtration, probably due to the pump effect and the stress this caused. This effect is observed in almost all cases where these species were used. But this aggregation effect does not occur with *D. tertiolecta*, where the particle size distributions are always similar. Nevertheless, a smooth shift of the unique existent peak occurs, indicating some mass aggregation as the microscopy images show. The mean size ranges from 3 μm to 30 μm, whereas the size of the nominal microalgae cell is around 15 μm. The sample regarding the filtration retentate is the one with a smaller mean particle size due to the disaggregating role of the pump. The samples related to steam explosion treatment performed with acid have mean particle sizes which are slightly smaller than those performed without acid. Concerning *D. tertiolecta*, it is interesting to note the ability of sonication to break the microalgae cells. This only happens with this species and is probably due to the fact that *D. tertiolecta* does not have a cell wall. With other species, sonication only breaks aggregates. This is only observed with the sample after being harvested, but not with samples after steam explosion and membrane filtration. The reason for this is that at those stages, cells are almost totally unstructured, in agreement with microscopy images.

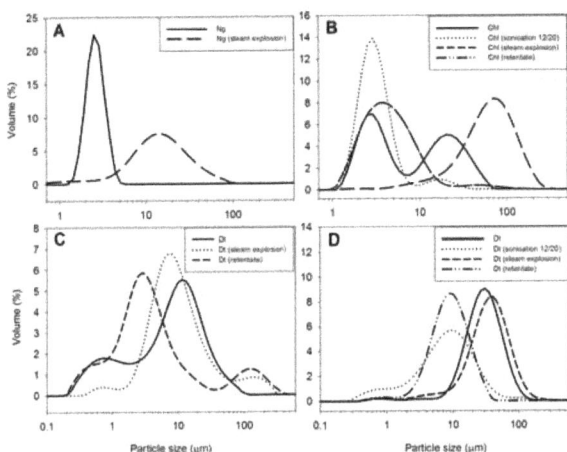

Figure 2. Particle size distribution results. (**A**) *Nannochloropsis gaditana* (**B**) *Chlorella sorokiniana* (**C**) *Dunaliella tertiolecta* (steam explosion with acid) (**D**) *D. tertiolecta* (steam explosion without acid). In all cases except those indicated, sonication was 0/12. All plots were obtained from an average of three measurements.

3.1.3. Lipid, Sugar, and Protein Contents

Table 1 shows the results of the steam explosion experiments. The amount of lipid extracted (by Bligh and Dyer and *n*-hexane), sugar, and protein contents are indicated. For the purpose of comparison, the values of lipid, total sugar content, and proteins from the fresh untreated samples are also included.

Table 1. Results of lipid, sugar, and protein analysis of steam explosion experiments (150 °C, 5 min and 5% *w/w*. H_2SO_4 except sample *D. tertiolecta* (II) with no acid). Values are expressed as the mean and the standard deviation is indicated in brackets.

		Lipids		Sugar	Protein	
		Bligh & Dyer	Hexane		Hot NaOH	Solubilization
Nannochloropsis gaditana	Untreated	22.2% (0.4)	2.1% (0.3)	18.8% (0.8)	17.3% (0.8)	1.4% (0.1)
	Steam exploded	22.3% (0.1)	17.6%(0.2)	12.9% (0.6)	8.4% (0.6)	9.1% (0.4)
Chlorella sorokiniana	Untreated	13.0% (0.2)	0.6% (0.0)	23.5% (1.3)	19.2% (0.3)	2.2% (0.0)
	Steam exploded	11.8% (0.1)	4.8% (0.2)	18.6% (0.9)	9.2% (0.1)	10.7% (0.1)
Dunaliella tertiolecta (I)	Untreated	26.6% (0.8)	2.8% (0.7)	26.1% (2.2)	14.5% (0.5)	12.0% (0.3)
	Steam exploded	29.7% (3.2)	10.6% (0.1)	19.2% (0.8)	2.6% (0.0)	5.1% (0.3)
Dunaliella tertiolecta (II)	Untreated	11.4% (1.2)	1.6% (0.1)	25.8% (2.4)	10.5% (0.0)	5.9% (0.1)
	Steam exploded No acid	11.9% (0.1)	2.1% (0.0)	8.6% (0.6)	4.8% (0.1)	4.4% (0.4)

By comparing the total lipid contents, as determined by the Bligh and Dyer method, of the untreated and steam exploded samples, we can observe that similar values are obtained in all the cases. This is because the Bligh and Dyer method yields the highest lipid recoveries, because it is a stronger method. But the use of *n*-hexane was considered as organic solvent for lipid isolation from microalgae to avoid the use of chloroform, which presents environmental and health risks, especially when it is used at an industrial scale. The experiments performed with the untreated microalgae samples showed the low extraction capability of *n*-hexane, with a maximum of 2.8% (*w/w*, DAF basis) lipid yield in the case of *D. tertiolecta*. But the amount of lipid extracted with *n*-hexane improved with the application of the steam explosion technique. Among the three microalgae species studied, *N. gaditana* yielded the maximum amount of lipid recovery of the steam exploded sample (at 150 °C, with 5% sulfuric acid), with 17.6% (*w/w*, DAF of untreated microalga basis). It signifies 79% of the total lipid as obtained by the Bligh and Dyer method. For *C. sorokiniana*, the amount of lipid extracted after steam explosion (at 150 °C, with 5% sulfuric acid) was only 4.8% (*w/w*, DAF of untreated microalga basis), representing 41% of the total amount of lipids of this microalga. In the case of *D. tertiolecta*, the extraction of lipids with *n*-hexane greatly enhanced due to the use of acid in the steam explosion process. A lipid yield of 2.1% (*w/w*, DAF of untreated microalga basis) was obtained when steam explosion was applied without acid impregnation, whereas this value increased to 10.6% (*w/w*, DAF of untreated microalga basis), as a consequence of using 5% sulphuric acid in the steam explosion experiment. This result is in agreement with our previous study [29], and shows the importance of carbohydrate hydrolysis to achieve a higher lipid extraction yield from microalgal sludge, using *n*-hexane as the solvent.

Concerning carbohydrates, the total sugar content of the untreated microalga, obtained by analytical acid hydrolysis, was determined for each microalgae species and the specific values are presented in Table 1. These values can be compared with the measured concentration of sugar in the solution of the steam exploded samples, which are also included in Table 1. For the steam explosion experiments performed with acid impregnation, a high percentage, between 70% and 80%, of the total sugar content of the microalga was found in solution after steam explosion. Contrary to this, the experiment performed with *D. tertiolecta* without the use of acid resulted in a low sugar concentration, representing 33% of the total sugar content of the untreated sample.

The protein concentration of the untreated microalgal samples ranged between 10% and 19% of DAF in the three species (Table 1). These values are in the range reported for species of the same genera in the stationary phase of culture.

The protein contents of *D. tertiolecta* detected after solubilization with dilute NaOH or after extraction at a high temperature were similar (Table 1). Thus, proteins were already available for solubilization in the harvested cultures of this naked microalgae species, meaning that it was not necessary to apply a disruption treatment. On the other hand, the protein contents detected after solubilization with dilute NaOH of both *N. gaditana* and *C. sorokiniana* were much higher after steam explosion. This rise in the detected protein revealed the cell disruption effect of steam explosion. However, the number of proteins detected after extraction at high temperature was lower in the steam exploded material than in the untreated sample for the three species. This protein loss may be explained by the occurrence of protein hydrolysis during steam explosion. The color reaction that is measured in the bicinchoninic acid assay is due to the reduction of Cu^{2+} to Cu^{+} by the oxidation of aromatic residues and peptide bonds in the protein in the reaction solution. Therefore, a lighter coloration may evidence a reduction in the number of peptide bonds due to protein hydrolysis.

3.2. Fractionation of Steam Exploded Samples by Means of Membrane Filtration

According to the results of a previous study [24], the fractionation strategy followed in the present work consists of filtrating the exploded sample with a membrane set-up and then extracting the retentate and permeate streams with solvent. The filtration was performed with dynamic filtration, which allowed for a much better permeability with just a little more energy compared to conventional cross-flow filtration. This was because fouling is highly reduced. Not only are less pores blocked, but, primarily, the cake molding over the surface of the membrane that occurs in conventional filtration is hardly produced in dynamic filtration. Therefore, vibrating filtration highly reduces microalgae attachment on the membrane surface. A PE5 membrane (MWCO = 5000 Da) was used, since it exhibited the best performance in the filtration experiments regarding permeability and irreversible fouling.

3.2.1. Rejection

Table 2 presents the results of the filtration experiments including the total weight and DAF percentage and the lipids, sugars, and protein content of each of the different streams. From the values of the DAF percentages, it can be observed that different concentrations of the retentate streams were attained (from 3% to 10% DAF). This mainly depended on the concentration of the starting material.

Table 2. Results of total mass balance, and lipid, sugar, and protein analysis of filtration experiments.

	Nannochloropsis gaditana			*Chlorella sorokiniana*		
	Steam Exploded Sample	**Retentate**	**Permeate**	**Steam Exploded Sample**	**Retentate**	**Permeate**
Total weight (g)	6000	2400	3600	6000	2240	3760
DAF percentage	5.1 (0.1)	10.1 (0.2)	1.8 (0.05)	2.7 (0.02)	5.8 (0.08)	0.9 (0.01)
Lipid (g/L)	9.2 (0.3)	22.7 (0.5)	0.07 (0.01)	1.3 (0.05)	3.9 (0.09)	0.05 (0.01)
Sugar (g/L)	6.8 (0.3)	5.9 (0.2)	6.0 (0.2)	5.1 (0.1)	5.2 (0.2)	4.9 (0.1)
Protein (g/L)	4.7 (0.2)	5.65 (0.15)	n.d.	2.92 (0.04)	4.8 (0.14)	n.d.
	Dunaliella tertiolecta			*Dunaliella tertiolecta* (No Acid)		
	Steam Exploded Sample	**Retentate**	**Permeate**	**Steam Exploded Sample**	**Retentate**	**Permeate**
Total weight (g)	6000	2290	3710	6000	2400	3600
DAF percentage	1.7 (0.01)	3.2 (0.04)	0.9 (0.01)	1.6 (0.01)	3.0 (0.02)	0.7 (0.01)
Lipid (g/L)	1.8 (0.02)	3.5 (0.08)	0.07 (0.01)	0.34 (0.03)	1.3 (0.01)	0.07 (0.01)
Sugar (g/L)	3.3 (0.2)	2.9 (0.1)	3.2 (0.2)	1.4 (0.1)	1.3 (0.1)	1.2 (0.1)
Protein (g/L)	0.89 (0.05)	1.46 (0.06)	n.d.	0.71 (0.06)	1.13 (0.07)	n.d.

The amount of lipid extracted with *n*-hexane and the proteins obtained with the solubilization method from the steam exploded and the permeate and retentate are included in Table 2. These values are expressed as a concentration of each stream, to allow for a better comparison. The permeate streams have a negligible content of lipids and no proteins. This result was also confirmed by optical density measurements. OD750 nm of permeates were like that of filtered (0.45 µm) seawater (Table 3). Therefore, it was assumed that lipid rejection was obtained in all the experiments. The absence of lipids and proteins in the permeate implies that the membrane PE5 is suitable for rejecting lipids and proteins from different microalgae species. On the other hand, the concentration of lipids and proteins in the retentate streams is much higher than that of the steam exploded sample before filtration.

Table 3. Optical density at 750 nm after filtration of steam exploded microalgae. Raw values are compared to filtered (0.45 µm) seawater (blank). Values are expressed as the mean and the standard deviation is provided in brackets.

	OD_{750nm}	
	Blank	**Permeate**
Nannochloropsis gaditana	0.081 (0.001)	0.091 (0.003)
Chlorella sorokiniana	0.081 (0.001)	0.101 (0.002)
Dunaliella tertiolecta	0.083 (0.001)	0.085 (0.001)
Dunaliella tertiolecta (no acid)	0.083 (0.001)	0.083 (0.000)

Concerning the sugar analysis, approximately the same values of concentration were obtained for the steam exploded sample and retentate and permeate streams, for the different microalgae species. This means that the employed membrane (PE5) is unable to retain sugars.

3.2.2. Permeability

Regarding the performance of the membrane in using dynamic filtration, Figure 3 shows membrane permeabilities including water permeability with the new (unused) membrane and after the experiment, for the different microalgae species studied. The permeabilities of steam-exploded biomass were measured. With them, the total fouling of materials was calculated. Concerning the permeability for the water of new PE5 membranes, the values between 30.4 L/h/m^2/bar (for *D. tertiolecta* exploded without acid) and 90.8 L/h/m^2/bar (for *N. gaditana*) were obtained. In an ideal system where a liquid that does not provide fouling is used and virgin membranes perfectly manufactured are used, the same permeabilities would be obtained. But in laboratory or pilot-scale scenarios, both conditions hardly occur. As checked earlier with the help of a scanning electron microscope, membrane thicknesses differ within the same sample. Following Darcy's law, this makes the permeability change accordingly.

If enough surface of membrane is used, a mean permeability value with a low deviation is normally obtained. But this is not the case with a pilot unit as the one used in this work.

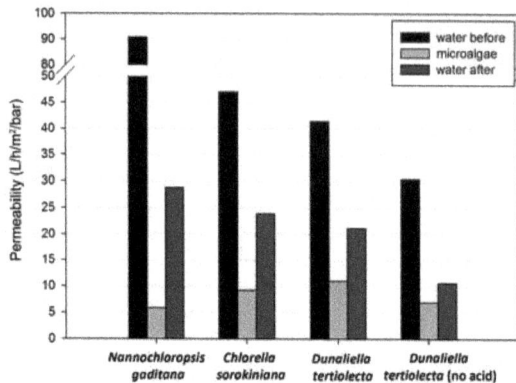

Figure 3. Water and sample permeabilities for the different microalgae samples.

In terms of pretreated microalgae filtration, the *N. gaditana* sample resulted in a microalgae permeability of 5.84 L/h/m^2/bar, the lowest value among the samples. With *D. tertiolecta* exploded without acid, a microalgae permeability of 6.93 L/h/m^2/bar was obtained, and with *C. sorokiniana*, a permeability of 9.18 L/h/m^2/bar was reached. The best membrane performance was obtained when filtrating the sample of *D. tertiolecta* exploded with acid, with the permeability value of 10.97 L/h/m^2/bar.

The total fouling factor (TF) of PE5 was the highest for *N. gaditana*, with the value of 15.55. In the case of *C. sorokiniana*, TF was lower with the value of 5.12 and with *D. tertiolecta* exploded without acid, where TF = 4.39. The best performance in terms of TF was obtained with *D. tertiolecta* exploded with acid, where the value of 3.77 was given.

3.2.3. Irreversible Fouling

To calculate irreversible fouling, membrane permeability with water before and after the experiment was measured (the system was cleaned before performing the water permeability measurements after the experiment). PE5 with *N. gaditana* resulted in the permeability of 28.7 L/h/m^2/bar, *C. sorokiniana* performed with the value of 23.85 L/h/m^2/bar, *D. tertiolecta* exploded with acid gave the value of 21.07 L/h/m^2/bar, and finally, *D. tertiolecta* exploded without acid performed with the value of 10.61 L/h/m^2/bar. Therefore, the experiment with the lowest irreversible fouling factor of 1.96 was *D. tertiolecta* exploded with acid, while *N. gaditana*, *C. sorokiniana*, and *D. tertiolecta* exploded without acid resulted in IF = 3.16, IF = 2.86, and IF = 1.97, respectively.

Figure 4 presents the exploded microalgae permeability profiles vs. time for dynamic filtration with *N. gaditana* and *C. sorokiniana*. In the filtration of *C. sorokiniana*, a steady state was reached after 30 min of the experiment with the permeability value of 9.5 L/h/m^2/bar. On the contrary, in the filtration of *N. gaditana*, the plateau was not reached, even though the experiment lasted longer than *C. sorokiniana*. After 130 min of filtrating, the value of permeability with *N. gaditana* was 4.2 L/h/m^2/bar and still decreasing.

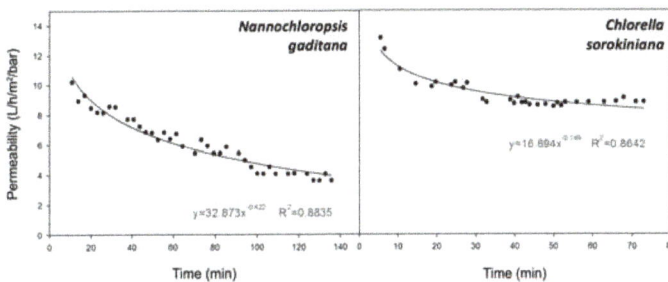

Figure 4. Permeability profiles vs. time of filtration experiments performed with (**a**) *Nannochloropsis gaditana* and (**b**) *Chlorella sorokiniana*.

4. Process Costs

One of the biggest drawbacks in the profitability of a bio-refinery of microalgae is to reduce harvest and fractionation costs [43,44]. It is necessary to have information that allows a careful calculation of costs to evaluate the different stages of a given process and that is what is discussed below.

4.1. Steam Explosion Costs

The energy consumption associated with the steam explosion operation was calculated by using the analysis provided by Sui and Chen [45]. Once a continuous stable operation is reached (e.g., the reactor steel has been heated to the operating temperature), the different energy topics considered are heating the steam, the air in the reactor, and the microalgal sludge. The other energy losses considering the expansion to atmospheric pressure are those from steam, water, air, and dry materials. We considered that the good thermal insulation of the reactor allowed us to ignore other heat losses. The experiments performed with microalgae sludge show that the humidity of the sample is the factor that most influences the steam consumption, in complete agreement with the conclusions of the above-mentioned paper. Considering a humidity of 80% in the sample, the required steam is 1.2 kg/kg dry sample. If the humidity is 85%, the steam needed is 1.6 kg/kg dry sample.

The cost of the steam strongly depends on the cost of the fuel used to produce it (TLV 2018), but it can be greatly reduced if the heat needed is produced on-site [46]. The vaporization enthalpy experiences a very minor change with the steam pressure.

To proceed with the first stage of fractionation, we consider the production of saturated steam at 5 barg (159.2 °C) from water at 20 °C. If the cost of fuel oscillates from 1.65 $/GJ to 4.5 $/GJ, the cost of the steam oscillates from 4.4 $/metric Ton to 12.2 $/metric Ton. Assuming that it is needed, for 1.2 kg of steam/kg of dry microalgae sample, the cost of the energy associated with the steam explosion operation oscillates between 0.005 $/kg of dry sample to 0.014 $/kg of dry sample.

A comparison with other techniques such as High Pressure Homogenization, Ultrasonication, Microwave treatment, Laser treatment, High Speed Homogenization, Bead Milling, and Pulse Electric Field gives a range of disruption cost between 14.68 and 0.006 $/kg dry sample [47]. The high range of cost is also due to different levels of humidity in the sample to be treated. Another aspect is that all the energy involved in those technologies is electrical, which implies a higher energy cost compared to the cost of steam production. A good energy integration in the plant using residual heat will allow for the reduction of the initial sample humidity and help reduce the cost of steam production.

4.2. Dynamic Membrane Filtration Operational Costs

Vibrational membrane filtration lowers the separation operational cost compared to conventional filtration or centrifugation. The reduction of fouling causes high flux rates that means more throughput capacity per monetary unit of capital invested. For the same reason, the membranes need less replacement and less cleaning and maintenance, which also reduces the operational cost.

In membrane filtration, microalgae can be almost fully rejected, whereas centrifugation only recovers between 80% and 90% of the biomass [48].

Considering an available commercial dynamic membrane filtration setup with a membrane area of 140 m² [49], a microalgae sludge permeability of 2 L/h/m²/bar, one year of operation (1800 h), and a final stream with a lipid concentration of 100 g/L, a yield of 6 tones/year of lipids (dry basis) could be produced. The permeability value used corresponds to 25% of the mean value of the experimental permeability measured. It is a conservative value considering that the final concentration is higher than the one reached in the experiments. The required power of the equipment is about 47 kW [49], which means that around 84,000 kWh/year would be consumed. Considering an energy cost basis in Spain of 0.097 $/kWh [50], an electrical cost of 8100 $/year would be necessary. It must be noted that the power cost in Spain is quite high compared to other neighboring countries. If this cost is normalized per unit of product, the cost would be about 0.08 $/kg microalgae (dry basis). Other operational costs are membrane replacement, maintenance, and cleaning. With current research results demonstrating that much cheaper membranes can be used for this application [17], the membrane replacement cost can be assumed to be half the electricity cost [51], and the total operation cost should be around 0.12 $/kg microalgae (dry basis).

Bibliographical data estimates that electrical centrifugation costs, with a sludge in similar conditions and with a similar yield, are higher than 0.6 $/kg microalgae (dry basis) [52].

To sum up, accumulated operating costs from cell disruption and membrane filtration are around 0.13 $/kg dry microalgae (0.01 steam explosion +0.12 membrane filtration). Filtration contributes to more than 90% of the cost and it is where the research attention should be focused.

5. Conclusions

Steam explosion has the potential to become a broad-spectrum microalgae cell disruption, as well as pre-fractionation, treatment. It provided proper availability of organic compounds and carbohydrate hydrolysis into sugars with all the various kinds of used microalgae and it is particularly effective when the strains have recalcitrant cell walls.

The use of steam explosion, besides breaking the cell wall, partially hydrolyzes proteins.

With all the tested strains, dynamic membrane filtration offers an excellent performance regarding permeability by rejecting lipids.

The sequence of steam explosion, dynamic membrane filtration, and solvent extraction as downstream unit operations in a microalgae biorefinery clearly allows for the reduction of process costs. All the mentioned technologies for all the stages are already commercially available.

Acknowledgments: This study was funded by the Spanish Ministry of Economy and Competitiveness (grant number CTQ2014-56285-R). The research was also supported by the European Regional Development Funds (ERDF, FEDER Programa Competitividad de Catalunya 2007–2013) and by the project "Fuels from Biomass" funded by Excma. Diputació de Tarragona. M. Hapońska is grateful to the Universitat Rovira i Virgili (URV) for her PhD scholarship.

Author Contributions: E.L. designed part of the experiments, performed the experiments, analyzed the data, and wrote the paper. M.H. designed part of the experiments, performed the experiments, analyzed the data, and wrote the paper. E.C. designed part of the experiments performed part of the experiments, analyzed the data, and wrote the paper. C.T. conceived and designed part of the experiments, performed part of the experiments, analyzed the data, and wrote the paper. J.S. conceived and designed part of the experiments, analyzed the data, and wrote the paper.

Conflicts of Interest: The authors declare no conflict of interest.

References

1. Chisti, Y. Biodiesel from microalgae. *Biotechnol. Adv.* **2007**, *25*, 294–306. [CrossRef] [PubMed]

2. Sheehan, J.; Dunahay, T.; Benemann, J.; Roessler, P. *A Look Back at the US Department of ENERGY'S Aquatic Species, Program: Biodiesel from Algae*; National Renewable Energy Laboratory (NREL): Golden, CO, USA, 1998.

3. Zhu, L.; Ketola, T. Microalgae production as a biofuel feedstock: Risks and challenges. *Int. J. Sustain. Dev. World* **2012**, *19*, 268–274. [CrossRef]

4. Zhu, L. Biorefinery as a promising approach to promote microalgae industry: An innovative framework. *Renew. Sustain. Energy Rev.* **2015**, *41*, 1376–1384. [CrossRef]

5. Wijffels, R.H.; Barbosa, M.J.; Eppink, M.H.M. Microalgae for the production of bulk chemicals and biofuels. *Biofuels Bioprod. Biorefin.* **2010**, *4*, 287–295. [CrossRef]

6. Aslam, A.; Thomas-Hall, S.R.; Mughal, T.A.; Schenk, P.M. Selection and adaptation of microalgae to growth in 100% unfiltered coal-fired flue gas. *Bioresour. Technol.* **2017**, *233*, 271–283. [CrossRef] [PubMed]

7. Zhao, B.; Su, Y. Process effect of microalgal-carbon dioxide fixation and biomass production: A review. *Renew. Sustain. Energy Rev.* **2014**, *31*, 121–132. [CrossRef]

8. Gupta, P.L.; Lee, S.M.; Choi, H.J. Integration of microalgal cultivation system for wastewater remediation and sustainable biomass production. *World J. Microbiol. Biotechnol.* **2016**, *32*, 139. [CrossRef] [PubMed]

9. Suresh Kumar, K.; Dahms, H.U.; Won, E.J.; Lee, J.S.; Shin, K.H. Microalgae—A promising tool for heavy metal remediation. *Ecotoxicol. Environ. Saf.* **2015**, *113*, 329–352. [CrossRef] [PubMed]

10. Lupatini, A.L.; Colla, L.M.; Canan, C.; Colla, E. Potential application of microalga spirulina platensis as a protein source. *J. Sci. Food Agric.* **2017**, *97*, 724–732. [CrossRef] [PubMed]

11. Sanchez Rizza, L.; Sanz Smachetti, M.E.; Do Nascimento, M.; Salerno, G.L.; Curatti, L. Bioprospecting for native microalgae as an alternative source of sugars for the production of bioethanol. *Algal Res.* **2017**, *22*, 140–147. [CrossRef]

12. Reyes, F.A.; Mendiola, J.A.; Ibañez, E.; Del Valle, J.M. Astaxanthin extraction from haematococcus pluvialis using CO_2-expanded ethanol. *J. Supercrit. Fluids* **2014**, *92*, 75–83. [CrossRef]

13. Uquiche, E.; Antilaf, I.; Millao, S. Enhancement of pigment extraction from *B. braunii* pretreated using CO_2 rapid depressurization. *Braz. J. Microbiol.* **2016**, *47*, 497–505. [CrossRef] [PubMed]

14. Cuello, J.L.; Hoshino, T.; Kuwahara, S.; Brown, C.L. Scale-up-bioreactor design and culture optimization. In *Biotechnology for Biofuel Production and Optimization*; Elsevier: Amsterdam, The Netherlands, 2016; pp. 497–511.

15. Thomassen, G.; Egiguren Vila, U.; Van Dael, M.; Lemmens, B.; Van Passel, S. A techno-economic assessment of an algal-based biorefinery. *Clean Technol. Environ. Policy* **2016**, *18*, 1849–1862. [CrossRef]

16. Ersahin, M.E.; Ozgun, H.; Dereli, R.K.; Ozturk, I.; Roest, K.; van Lier, J.B. A review on dynamic membrane filtration: Materials, applications and future perspectives. *Bioresour. Technol.* **2012**, *122*, 196–206. [CrossRef] [PubMed]

17. Hapońska, M.; Clavero, E.; Salvadó, J.; Torras, C. Application of ABS membranes in dynamic filtration for *Chlorella sorokiniana* dewatering. *Biomass Bioenergy* **2016**, *111*, 224–231. [CrossRef]

18. Kim, J.; Yoo, G.; Lee, H.; Lim, J.; Kim, K.; Kim, C.W.; Park, M.S.; Yang, J.W. Methods of downstream processing for the production of biodiesel from microalgae. *Biotechnol. Adv.* **2013**, *31*, 862–876. [CrossRef] [PubMed]

19. Nurra, C.; Clavero, E.; Salvadó, J.; Torras, C. Vibrating membrane filtration as improved technology for microalgae dewatering. *Bioresour. Technol.* **2014**, *157*, 247–253. [CrossRef] [PubMed]

20. Rossignol, N.; Vandanjon, L.; Jaouen, P.; Quéméneur, F. Membrane technology for the continuous separation microalgae/culture medium: Compared performances of cross-flow microfiltration and ultrafiltration. *Aquac. Eng.* **1999**, *20*, 191–208. [CrossRef]

21. Zhao, F.; Chu, H.; Zhang, Y.; Jiang, S.; Yu, Z.; Zhou, X.; Zhao, J. Increasing the vibration frequency to mitigate reversible and irreversible membrane fouling using an axial vibration membrane in microalgae harvesting. *J. Membr. Sci.* **2017**, *529*, 215–223. [CrossRef]

22. Cheng, J.; Huang, R.; Li, T.; Zhou, J.; Cen, K. Physicochemical characterization of wet microalgal cells disrupted with instant catapult steam explosion for lipid extraction. *Bioresour. Technol.* **2015**, *191*, 66–72. [CrossRef] [PubMed]

23. Grimi, N.; Dubois, A.; Marchal, L.; Jubeau, S.; Lebovka, N.I.; Vorobiev, E. Selective extraction from *Microalgae nannochloropsis* sp. Using different methods of cell disruption. *Bioresour. Technol.* **2014**, *153*, 254–259. [CrossRef] [PubMed]

24. Lorente, E.; Hapońska, M.; Clavero, E.; Torras, C.; Salvadó, J. Microalgae fractionation using steam explosion, dynamic and tangential cross-flow membrane filtration. *Bioresour. Technol.* **2017**, *237*, 3–10. [CrossRef] [PubMed]

25. Mendes-Pinto, M.M.; Raposo, M.F.J.; Bowen, J.; Young, A.J.; Morais, R. Evaluation of different cell disruption processes on encysted cells of haematococcus pluvialis: Effects on astaxanthin recovery and implications for bio-availability. *J. Appl. Phycol.* **2001**, *13*, 19–24. [CrossRef]
26. Nurra, C.; Torras, C.; Clavero, E.; Ríos, S.; Rey, M.; Lorente, E.; Farriol, X.; Salvadó, J. Biorefinery concept in a microalgae pilot plant. Culturing, dynamic filtration and steam explosion fractionation. *Bioresour. Technol.* **2014**, *163*, 136–142. [CrossRef] [PubMed]
27. Yap, B.H.J.; Dumsday, G.J.; Scales, P.J.; Martin, G.J.O. Energy evaluation of algal cell disruption by high pressure homogenisation. *Bioresour. Technol.* **2015**, *184*, 280–285. [CrossRef] [PubMed]
28. Safi, C. *Microalgae Biorefinery: Proposition of a Fractionation Process*; Université de Toulouse: Toulouse, France, 2013.
29. Lorente, E.; Farriol, X.; Salvadó, J. Steam explosion as a fractionation step in biofuel production from microalgae. *Fuel Process. Technol.* **2015**, *131*, 93–98. [CrossRef]
30. Lee, A.K.; Lewis, D.M.; Ashman, P.J. Disruption of microalgal cells for the extraction of lipids for biofuels: Processes and specific energy requirements. *Biomass Bioenergy* **2012**, *46*, 89–101. [CrossRef]
31. Heitz, M.; Capek-Ménard, E.; Koeberle, P.G.; Gagné, J.; Chornet, E.; Overend, R.P.; Taylor, J.D.; Yu, E. Fractionation of populus tremuloides at the pilot plant scale: Optimization of steam pretreatment conditions using the stake II technology. *Bioresour. Technol.* **1991**, *35*, 23–32. [CrossRef]
32. Verwijst, T.; Baggerman, J.; Liebermann, F.; van Rijn, C.J.M. High-frequency flow reversal for continuous microfiltration of milk with microsieves. *J. Membr. Sci.* **2015**, *494*, 121–129. [CrossRef]
33. Malmali, M.; Stickel, J.J.; Wickramasinghe, S.R. Sugar concentration and detoxification of clarified biomass hydrolysate by nanofiltration. *Sep. Purif. Technol.* **2014**, *132*, 655–665. [CrossRef]
34. Kumar, K.; Das, D. Growth characteristics of chlorella sorokiniana in airlift and bubble column photobioreactors. *Bioresour. Technol.* **2012**, *116*, 307–313. [CrossRef] [PubMed]
35. Chua, E.T.; Schenk, P.M. A biorefinery for nannochloropsis: Induction, harvesting, and extraction of epa-rich oil and high-value protein. *Bioresour. Technol.* **2017**, *244*, 1416–1424. [CrossRef] [PubMed]
36. Francavilla, M.; Kamaterou, P.; Intini, S.; Monteleone, M.; Zabaniotou, A. Cascading microalgae biorefinery: Fast pyrolysis of dunaliella tertiolecta lipid extracted-residue. *Algal Res.* **2015**, *11*, 184–193. [CrossRef]
37. Scholz, M.J.; Weiss, T.L.; Jinkerson, R.E.; Jing, J.; Roth, R.; Goodenough, U.; Posewitz, M.C.; Gerken, H.G. Ultrastructure and composition of the nannochloropsis gaditana cell wall. *Eukaryot. Cell* **2014**, *13*, 1450–1464. [CrossRef] [PubMed]
38. Takeda, H. Taxonomical assignment of chlorococal algae from their cell wall composition. *Phytochemistry* **1993**, *34*, 1053–1055. [CrossRef]
39. Kodner, R.B.; Summons, R.E.; Knoll, A.H. Phylogenetic investigation of the aliphatic, non-hydrolyzable biopolymer algaenan, with a focus on green algae. *Org. Geochem.* **2009**, *40*, 854–862. [CrossRef]
40. Bligh, E.G.; Dyer, W.J. A rapid method of total lipid extraction and purification. *Can. J. Biochem. Phys.* **1959**, *37*, 911–917. [CrossRef]
41. Templeton, D.W.; Quinn, M.; Van Wychen, S.; Hyman, D.; Laurens, L.M.L. Separation and quantification of microalgal carbohydrates. *J. Chromatogr. A* **2012**, *1270*, 225–234. [CrossRef] [PubMed]
42. Barbarino, E.; Lourenço, S.O. An evaluation of methods for extraction and quantification of protein from marine macro- and microalgae. *J. Appl. Phycol.* **2005**, *17*, 447–460. [CrossRef]
43. Hoffman, J.; Pate, R.C.; Drennen, T.; Quinn, J.C. Techno-economic assessment of open microalgae production systems. *Algal Res.* **2017**, *23*, 51–57. [CrossRef]
44. t Lam, G.P.; Vermuë, M.H.; Eppink, M.H.M.; Wijffels, R.H.; van den Berg, C. Multi-product microalgae biorefineries: From concept towards reality. *Trends Biotechnol.* **2017**, *36*, 216–277. [CrossRef] [PubMed]
45. Sui, W.; Chen, H. Multi-stage energy analysis of steam explosion process. *Chem. Eng. Sci.* **2014**, *116*, 254–262. [CrossRef]
46. Carrere, H.; Antonopoulou, G.; Affes, R.; Passos, F.; Battimelli, A.; Lyberatos, G.; Ferrer, I. Review of feedstock pretreatment strategies for improved anaerobic digestion: From lab-scale research to full-scale application. *Bioresour. Technol.* **2016**, *199*, 386–397. [CrossRef] [PubMed]
47. Günerken, E.; D'Hondt, E.; Eppink, M.H.M.; Garcia-Gonzalez, L.; Elst, K.; Wijffels, R.H. Cell disruption for microalgae biorefineries. *Biotechnol. Adv.* **2015**, *33*, 243–260. [CrossRef] [PubMed]
48. Chen, C.-Y.; Yeh, K.-L.; Aisyah, R.; Lee, D.-J.; Chang, J.-S. Cultivation, photobioreactor design and harvesting of microalgae for biodiesel production: A critical review. *Bioresour. Technol.* **2011**, *102*, 71–81. [CrossRef] [PubMed]

49. New Logic Research, Inc. Vsep Products. Available online: http://www.vsep.com/products/index.html (accessed on 23 October 2017).

50. Ministerio de Energía, Turismo y Agenda Digital de España. Energía y Emisiones. Available online: http://www.minetad.gob.es/es-ES/IndicadoresyEstadisticas/DatosEstadisticos/IV.%20Energ%C3%ADa%20y%20emisiones/IV_12.pdf (accessed on 23 October 2017).

51. New Logic Research, Inc. Vsep Case Studies and Application Notes. Available online: http://www.vsep.com/downloads/case_studies_application_notes.html (accessed on 23 October 2017).

52. Dassey, A.J.; Theegala, C.S. Harvesting economics and strategies using centrifugation for cost effective separation of microalgae cells for biodiesel applications. *Bioresour. Technol.* **2013**, *128*, 241–245. [CrossRef] [PubMed]

processes

MDPI

Article

Supported Ionic Liquid Membranes for Separation of Lignin Aqueous Solutions

Ricardo Abejón * [ID], Javier Rabadán, Silvia Lanza, Azucena Abejón, Aurora Garea and Angel Irabien [ID]

Chemical and Biomolecular Engineering Department, University of Cantabria. Avda. Los Castros s/n, 39005 Santander, Spain; javier.rabadan.mtz@gmail.com (J.R.); silvialan19@gmail.com (S.L.); azucena.abejon@unican.es (A.A.); aurora.garea@unican.es (A.G.); angel.irabien@unican.es (A.I.)
* Correspondence: abejonr@unican.es; Tel.: +34-9-4220-1579; Fax: +34-9-4220-1591

Received: 3 August 2018; Accepted: 28 August 2018; Published: 1 September 2018

Abstract: Lignin valorization is a key aspect to design sustainable management systems for lignocellulosic biomass. The successful implementation of bio-refineries requires high value added applications for the chemicals derived from lignin. Without effective separation processes, the achievement of this purpose is difficult. Supported ionic liquid membranes can play a relevant role in the separation and purification of lignocellulosic components. This work investigated different supported ionic liquid membranes for selective transport of two different types of technical lignins (Kraft lignin and lignosulphonate) and monosaccharides (xylose and glucose) in aqueous solution. Although five different membrane supports and nine ionic liquids were tested, only the system composed by [BMIM][DBP] as an ionic liquid and polytetrafluoroethylene (PTFE) as a membrane support allowed the selective transport of the tested solutes. The results obtained with this selective membrane demonstrated that lignins were more slowly transferred from the feed compartment to the stripping compartment through the membrane than the monosaccharides. A model was proposed to calculate the effective mass transfer constants of the solutes through the membrane (values in the range $0.5–2.0 \times 10^{-3}$ m/h). Nevertheless, the stability of this identified selective membrane and its potential to be implemented in effective separation processes must be further analyzed.

Keywords: supported ionic liquid membranes; separation; lignin; glucose; xylose

---

1. Introduction

Among renewable raw materials, wood must be highlighted, because more effective, cost-competitive and sustainable alternatives have not been identified for some applications. Forest exploitation provides economic and social values from this natural resource and promotes a sustainable development chance for rural areas [1]. The wood processing industrial sector obtains forest products such as lumber, engineered wood, and pulp. Traditional wood pulping has been focused on the fractionation of the main lignocellulosic components (cellulose, hemicellulose, and lignin), but paying special attention to the cellulosic fraction, which is the relevant one for the production of paper. In this framework, hemicellulose and lignin have been only employed for energy recovery by direct combustion [2]. However, recent research interest is being targeted to hemicellulose and lignin as raw materials for renewable chemicals to replace those derived from petroleum. Therefore, the integral use of the lignocellulosic biomass must take into account the valorization of hemicellulose and lignin [3–5]. For example, lignin must be considered the most promising renewable source to produce aromatic chemicals at a real industrial scale because of its structure (Figure 1) and its abundance in nature [6].

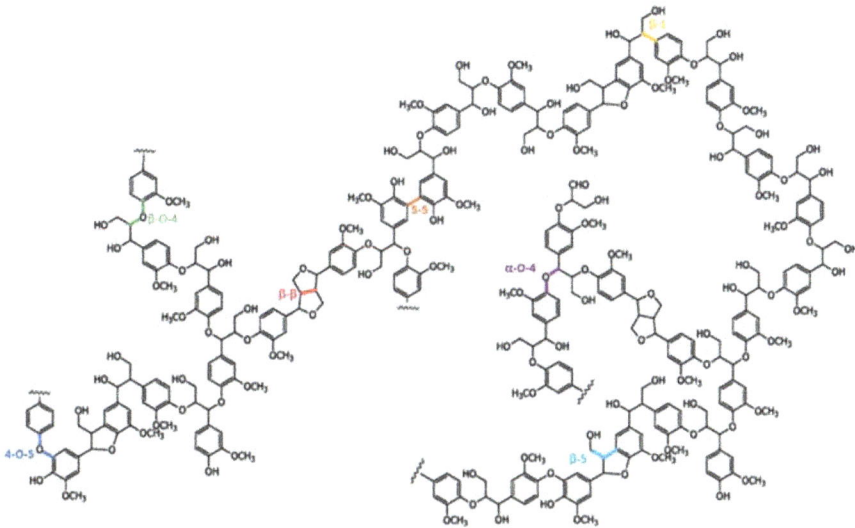

Figure 1. Example of lignin structure. Reproduced with permission from Chávez-Sifontes, Lignina, estructura y aplicaciones: métodos de despolimerización para la obtención de derivados aromáticos de interés industrial; published by Avances en Ciencias e Ingeniería, 2013.

In this new scenario, biorefineries have been introduced to provide an alternative to traditional petroleum refineries. A biorefinery is a facility that integrates the biomass conversion processes to produce bioenergy, biofuels, and bio-based chemicals from biomass [7]. While the valorization of cellulose and hemicellulose has been successfully implemented in biorefineries [8], the optimal valorization of lignin remains as a great challenge to be solved. Enzymatic hydrolysis of cellulose and hemicellulose results in fermentable sugars, which can be easily transformed to biofuels (bioethanol) or precursors for production of valuable bio-based chemicals. Delignification is a necessary prerequisite for enzymatic hydrolysis, as lignin interferes the reaction and blocks the process [9,10]. Therefore, lignin must be separated during the pretreatment of lignocellulosic biomass, which facilitates its posterior valorization.

The production of commercially available lignin-derived chemicals has been very limited until recent days: only dispersing and emulsifying agents obtained from lignosulphonates and plywood panels can be mentioned. This lack of commercial application can be justified by the heterogeneous structure of lignin, which, unlike cellulose or hemicellulose, is not formed by the systematic series of regular monomers. Despite this irregular and complex structure, research efforts have been applied to find the most suitable options for lignin conversion to valuable products [11–16]. Although some possibilities have been identified for direct valorization of raw lignin, lignin depolymerization is a more promising route. On the one hand, aggressive unselective depolymerization to break C-C and C-O linkages results in aromatic compounds mixtures like benzene, toluene, xylene, and phenol [17,18]. In addition, some short aliphatic (C1-C3) and, in less extent, longer cycloaliphatic (C6-C7) hydrocarbons can be obtained. On the other hand, highly selective depolymerization processes are based on the cleavage of only determined links. This way, products that are not easily produced by traditional petrochemical routes can be obtained, like substituted coniferols, aromatic polyols, or oxidized monomers [19].

However, classic fractionation processes for lignocellulosic biomass (Kraft process, Organosolv process, alkaline treatment, steam explosion ...) result in low-purity lignin, mostly because of the presence of impurities derived from cellulose and hemicellulose. Consequently, new research efforts must be applied to develop an efficient and selective process to obtain a purified lignin fraction. Membrane separation technologies have been implemented in biorefineries because they show very advantageous properties: no phase change, no heat requirements, low energy consumption, and compact and easily scalable design [20]. Research works have investigated the potentiality of membrane techniques for lignin separation and purification. Lignin-rich liquors can be treated with ultrafiltration and nanofiltration membranes for purification of lignin and separation of other inorganic compounds [21–26]. Other authors have applied membrane separations for different tasks aimed to lignin valorization, such as concentration of lignin solutions and elimination of lower molecular weight impurities [27], fractionation of lignin fragments according to their molecular weight [28,29], or separation of different lignin derivatives [30–32].

Since lignin is not easily solubilized in conventional solvents, the use of ionic liquids (ILs) for fractionation of lignocellulosic biomass has been deeply investigated. ILs are organic salts formed by high-volume organic cations and smaller organic or inorganic anions. Some of the most interesting ILs have melting points below 100 °C, so they are liquid at room temperature. These ILs present some common characteristics, like negligible vapor pressure or high thermal and mechanic stabilities [33]. The physicochemical properties can be customized by an optimal combination of the most convenient cations and anions for each application. The viability of ILs for dissolution, separation, and recovery of the main components of lignocellulosic biomass has been investigated [34–42]. Imidazolium based ILs (with different radicals joined to the central ring and combined with simple inorganic and more complex organic anions) have been deeply investigated for the selective dissolution of lignin, since they are not good solvents for cellulose or hemicellulose [43–45]. The ILs based on 1,3-dialkyl-imidazolium have been object of most research works. However, the limited stability of these ILs in alkaline and oxidant media must be taken into account because posterior treatment of lignin can take place under these conditions [12]. Therefore, alternative ILs non-based on imidazolium have been tested for lignin processing, paying special attention to ILs based on ammonium, phosphonium, pyridinium, and pyrrolidinium [46–49]. Nevertheless, the high economic costs derived from the employment of this type of IL remains the main drawback for real-scale implementation in the biorefinery processes [50].

Supported liquid membranes (SLMs) consist of porous supports that have been impregnated with a specific solvent to get it imbedded in the pores. The solvent is kept there by capillary forces and forms a three-phase system, since it separates the feed and stripping phases [51,52]. When this solvent is an IL, a supported ionic liquid membrane (SILM) is obtained (Figure 2) [53]. SILMs require the existence of three simultaneous processes to be applied for effective separation of solutes: the extraction from the feed solution to the SILM, the diffusion through the SILM and the re-extraction from the SILM to the stripping solution. SILMs present some advantages over SLMs, mainly because of their improved stability, since the use of ILs reduces the solvent losses from the support by evaporation or dispersion in the feed and stripping phases [54]. Moreover, ILs can provide very high specificity for the solutes to be separated and the small amount of IL required in a SILM can reduce the economic costs significantly.

Although scientific information about the use of membranes and ILs for lignocellulosic biomass fractionation and further processing is abundant, the integration of both tools as SILMs has not been deeply investigated. SILMs have been successfully applied for extraction of minor components that appear during vegetal biomass fractionation (for example, lipophilic compounds linked to resin, such as fatty acids or sterols) for analytical purposes [55]. The potential of these systems for extraction and purification of lignin must be studied, since SILMs can be preferred over other technologies for extraction and purification of lignin. Since the solute extraction from the feed phase and the re-extraction to the stripping phase occur in a unique stage, very simple designs are possible, avoiding complex configurations or high energy requirements (separation can be carried out without heat or pressure application). The employment of highly porous materials to support the SILMs provides a very high interfacial area for mass transfer, which allows very compact equipment. Moreover, the coupling of the extraction and stripping results in mass transport without limitations due to the solubility limits [56]. The small amount of IL required to implement a SILM allows the selection of non-traditional ILs, which can offer better permeability and specificity for lignin without compromising the chemical structure and physicochemical properties of lignin or interfere in the characteristics of the rest of the lignocellulosic components. Lastly, the most relevant disadvantage of the use of ILs (their high economic costs) is minimized when SILMs are implemented, since the total volume of ILs required is greatly reduced [51].

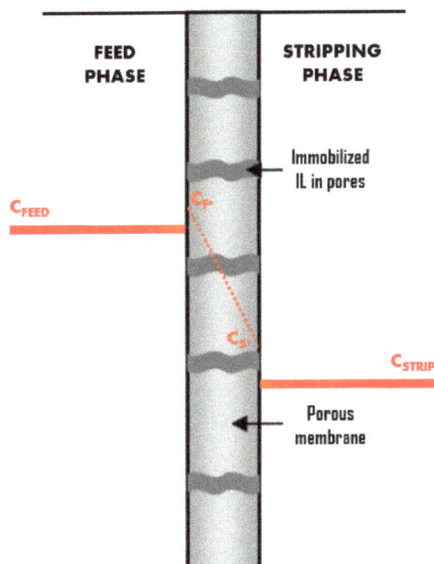

Figure 2. Schematic representation of transport through a SILM (Supported Ionic Liquid Membrane).

A previous work began the analysis of the application of SILMs for lignin separation, but the preliminary results revealed that the selective transport of lignin was not easily obtained [57]. The main objective of this work was a complete study of the potentiality of different SILMs for fractionation and separation of lignocellulosic biomass components, with special attention to lignin extraction and purification. The flow of two lignin types (Kraft lignin and lignosulphonates) from the feed compartment to the stripping one was characterized and compared with the flow of monosaccharides (glucose and xylose were selected because cellulose is made with repeated glucose units and xylose is the main sugar monomer in the structure of hemicellulose) to determine their potential for selective transport of the developed SILMs.

2. Experimental

2.1. Chemicals and Materials

Nine different ILs were selected to prepare SILMs (Figure 3). Six imidazolium-based ILs ([BMIM]MeSO₄, [BMIM][DBP], [BMIM][OTf] and [HMIM][OTf] from Iolitec, and [EMIM]EtSO₄ and [EMIM]Ac from Sigma-Aldrich, Munich, Germany), two phosphonium-based ILs (CYPHOS 101 and CYPHOS 108 from Cytec, Woodland Park, NJ, USA) and a mixture of quaternary ammonium salts (Aliquat 336 from Sigma-Aldrich, Munich, Germany) were used as supplied. Kraft lignin (low sulfonate content), D-(+)-xylose (>99%) and D-(+)-glucose (>99.5%) were provided by Sigma-Aldrich, Munich, Germany, while sodium lignosulphonate was purchased from TCI Chemicals, Tokyo, Japan. The employed water was obtained by an Elix purification system (Millipore, Darmstadt, Germany).

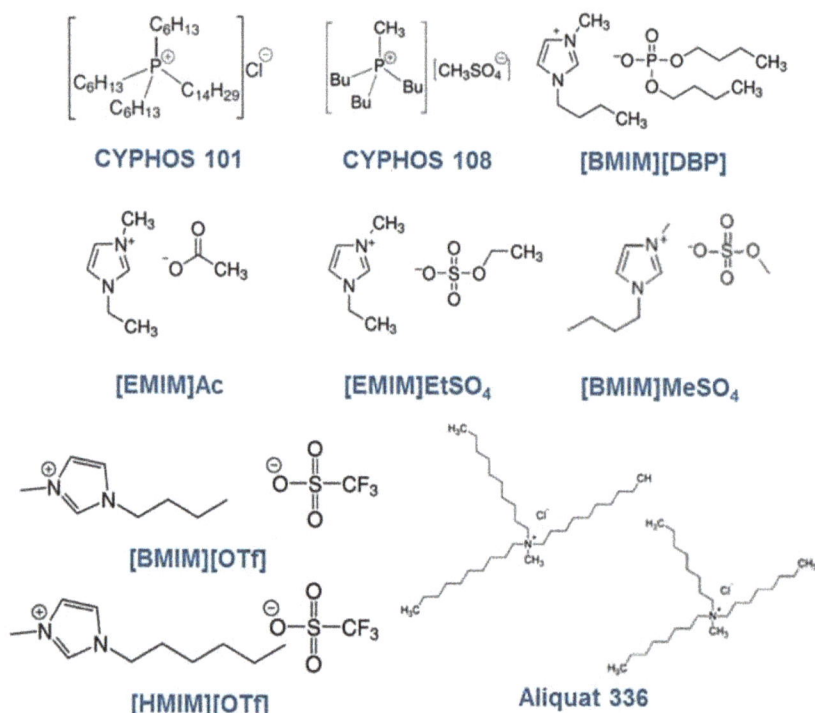

Figure 3. Structure of the selected ILs (Ionic Liquids).

As membrane supports, membrane disc filters were employed. Five different polymeric materials were tested: PP (polypropylene) and PTFE (polytetrafluoroethylene) from Filter-Lab, Barcelona, Spain, PCTE (polycarbonate) from Sterlitech, Kent, WA, USA, PVDF (hydrophobic polyvinylidene fluoride), and HPVDF (hydrophilic polyvinylidene fluoride) from Millipore, Darmstadt, Germany. All the membranes had the same diameter (47 mm) and pore diameter (0.45 μm), except PCTE (0.40 μm).

2.2. SILMs Preparation

The SILMs were prepared using the different polymeric membranes and ILs. Firstly, the corresponding membrane and IL were introduced in a vacuum oven (<35 mbar and 70 °C) within separate Petri dishes to remove the humidity, gases, and any other traces of volatile compounds. Later, the membrane was soaked in the IL, keeping the vacuum for 24 h to promote a proper impregnation by removal of air from the membrane pores. Finally, the liquid excess over the membrane surface was removed by allowing dripping overnight. This way, the SILM was ready to be employed.

2.3. Installation and Analytical Procedures

The experimental tests were carried out in a membrane cell designed for this specific purpose (Figure 4). The glass cell was composed of two identical compartments (volume lower than 150 mL each compartment), one for the feed solution and the other for the stripping one, which were separated by the SILM. The feed and stripping solutions were poured into the cell at the same time and the compartments were closed without inlet or outlet streams in the system. Samples were taken at regular time intervals from both compartments. All the experiments were carried out at room temperature. It was decided to stop the experiments after the capture of enough samples to characterize the transport.

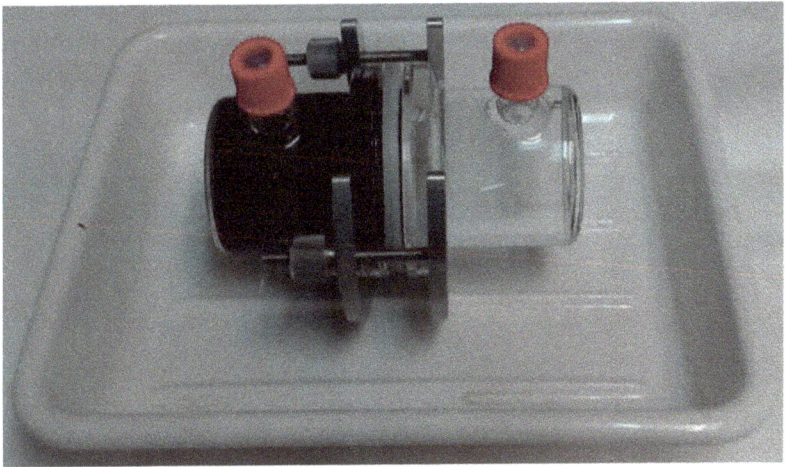

Figure 4. Image of the experimental cell selected for testing SILMs.

Kraft lignin, lignosulphonate, glucose and xylose concentrations were determined by a ultraviolet–visible (UV-VIS) spectrophotometer DR 5000 (Hach, Düsseldorf, Germany), using a wavelength of 280 nm for Kraft lignin and lignosulphonate [58,59] and of 575 nm for monosaccharides, according to the dinitrosalicylic acid method for determination of reducing sugars [60,61]. The calibration curves for the different solutes are compiled in Figure 5.

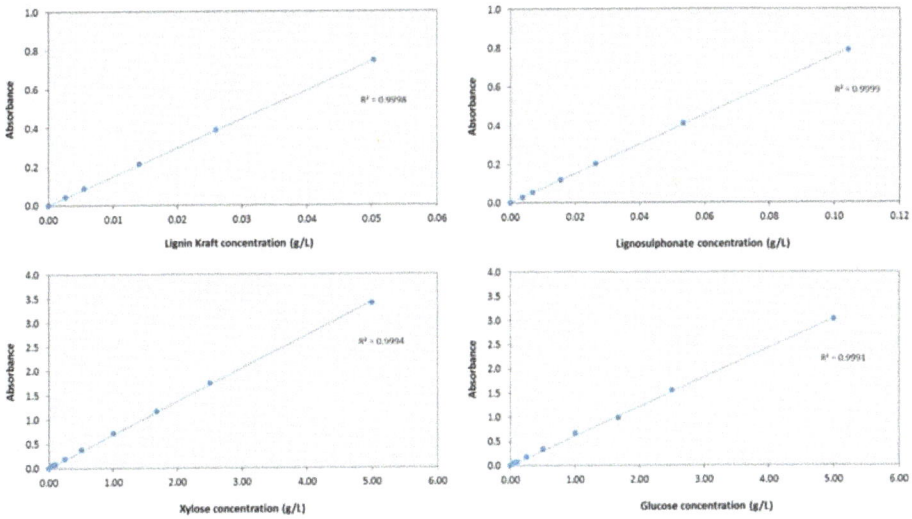

Figure 5. Calibration curves for the determination by ultraviolet–visible (UV-VIS) spectrophotometry of the concentration of the different tested solutes.

The calibration curves showed good linearity in the defined intervals (until 0.05 g/L for Kraft lignin, 0.10 g/L for lignosulphonate and 5.00 g/L for monosaccharides). From the obtained lines, the corresponding absorptivity coefficients a_λ were calculated according to the Beer–Lambert law:

$$A = a_\lambda \cdot b \cdot C \tag{1}$$

where A is the absorbance, b is the length of the path and C the solute concentration. The obtained absorptivity coefficients are compiled in Table 1.

Table 1. The absorptivity coefficients for determination by UV-VIS spectrophotometry of the concentration of the different tested solutes.

	Absorptivity Coefficients (L/g·cm)	
	a_{280}	a_{575}
Kraft lignin	14.706	-
Lignosulphonate	7.604	-
Xylose	-	0.687
Glucose	-	0.608

2.4. Membrane Transport Characterization

The previously published paper with preliminary results presented a transport model for the selective transport of solutes through the prepared SILMs [57]. However, the preliminary results revealed that the transport through the SILMs was not selective. Therefore, in this section the previously prepared model for the selective transport is summarized and the adaptations required to consider non-selective transport are added.

The amount of a solute that passes by selective transport through a SILM per unit of time and surface is called flux J and it is proportional to the gradient of concentration C between both solutions:

$$J = k \cdot \Delta C \tag{2}$$

where k is the proportionality constant that can be defined as the permeability. To model the evolution of the concentration of the solute in the feed compartment C_F, a mass balance to the feed compartment can be applied:

$$V \cdot \frac{dC_F}{dt} = -k \cdot \Delta C \cdot A_M \tag{3}$$

where V is the volume of the feed solution and A_M the active surface of the SILM. The equation can be reorder and modified considering the gradient between the feed (F) and stripping (S) compartments, and the total mass balance with C_0 as initial feed concentration

$$\Delta C = C_F - C_S \tag{4}$$

$$C_0 = C_F + C_S \tag{5}$$

$$\frac{dC_F}{2C_F - C_0} = -\frac{k \cdot A_M \cdot dt}{V} \tag{6}$$

All the constant can be grouped into a unique constant K, called effective mass transfer constant:

$$\frac{dC_F}{C_F - \frac{C_0}{2}} = -\frac{2 \cdot k \cdot A_M \cdot dt}{V} = -K \cdot dt \tag{7}$$

Integration can be applied and the evolution of the concentration in the feed compartment C_F can be assessed:

$$C_F = \frac{C_0}{2} \cdot \left(1 + e^{-K \cdot t}\right) \tag{8}$$

In a totally equivalent way, the evolution of the concentration in the stripping compartment C_S can be obtained:

$$C_S = \frac{C_0}{2} \cdot \left(1 - e^{-K \cdot t}\right) \tag{9}$$

For the case of non-selective transport, the corresponding mass balance can be formulated, just by replacement of the proportionality constant k by the permeate flux F_P:

$$V \cdot \frac{dC_F}{dt} = -F_P \cdot \Delta C \cdot A_M \tag{10}$$

This way, the evolution of the concentrations in both compartments can be derived, defining a new effective mass transfer constant K_P:

$$C_F = \frac{C_0}{2} \cdot \left(1 + e^{-K_P \cdot t}\right) \tag{11}$$

$$C_S = \frac{C_0}{2} \cdot \left(1 - e^{-K_P \cdot t}\right) \tag{12}$$

$$K_P = \frac{2 \cdot F_P \cdot A_M}{V} \tag{13}$$

3. Results and Discussion

During the preparation of the SILMs, some experimental drawbacks were identified. For example, when lignosulphonate solutions were mixed with Aliquat 336 or CYPHOS 101, unstable systems were obtained, since sticky precipitated solids appeared (Figure 6a,b). Under these conditions, a SILM with CYPHOS 101 in PVDF support did not result in effective results for Kraft lignin transport, because of the precipitation of solids on the membrane surface (Figure 6c).

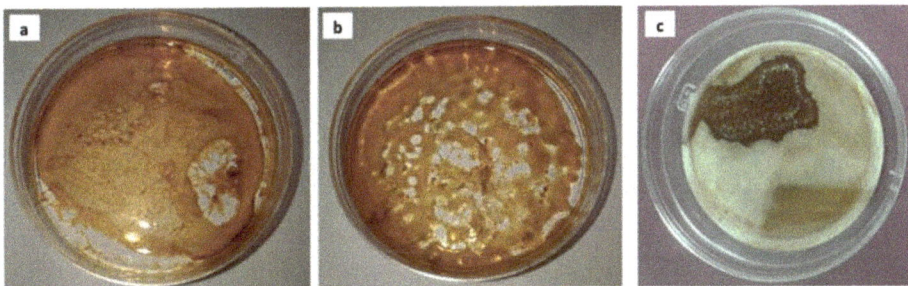

Figure 6. Identification of unstable IL and lignin systems: (**a**) 1.4 g Aliquat + 1.6 g lignosulphonate solution (40 g/L). (**b**) 1.6 g CYPHOS 108 + 1.9 g lignosulphonate solution (40 g/L). (**c**) CYPHOS 108 in PVDF support after contact with Kraft lignin solution (40 g/L).

Another identified problematic issue was the employment of [EMIM]Ac for the preparation of SILMs. After the contact between [EMIM]Ac and the PCTE support for 24 h at 70 °C in the vacuum oven, the membrane was totally dissolved in the IL (Figure 7a). Even when the temperature in the vacuum oven was reduced to 30 °C, the PCTE membrane suffered severe degradation after contact with this IL (Figure 7b). Other membranes, like PVDF or HPVDF, were highly modified (color changed) after impregnation with [EMIM]Ac at 70 °C (Figure 7c), but a decreased value of the temperature in the vacuum oven (30 °C) resulted in stable SILMs.

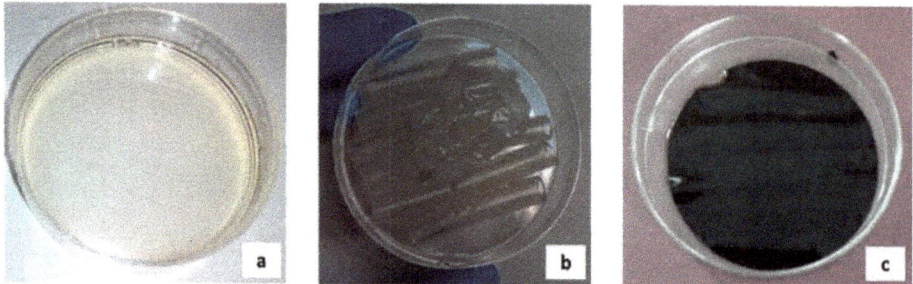

Figure 7. Identification of unstable SILMs with [EMIM]Ac: (**a**) PCTE in [EMIM]Ac (oven at 70 °C). (**b**) PCTE in [EMIM]Ac (oven at 30 °C). (**c**) HPVDF in [EMIM]Ac (oven at 70 °C).

The only virgin membrane that was permeable to Kraft lignin solution and allowed the transport of the solute from the feed to the stripping compartment was the HPVDF membrane. The evolution of the normalized concentration of Kraft lignin in the stripping compartment of the cell with respect to time is shown in Figure 8, where the transport through the virgin membrane was compared to the transport through some SILMs based on HPVDF. As can be observed in the graph, the virgin membrane showed the fastest transport, while the performance of the permeable SILMs was very similar and only minor differences could be identified among the different ILs.

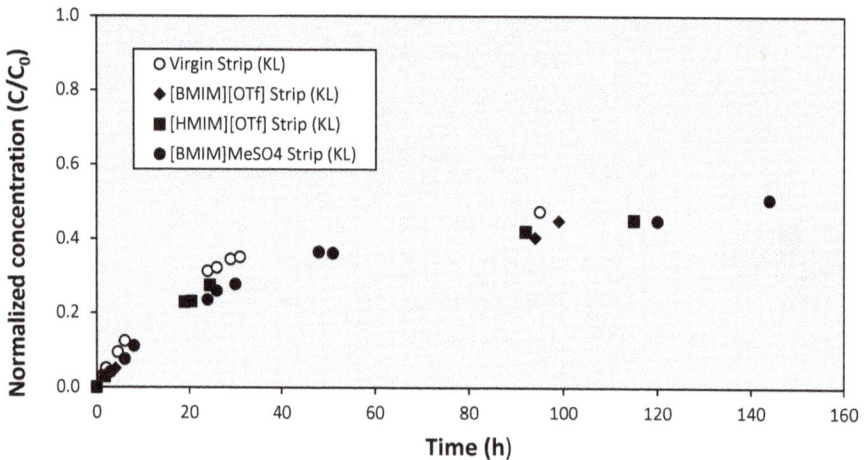

Figure 8. Evolution of the concentration of Kraft lignin in the stripping compartment with virgin HPVDF membrane and three SILMs supported in HPVDF.

Higher permeation of Kraft lignin occurred during the initial experimental phase, when the solute concentration gradient was maximum between both compartments of the cell. Then, the transport was slowed down due to reduced gradient for concentrations getting closer to the equilibrium. These experimental results fitted satisfactorily with the proposed model for non-selective transport through permeable membranes. The corresponding equations were employed for direct assessment of the effective mass transfer constants K_P (values compiled in Table 2). The R^2 values of the fitting of the experimental results ranged from 0.936 to 0.982. As expected from the experimental results in Figure 8, the maximal K_P value corresponded to the virgin HPVDF membrane (0.0334 h^{-1}), with lower values for the HPVDF-based SILMs. Taking into account the effective membrane area in the cell (14.6 cm^2) and the volume of each cell compartment (120 mL), the corresponding permeate flux F$_P$ were calculated from the effective mass transfer constants K_P (Table 2).

Table 2. Values of effective mass transfer constants K_P, permeate fluxes F_P, and resistances (attributable to the membrane support and the IL) for virgin HPVDF membranes and SILMs based on these membranes (viscosity and density values of the ILs are included).

	K_P (h^{-1})	F_P (m/h)	R_{MEMB} (h/m)	R_{IL} (h/m)	Viscosity [1] (mPa·s)	Density (g/cm^3)
Virgin HPVDF	0.0334	1.37×10^{-3}	729	-	-	-
HPVDF + [EMIM]EtSO$_4$	0.0263	1.08×10^{-3}	-	197	122	1.20 (80 °C)
HPVDF +[BMIM]MeSO$_4$	0.0216	0.89×10^{-3}	-	398	214	1.17 (80 °C)
HPVDF + CYPHOS 108	0.0238	0.98×10^{-3}	-	294	409	1.07 (25 °C)
HPVDF + [EMIM]Ac	0.0248	1.02×10^{-3}	-	253	93	1.07 (80 °C)
HPVDF + [BMIM][DBP]	0.0141	0.58×10^{-3}	-	997	1539	1.04 (40 °C)
HPVDF + [HMIM][OTf]	0.0157	0.65×10^{-3}	-	821	135	1.24 (29 °C)
HPVDF + [BMIM][OTf]	0.0225	0.92×10^{-3}	-	353	75	1.30 (25 °C)

[1] Measured at room temperature.

The maximal permeate flux of the virgin membrane can be directly justified taking into consideration the resistances in the series model, since the ILs supported in the membranes provide an additional resistance to the permeate transport:

$$F_P = \frac{1}{R_{TOT}} = \frac{1}{R_{MEMB} + R_{IL}} \tag{14}$$

The value of the membrane resistance R_{MEMB} was calculated from the F_P value of the virgin membrane and the resistance values attributable to the ILs were derived from the corresponding F_P values once R_{MEMB} was known. According to the figures in Table 2, the R_{IL} values were lower than the R_{MEMB} value for all the SILMs but the ones with [BMIM][DBP] and [HMIM][OTf]. The high resistance exhibited by [BMIM][DBP] can be directly attributed to its high viscosity, but the value corresponding to [HMIM][OTf] was not easily related to the viscosity or density of the IL because other more viscous or denser ILs showed lower resistance [62–67].

Apart from HPVDF, the rest of the virgin membranes were impermeable to water or aqueous solutions. The SILMs prepared with these impermeable membranes were initially tested with Kraft lignin and lignosulphonate solutions. The results obtained with some of these SILMs are graphed in Figure 9, where the evolution of the solute concentration in both compartments is presented. The differences among the experimental data for SILMs with different ILs and solutes were not obvious and very similar results were obtained for all the SILMs that exhibited solute transport. When additional experiments were carried out with monosaccharides (glucose and xylose) as solutes, the performance of the tested SILMs was comparable and the evolution of the monosaccharides concentration in the cell compartments followed a similar trend to the identified one for Kraft lignin or lignosulphonates (Figure 10). A further analysis of the experimental results shown in Figures 8–10 suggested non-selective transport through the SILMs, as the different investigated membranes, ILs or solutes had not relevant influence on the performance of the SILMs. For example, the performance of the SILMs based on impermeable membranes could be compared to the SILMs based on permeable HPVDF.

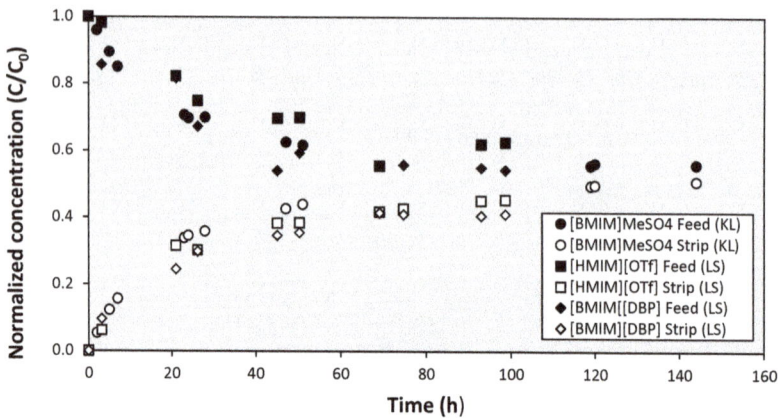

Figure 9. Evolution of the concentration of Kraft lignin (KL) or lignosulphonate (LS) in the feed and stripping compartments with three different SILMs supported in PCTE.

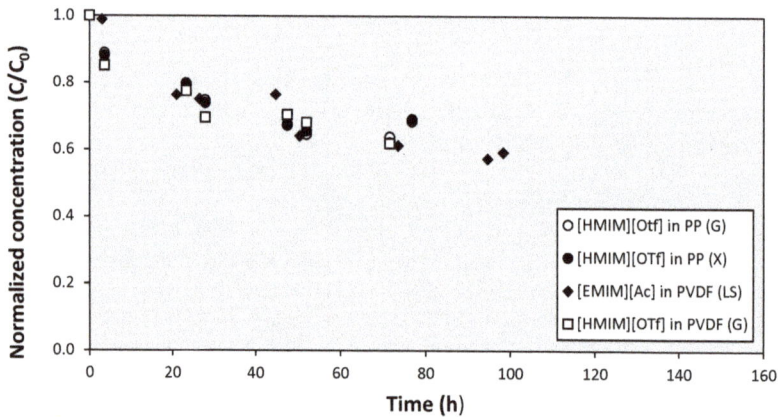

Figure 10. Evolution of the concentration of lignosulphonate (LS), glucose (G), or xylose (X) in the feed compartment with three different SILMs.

The lack of selectivity of the studied SILMs was previously reported during the preliminary works of this research group [57]. Those results demonstrated that the performance of the freshly prepared SILMs was totally similar to SILMs used repeatedly in consecutive cycles, even when the IL had been apparently lost from the membrane support. The mass loss of the SILMs once the first cycle was finished corresponded to the amount of IL previously immobilized in the membrane. Scanning electron microscope (SEM) images confirmed this fact, since the appearance of the membrane before the second cycle was much more similar to the virgin membrane than the SILM before the first cycle (Figure 11). Therefore, the transport through the SILMs seemed to be more closely related to the modification of the membrane supports than the presence of ILs in the membranes. Further experiments were carried out to understand the transport through the SILMs and tests to analyze the permeability of the SILMs to water and aqueous solutions were proposed. While the feed compartment was filled with water or an aqueous solution of Kraft lignin, the stripping compartment remained empty. Surprisingly, nearly all the tested SILMs that exhibited solute transport were identified as permeable and the only SILM that maintained its impermeability was that one based on [BMIM][DBP] supported in PTFE (Figure 12).

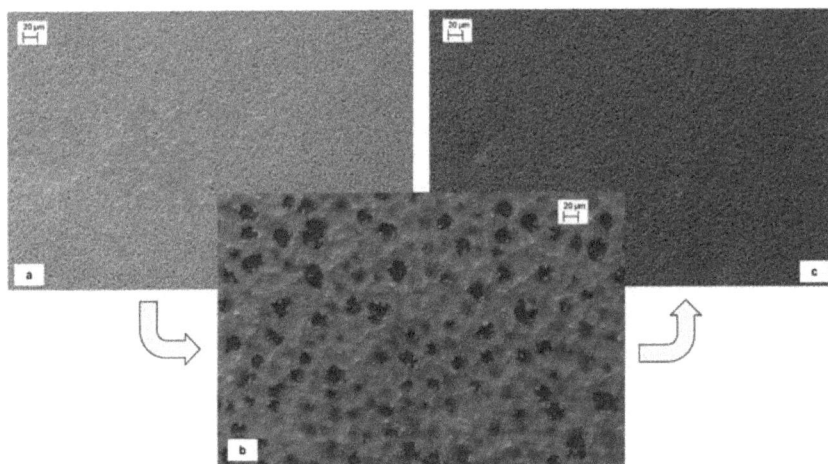

Figure 11. Scanning electron microscope (SEM) microphotographs of virgin PVDF membrane (**a**), SILM with CYPHOS 108 in PVDF before the transport test (**b**) and SILM with CYPHOS 108 in PVDF after the transport test (**c**).

Figure 12. Images of the permeability tests of virgin PVDF membrane (**a**), SILM with CYPHOS 108 in PVDF (**b**), and SILM with [BMIM][DBP] in PTFE (**c**).

As a summary, Table 3 compiles all the calculated K_P values of the tested SILMs. Moreover, the same table identifies the SILMs that were not tested because of the technical drawbacks during their preparation and the SILMS that did not show any solute transport. The reported K_P values ranged between 0.0139 and 0.0284 h^{-1} for [BMIM][DBP] in PP and [BMIM]MeSO$_4$ in PCTE, respectively. These figures fell in the range of the SILMS prepared with HPVDF (0.0141–0.0263 h^{-1}). This concordance once again confirmed the common transport mechanisms between both types of SILMs: convective transport of the solute as consequence of the permeation of the solutions.

The SILM with [BMIM][DBP] supported in PTFE membrane, which was the only one that showed impermeability to the tested solutions, was the only case that exhibited selective solute transport. As can be observed in Figure 13, the evolution of the concentration in the stripping compartment was very different for the two graphed solutes: the transport of glucose is much faster than the transport of lignosulphonate. While in the case of glucose, 70 h was time enough to achieve a concentration that could be considered an equilibrium situation; after 100 h the normalized concentration of lignosulphonate in the stripping compartment was still below 0.38.

Table 3. Summary table with the values of effective mass transfer constants K_P of the different tested SILMs.

Membrane Support	K_P (h^{-1})								
	CYPHOS 101	CYPHOS 108	ALIQUAT	[BMIM][MeSO$_4$]	[BMIM][DBP]	[BMIM][OTf]	[HMIM][OTf]	[EMIM][EtSO$_4$]	[EMIM]Ac
HPVDF	No tested	0.0238	No tested	0.0216	0.0141	0.0225	0.0157	0.0263	0.0248
PVDF	No tested	0.0254	No tested	0.0238	0.0247	0.0221	0.0171	0.0237	0.0186
PCTE	No tested	No tested	No tested	0.0284	0.0225	No tested	0.0227	No tested	No tested
PP	No tested	0.0248	No tested	No transport	0.0139	0.0179	0.0182	No transport	No transport
PTFE	No tested	No transport	No tested	No transport	Selective transport	No transport	No transport	No transport	No transport

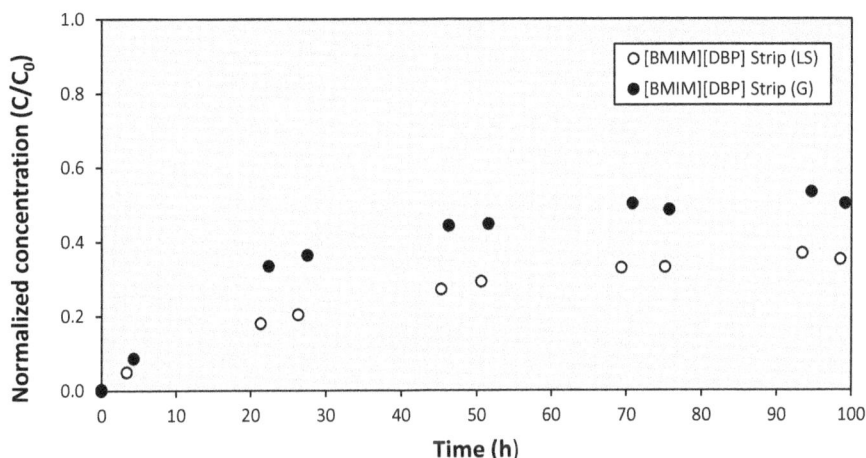

Figure 13. Evolution of the concentration of lignosulphonate (LS), glucose (G) in the stripping compartment with SILM with [BMIM][DBP] in PTFE.

When Kraft lignin and xylose were selected as solutes, the results were equivalent. The transport of xylose was much faster, with very similar results to glucose, and the results of Kraft lignin and lignosulphonates were comparable. The corresponding effective mass transfer constants were calculated (Table 4). These values revealed that glucose was the most easily transported solute, followed by xylose. The lignin compounds showed lower values: 0.67×10^{-3} m/h for lignosulphonate and 0.71×10^{-3} m/h for Kraft lignin. Therefore, the transport of monosaccharides was favored against lignin compounds. Further work will be carried out in order to improve the understanding of the mechanisms of the selective transport through the SILM and its stability. Furthermore, the design of effective separation processes based on this SILM will be investigated.

Table 4. Values of effective mass transfer constants K, and permeabilities k of the SILMs based on [BMIM][DBP] supported in PTFE membranes.

	K (h^{-1})	k (m/h)
Kraft lignin (KL)	0.0172	0.71×10^{-3}
Lignosulphonate (LS)	0.0162	0.67×10^{-3}
Glucose (G)	0.0453	1.86×10^{-3}
Xylose (X)	0.0429	1.76×10^{-3}

4. Conclusions

This study investigated the potential of SILMs for selective transport of two different types of technical lignins (Kraft lignin and lignosulphonate) and monosaccharides (xylose and glucose) in an aqueous solution. The SILMs obtained by the combination of five different membrane supports and nine ILs were tested. Some ILs (CYPHOS 101 and Aliquat 336) were not useful, since they resulted in unstable SILMs because of precipitation problems. [EMIM]Ac was another problematic IL as it degraded some of the membrane supports. Although the virgin hydrophobic membranes did not allow the permeation of the aqueous solutions, most membranes became permeable after the impregnation with the ILs. Therefore, the solutes were able to cross these SILMs by non-selective transport. However, the SILM based on [BMIM][DBP] as IL and PTFE as membrane support maintained its hydrophobicity and allowed selective transport of the tested solutes. The effective mass transfer constants of the solutes were determined according to the proposed transport model: lignosulphonate was the least easily transported solute (0.67×10^{-3} m/h), while glucose was the most easily transported one (1.86×10^{-3} m/h). Nevertheless, the stability of this identified selective SILM and its applicability to separation processes must be investigated more deeply and further work will be carried out.

Author Contributions: R.A. conceived and designed the experiments; J.R., S.L. and A.A. performed the experiments; R.A., J.R., S.L. and A.A. analyzed the data; A.G. and A.I. supervised the investigation; R.A. prepared the original draft of the manuscript and all authors revised the several versions of the manuscript.

Funding: This research has been financially supported by the Spanish Ministry of Economy and Competitiveness (MINECO) through CTQ2014-56820-JIN Project, co-financed by FEDER funds from European Union.

Conflicts of Interest: The authors declare no conflict of interest.

References

1. Bjärstig, T.; Sténs, A. Social Values of Forests and Production of New Goods and Services: The Views of Swedish Family Forest Owners. *Small-Scale For.* **2018**, *17*, 125–146. [CrossRef]
2. Hamaguchi, M.; Kautto, J.; Vakkilainen, E. Effects of hemicellulose extraction on the kraft pulp mill operation and energy use: Review and case study with lignin removal. *Chem. Eng. Res. Des.* **2013**, *91*, 1284–1291. [CrossRef]
3. Abejón, R. A Bibliometric Study of Scientific Publications regarding Hemicellulose Valorization during the 2000–2016 Period: Identification of Alternatives and Hot Topics. *ChemEngineering* **2018**, *2*, 7. [CrossRef]
4. Abejón, R.; Pérez-Acebo, H.; Clavijo, L. Alternatives for Chemical and Biochemical Lignin Valorization: Hot Topics from a Bibliometric Analysis of the Research Published During the 2000–2016 Period. *Processes* **2018**, *6*, 98. [CrossRef]
5. Li, T.; Takkellapati, S. The current and emerging sources of technical lignins and their applications. *Biofuels Bioprod. Biorefin.* **2018**, 1–32. [CrossRef]
6. Lee, H.S.; Jae, J.; Ha, J.M.; Suh, D.J. Hydro- and solvothermolysis of kraft lignin for maximizing production of monomeric aromatic chemicals. *Bioresour. Technol.* **2016**, *203*, 142–149. [CrossRef] [PubMed]
7. Ceapraz, I.L.; Kotbi, G.; Sauvee, L. The territorial biorefinery as a new business model. *Bio-Based Appl. Econ.* **2016**, *5*, 47–62. [CrossRef]
8. De Bhowmick, G.; Sarmah, A.K.; Sen, R. Lignocellulosic biorefinery as a model for sustainable development of biofuels and value added products. *Bioresour. Technol.* **2018**, *247*, 1144–1154. [CrossRef] [PubMed]
9. Barcelos, C.A.; Maeda, R.N.; Betancur, G.J.V.; Pereira, N. The essentialness of delignification on enzymatic hydrolysis of sugar cane bagasse cellulignin for second generation ethanol production. *Waste Biomass Valorization* **2013**, *4*, 341–346. [CrossRef]
10. Pihlajaniemi, V.; Sipponen, M.H.; Pastinen, O.; Nyyssölä, A.; Laakso, S. The effect of direct and counter-current flow-through delignification on enzymatic hydrolysis of wheat straw, and flow limits due to compressibility. *Biotechnol. Bioeng.* **2016**, *113*, 2605–2613. [CrossRef] [PubMed]
11. Ragauskas, A.J.; Beckham, G.T.; Biddy, M.J.; Chandra, R.; Chen, F.; Davis, M.F.; Davison, B.H.; Dixon, R.A.; Gilna, P.; Keller, M.; et al. Lignin valorization: Improving lignin processing in the biorefinery. *Science* **2014**, *344*. [CrossRef] [PubMed]
12. Chatel, G.; Rogers, R.D. Review: Oxidation of lignin using ionic liquids-an innovative strategy to produce renewable chemicals. *ACS Sustain. Chem. Eng.* **2014**, *2*, 322–339. [CrossRef]
13. Davis, K.M.; Rover, M.; Brown, R.C.; Bai, X.; Wen, Z.; Jarboe, L.R. Recovery and utilization of lignin monomers as part of the biorefinery approach. *Energies* **2016**, *9*, 808. [CrossRef]
14. Beckham, G.T.; Johnson, C.W.; Karp, E.M.; Salvachúa, D.; Vardon, D.R. Opportunities and challenges in biological lignin valorization. *Curr. Opin. Biotechnol.* **2016**, *42*, 40–53. [CrossRef] [PubMed]
15. Rinaldi, R.; Jastrzebski, R.; Clough, M.T.; Ralph, J.; Kennema, M.; Bruijnincx, P.C.A.; Weckhuysen, B.M. Paving the Way for Lignin Valorisation: Recent Advances in Bioengineering, Biorefining and Catalysis. *Angew. Chem. Int. Ed.* **2016**, *55*, 8164–8215. [CrossRef] [PubMed]
16. Wang, W.; Zhang, C.; Sun, X.; Su, S.; Li, Q.; Linhardt, R.J. Efficient, environmentally-friendly and specific valorization of lignin: Promising role of non-radical lignolytic enzymes. *World J. Microbiol. Biotechnol.* **2017**, *33*, 1–14. [CrossRef] [PubMed]

17. Fan, M.H.; Deng, S.M.; Wang, T.J.; Li, Q.X. Production of BTX through catalytic depolymerization of lignin. *Chin. J. Chem. Phys.* **2014**, *27*, 221–226. [CrossRef]

18. Elfadly, A.M.; Zeid, I.F.; Yehia, F.Z.; Abouelela, M.M.; Rabie, A.M. Production of aromatic hydrocarbons from catalytic pyrolysis of lignin over acid-activated bentonite clay. *Fuel Process. Technol.* **2017**, *163*, 1–7. [CrossRef]

19. Holladay, J.E.; White, J.F.; Bozell, J.J.; Johnson, D. *Top Value-Added Chemicals from Biomass Volume II—Results of Screening for Potential Candidates from Biorefinery Lignin*; DOE Scientific and Technical Information; US Department of Energy: Oak Ridge, TN, USA, 2007; ISBN PNNL-16983.

20. Abels, C.; Carstensen, F.; Wessling, M. Membrane processes in biorefinery applications. *J. Memb. Sci.* **2013**, *444*, 285–317. [CrossRef]

21. Dafinov, A.; Font, J.; Garcia-Valls, R. Processing of black liquors by UF/NF ceramic membranes. *Desalination* **2005**, *173*, 83–90. [CrossRef]

22. Bhattacharya, P.K.; Todi, R.K.; Tiwari, M.; Bhattacharjee, C.; Bhattacharjee, S.; Datta, S. Studies on ultrafiltration of spent sulfite liquor using various membranes for the recovery of lignosulphonates. *Desalination* **2005**, *174*, 287–297. [CrossRef]

23. Moniz, P.; Serralheiro, C.; Matos, C.T.; Boeriu, C.G.; Frissen, A.E.; Duarte, L.C.; Roseiro, L.B.; Pereira, H.; Carvalheiro, F. Membrane separation and characterisation of lignin and its derived products obtained by a mild ethanol organosolv treatment of rice straw. *Process Biochem.* **2018**, *65*, 136–145. [CrossRef]

24. Mota, I.F.; Pinto, P.R.; Ribeiro, A.M.; Loureiro, J.M.; Rodrigues, A.E. Downstream processing of an oxidized industrial kraft liquor by membrane fractionation for vanillin and syringaldehyde recovery. *Sep. Purif. Technol.* **2018**, *197*, 360–371. [CrossRef]

25. Servaes, K.; Varhimo, A.; Dubreuil, M.; Bulut, M.; Vandezande, P.; Siika-aho, M.; Sirviö, J.; Kruus, K.; Porto-Carrero, W.; Bongers, B. Purification and concentration of lignin from the spent liquor of the alkaline oxidation of woody biomass through membrane separation technology. *Ind. Crops Prod.* **2017**, *106*, 86–96. [CrossRef]

26. Weinwurm, F.; Drljo, A.; Waldmüller, W.; Fiala, B.; Niedermayer, J.; Friedl, A. Lignin concentration and fractionation from ethanol organosolv liquors by ultra- and nanofiltration. *J. Clean. Prod.* **2016**, *136*, 62–71. [CrossRef]

27. Jönsson, A.S.; Nordin, A.K.; Wallberg, O. Concentration and purification of lignin in hardwood kraft pulping liquor by ultrafiltration and nanofiltration. *Chem. Eng. Res. Des.* **2008**, *86*, 1271–1280. [CrossRef]

28. Toledano, A.; García, A.; Mondragon, I.; Labidi, J. Lignin separation and fractionation by ultrafiltration. *Sep. Purif. Technol.* **2010**, *71*, 38–43. [CrossRef]

29. Sevastyanova, O.; Helander, M.; Chowdhury, S.; Lange, H.; Wedin, H.; Zhang, L.; Ek, M.; Kadla, J.F.; Crestini, C.; Lindström, M.E. Tailoring the molecular and thermo-mechanical properties of kraft lignin by ultrafiltration. *J. Appl. Polym. Sci.* **2014**, *131*, 9505–9515. [CrossRef]

30. Žabková, M.; da Silva, E.A.B.; Rodrigues, A.E. Recovery of vanillin from lignin/vanillin mixture by using tubular ceramic ultrafiltration membranes. *J. Memb. Sci.* **2007**, *301*, 221–237. [CrossRef]

31. Dubreuil, M.F.S.; Servaes, K.; Ormerod, D.; Van Houtven, D.; Porto-Carrero, W.; Vandezande, P.; Vanermen, G.; Buekenhoudt, A. Selective membrane separation technology for biomass valorization towards bio-aromatics. *Sep. Purif. Technol.* **2017**, *178*, 56–65. [CrossRef]

32. Werhan, H.; Farshori, A.; von Rohr, P.R. Separation of lignin oxidation products by organic solvent nanofiltration. *J. Memb. Sci.* **2012**, *423–424*, 404–412. [CrossRef]

33. Mai, N.L.; Koo, Y.M. Computer-Aided Design of Ionic Liquids for High Cellulose Dissolution. *ACS Sustain. Chem. Eng.* **2016**, *4*, 541–547. [CrossRef]

34. Hossain, M.M.; Aldous, L. Ionic liquids for lignin processing: Dissolution, isolation, and conversion. *Aust. J. Chem.* **2012**, *65*, 1465–1477. [CrossRef]

35. Isik, M.; Sardon, H.; Mecerreyes, D. Ionic liquids and cellulose: Dissolution, chemical modification and preparation of new cellulosic materials. *Int. J. Mol. Sci.* **2014**, *15*, 11922–11940. [CrossRef] [PubMed]

36. Zhao, D.; Li, H.; Zhang, J.; Fu, L.; Liu, M.; Fu, J.; Ren, P. Dissolution of cellulose in phosphate-based ionic liquids. *Carbohydr. Polym.* **2012**, *87*, 1490–1494. [CrossRef]

37. Lan, W.; Liu, C.F.; Yue, F.X.; Sun, R.C.; Kennedy, J.F. Ultrasound-assisted dissolution of cellulose in ionic liquid. *Carbohydr. Polym.* **2011**, *86*, 672–677. [CrossRef]

38. Stolarska, O.; Pawlowska-Zygarowicz, A.; Soto, A.; Rodríguez, H.; Smiglak, M. Mixtures of ionic liquids as more efficient media for cellulose dissolution. *Carbohydr. Polym.* **2017**, *178*, 277–285. [CrossRef] [PubMed]

39. Liu, R.; Chen, Z.; Ren, H.; Duan, E. Synthesis and properties of non-aromatic ionic liquids and their role in cellulose dissolution. *BioResources* **2017**, *12*, 5407–5416. [CrossRef]

40. Roselli, A.; Asikainen, S.; Stepan, A.; Monshizadeh, A.; Von Weymarn, N.; Kovasin, K.; Wang, Y.; Xiong, H.; Turunen, O.; Hummel, M.; et al. Comparison of pulp species in IONCELL-P: Selective hemicellulose extraction method with ionic liquids. *Holzforschung* **2016**, *70*, 291–296. [CrossRef]

41. Heggset, E.B.; Syverud, K.; Øyaas, K. Novel pretreatment pathways for dissolution of lignocellulosic biomass based on ionic liquid and low temperature alkaline treatment. *Biomass Bioenergy* **2016**, *93*, 194–200. [CrossRef]

42. Carneiro, A.P.; Rodriguez, O.; Macedo, E.A. Dissolution and fractionation of nut shells in ionic liquids. *Bioresour. Technol.* **2017**, *227*, 188–196. [CrossRef] [PubMed]

43. Mäki-Arvela, P.; Anugwom, I.; Virtanen, P.; Sjöholm, R.; Mikkola, J.P. Dissolution of lignocellulosic materials and its constituents using ionic liquids-A review. *Ind. Crops Prod.* **2010**, *32*, 175–201. [CrossRef]

44. Lee, S.H.; Doherty, T.V.; Linhardt, R.J.; Dordick, J.S. Ionic liquid-mediated selective extraction of lignin from wood leading to enhanced enzymatic cellulose hydrolysis. *Biotechnol. Bioeng.* **2009**, *102*, 1368–1376. [CrossRef] [PubMed]

45. Tan, S.S.Y.; MacFarlane, D.R.; Upfal, J.; Edye, L.A.; Doherty, W.O.S.; Patti, A.F.; Pringle, J.M.; Scott, J.L. Extraction of lignin from lignocellulose at atmospheric pressure using alkylbenzenesulfonate ionic liquid. *Green Chem.* **2009**, *11*, 339. [CrossRef]

46. Balaji, C.; Banerjee, T.; Goud, V.V. COSMO-RS based predictions for the extraction of lignin from lignocellulosic biomass using ionic liquids: Effect of cation and anion combination. *J. Solut. Chem.* **2012**, *41*, 1610–1630. [CrossRef]

47. Hamada, Y.; Yoshida, K.; Asai, R.; Hayase, S.; Nokami, T.; Izumi, S.; Itoh, T. A possible means of realizing a sacrifice-free three component separation of lignocellulose from wood biomass using an amino acid ionic liquid. *Green Chem.* **2013**, *15*, 1863. [CrossRef]

48. Prado, R.; Erdocia, X.; Labidi, J. Lignin extraction and purification with ionic liquids. *J. Chem. Technol. Biotechnol.* **2013**, *88*, 1248–1257. [CrossRef]

49. Glas, D.; Van Doorslaer, C.; Depuydt, D.; Liebner, F.; Rosenau, T.; Binnemans, K.; De Vos, D.E. Lignin solubility in non-imidazolium ionic liquids. *J. Chem. Technol. Biotechnol.* **2015**, *90*, 1821–1826. [CrossRef]

50. Konda, N.M.; Shi, J.; Singh, S.; Blanch, H.W.; Simmons, B.A.; Klein-Marcuschamer, D. Understanding cost drivers and economic potential of two variants of ionic liquid pretreatment for cellulosic biofuel production. *Biotechnol. Biofuels* **2014**, *7*, 1–11. [CrossRef] [PubMed]

51. Parhi, P.K. Supported liquid membrane principle and its practices: A short review. *J. Chem.* **2013**, *2013*. [CrossRef]

52. Kocherginsky, N.M.; Yang, Q.; Seelam, L. Recent advances in supported liquid membrane technology. *Sep. Purif. Technol.* **2007**, *53*, 171–177. [CrossRef]

53. Abejón, R.; Pérez-Acebo, H.; Garea, A. A Bibliometric Analysis of Research on Supported Ionic Liquid Membranes during the 1995–2015 Period: Study of the Main Applications and Trending Topics. *Membranes* **2017**, *7*, 63. [CrossRef] [PubMed]

54. Lozano, L.J.; Godínez, C.; de los Ríos, A.P.; Hernández-Fernández, F.J.; Sánchez-Segado, S.; Alguacil, F.J. Recent advances in supported ionic liquid membrane technology. *J. Memb. Sci.* **2011**, *376*, 1–14. [CrossRef]

55. Kilulya, K.F.; Msagati, T.A.M.; Mamba, B.B.; Ngila, J.C.; Bush, T. Ionic liquid-liquid extraction and supported liquid membrane analysis of lipophilic wood extractives from dissolving-grade pulp. *Chromatographia* **2012**, *75*, 513–520. [CrossRef]

56. Venkateswaran, P. Di (2-ethylhexyl) phosphoric acid-coconut oil supported liquid membrane for the separation of copper ions from copper plating wastewater. *J. Environ. Sci.* **2007**, *19*, 1446–1453. [CrossRef]

57. Abejón, R.; Abejón, A.; Garea, A.; Irabien, A. Transport of lignin and other lignocellulosic components through supported ionic liquid membranes. *Chem. Eng. Trans.* **2017**, *57*, 1153–1158.

58. Alén, R.; Hartus, T. UV spectrophotometric determination of lignin from alkaline pulping liquors. *Cell. Chem. Technol.* **1987**, *618*, 613–618.

59. Lee, R.A.; Bédard, C.; Berberi, V.; Beauchet, R.; Lavoie, J.M. UV-Vis as quantification tool for solubilized lignin following a single-shot steam process. *Bioresour. Technol.* **2013**, *144*, 658–663. [CrossRef] [PubMed]

60. Miller, G.L. Use of Dinitrosalicylic Acid Reagent for Determination of Reducing Sugar. *Anal. Chem.* **1959**, *31*, 426–428. [CrossRef]

61. Oveissi, F.; Fatehi, P. Isolating lignin from spent liquor of thermomechanical pulping process via adsorption. *Environ. Technol.* **2014**, *35*, 2597–2603. [CrossRef] [PubMed]

62. Iolitec Ionic liquids: Catalogue. Available online: https://iolitec.de/index.php/en/products/ionic_liquids/catalogue (accessed on 6 June 2018).

63. Zhang, S.; Zhou, Q.; Lu, X.; Song, Y.; Wang, X. *Physicochemical Properties of Ionic Liquid Mixtures*; Springer: Dordrecht, The Netherlands, 2016; ISBN 9789401775717.

64. Palgunadi, J.; Kang, J.E.; Cheong, M.; Kim, H.; Lee, H.; Kim, H.S. Fluorine-free imidazolium-based ionic liquids with a phosphorous-containing anion as potential CO_2 absorbents. *Bull. Korean Chem. Soc.* **2009**, *30*, 1749–1754. [CrossRef]

65. Ge, M.-L.; Zhao, R.-S.; Yi, Y.-F.; Zhang, Q.; Wang, L.-S. Densities and Viscosities of 1-Butyl-3-methylimidazolium Tetrafluoroborate + H_2O Binary Mixtures at T = (303.15 to 343.15) K. *J. Chem. Eng. Data* **2008**, *53*, 2408–2411. [CrossRef]

66. Fredlake, C.P.; Crosthwaite, J.M.; Hert, D.G.; Aki, S.N.V.K.; Brennecke, J.F. Thermophysical Properties of Imidazolium-Based Ionic Liquids. *J. Chem. Eng. Data* **2004**, *49*, 954–964. [CrossRef]

67. Fraser, K.J.; MacFarlane, D.R. Phosphonium-based ionic liquids: An overview. *Aust. J. Chem.* **2009**, *62*, 309–321. [CrossRef]

processes

Review

Recent Advance on Draw Solutes Development in Forward Osmosis

Qingwu Long [1,2,3,*], Yongmei Jia [2,4], Jinping Li [3], Jiawei Yang [3], Fangmei Liu [3], Jian Zheng [1,2,3] and Biao Yu [2,4,*]

[1] Key Laboratory of Clean Energy and Materials Chemistry of Guangdong Province College (Lingnan Normal University), Department of Education of Guangdong Province, Zhanjiang 524048, China; zhengjianphd@hotmail.com
[2] Resource and Chemical Engineering Technology Research Center of Western Guangdong Province, Zhanjiang 524048, China; jiayongmei214@126.com
[3] School of Chemistry and Chemical Engineering, Lingnan Normal University, Zhanjiang 524048, China; p13420562529@163.com (J.L.); 18476092452@163.com (J.Y.); 13414854179@163.com (F.L.)
[4] Institute of New Materials, Lingnan Normal University, Zhanjiang 524048, China
*   Correspondence: wubi86@126.com (Q.L.); y.biao@lingnan.edu.cn (B.Y.); Tel.: +86-0759-3174029 (Q.L.)

Received: 31 July 2018; Accepted: 11 September 2018; Published: 13 September 2018

Abstract: In recent years, membrane technologies have been developed to address water shortage and energy crisis. Forward osmosis (FO), as an emerging membrane-based water treatment technology, employs an extremely concentrated draw solution (DS) to draw water pass through the semi-permeable membrane from a feed solution. DS as a critical material in FO process plays a key role in determining separation performance and energy cost. Most of existing DSs after FO still require a regeneration step making its return to initial state. Therefore, selecting suitable DS with low reverse solute, high flux, and easy regeneration is critical for improving FO energy efficiency. Numerous novel DSs with improved performance and lower regeneration cost have been developed. However, none reviews reported the categories of DS based on the energy used for recovery up to now, leading to the lack of enough awareness of energy consumption in DS regeneration. This review will give a comprehensive overview on the existing DSs based on the types of energy utilized for DS regeneration. DS categories based on different types of energy used for DS recovery, mainly including direct use based, chemical energy based, waste heat based, electric energy based, magnetic field energy based, and solar energy based are proposed. The respective benefits and detriments of the majority of DS are addressed respectively according to the current reported literatures. Finally, future directions of energy applied to DS recovery are also discussed.

Keywords: forward osmosis; draw solutes; membrane separation; regeneration; energy

1. Introduction

With the development of economy, population boom and increasing urbanization, water shortage has become a global problem. To address the water scarcity and prevent water pollution, the seawater desalination and wastewater reuse has been considered as a good solution to solving mentioned above [1–5]. In the past decade, numerous desalination technologies, including multistage flash evaporation, low-temperature multi-effect distillation, electrodialysis, reverse osmosis (RO) have been developed to desalinate seawater. Among them, RO is one of the most widely used desalination technology, due to its high salt rejection and easy large-scale operation. The energy consumption of 1 m³ fresh water produced by RO is still up to 1.5–2.5 kWh, which is 5~10 times lower than that of distillation [6]. However, the discharge of concentrated salt solution produced by RO may give rise to the secondary pollution [1,7–11].

Forward osmosis (FO) is an emerging water treatment technique [12]. In FO process, the net osmotic pressure difference generated by the draw and feed solution with different concentrations on both sides of the membrane impels water molecules to permeate from the side of feed solution to the draw solution (DS). Since water molecules spontaneously permeate without additional pressure, the energy consumption of seawater desalination by FO alone without DS recovery is around 0.84 kWh/m^3, which is 72.1% lower than that of RO [13]. Compared with RO, FO possesses a stronger antifouling capability [2,14–17], higher water recovery, and less salt discharge. Therefore, it has good prospect of application in such fields as seawater desalination [18,19], agricultural irrigation [20], anaerobic sludge disposal [8,21,22], osmotic power generation [23–25], wastewater treatment [19,26], protein enrichment [27,28], drug delivery and release [14,29], aerospace [30], and food industry [31].

Although FO is well considered as an energy-saving process by using some smart agents as DS, the result of thermodynamic calculations shows that reducing energy consumption in FO process isn't a "free lunch" [15]. In general, FO desalination process includes two steps: The concentrative DS is diluted, and the dilutive DS is re-concentrated. This means the use of higher osmotic pressure DS dewater from feed solution in the first step and the diluted DS should be regenerated to initial state by using other technologies in the second step. Especially, considering DS regeneration in seawater desalination, the energy consumption of a standalone FO process far surpasses than that of RO [15]. However, FO hybrid systems (i.e., FO-RO) are capable of desalinating the high-salinity waters, which RO process is limited, such as those highly fouled or specific waste stream or high osmotic pressure feeds [15]. In addition, FO as an immature technology suffers some limitations, such as severe DS leakage, which may decrease the quality of product water or reduced FO performance or increase DS recycling costs, leading to hinder its practical industrial application. However, FO still has the opportunities by employing the low-cost thermal energy for DS regeneration. For example, the use of thermolytic compounds as DSs in FO hybrid systems can reduce total energy requirement for desalting high-salinity waters, which is more energy-saving than other desalination solutions [15]. This case shows the advantages of FO over RO, such as low energy cost, considerably high water recovery, and minimal fouling. These advantages imply that FO is facing some accompanied challenges. In fact, FO could not be applied in widespread applications, mainly due to the limited choices of DSs.

DS as one of critical materials of FO has a great impact on the energy efficiency. Although FO is common an energy-efficient process without DS recovery, it, in most instances, still requires the secondary step to concentrate DS, which would directly increase the relevant energy cost. Therefore, it is necessary to develop suitable DS so as to break through the bottleneck of FO's development toward the practical application. An ideal DS should have some characteristics [2,32], including high solubility, high osmotic pressure, low molecular weight (MW), low-cost regeneration, good compatibility with the membrane, low reverse solute flux, nontoxicity, and so forth.

In the past few decades, plenty of materials have been investigated as DS in FO. Several review articles on traditional and advanced DS have been published [1,4,10,18,21–25,33–44], these DSs are shown in Table 1, which include (1) inorganic compounds, (2) organic compounds (e.g., polyacrylic acid sodium, methylimidazole-based compounds, hexavalent phosphazene salts, stimuli-responsive hydrogels), (3) functionalized nanoparticles (e.g., magnetic nanoparticles (MNPs)), Na$^+$-functionalized carbon quantum dots (Na-CQD). However, none has systematically addressed the energy type of DS in recovery process. Most of these reviews focused mainly on the synthesis, application, performance, and theories of DS in FO process. Specific discussion on the energy used to re-concentrate dilutive DS was rather brief. Moreover, lots of earlier research literatures were highlighting FO's feature of low energy consumption, and more discussion on DSs was focused on achieving superior FO performance, but mentioned lightly on the energy cost, resulting in a misplaced expectation of FO with higher energy efficiency than RO desalination. Therefore, in order to strengthen a comprehensive understanding on DS regeneration. A review that analyzes the types of energy applied in DS regeneration is crucially needed. Up to now, there are no relative reports on DS's classification based on the type of energy used for DS recovery [10,20].

Table 1. Overview of the traditional classification of draw solutions (DSs) in forward osmosis (FO) process.

Categories	Draw Solutes	Recovery Methods	Ref.
Inorganic compounds	NaCl	reverse osmosis (RO)	[33]
	inorganic fertilizer	direct use	[45]
	potassium sulfate (K_2SO_4)	RO	[33]
	sodium nitrate ($NaNO_3$)	direct use	[46]
	aluminum sulfate ($Al_2(SO_4)_3$)	precipitation	[47]
	magnesium sulfate ($MgSO_4$), copper sulfate ($CuSO_4$)	precipitation	[48,49]
Organic compounds	Switchable polarity solvent (SPS)	RO	[50]
	sodium polyacrylate (PAA-Na)	ultrafiltration (UF), membrane distillation (MD)	[51,52]
	CO_2-responsive polymers (PDMAEMA)	UF	[53]
	poly(sodium styrene-4-sulfonate-co-*N*-isopropylacrylamide) (PSSS-PNIPAM)	MD	[54]
	poly (aspartic acid sodium salt)	MD	[55]
	N,N-dimethylcyclohexylamine (N(Me)$_2$Cy)	heating	[56]
	1-Cyclohexylpiperidine (CHP)	heating	[57]
	Micellar solution	UF	[58]
	oxalic acid complexes with Fe/Cr/Na	nanofiltration (NF)	[59]
	2-Methylimidazole compounds	MD	[60]
	trimethylamine–carbon dioxide	heating	[61]
	glucose, fructose	RO	[62–64]
	polyelectrolyte incorporated with triton-x114	MD	[65]
	dimethyl ether	heating with solar energy	[66]
	poly(4-styrenesulfonic acid-co-maleic acid)	NF	[67]
Functional nanoparticles	Super hydrophilic nanoparticles	UF	[40]
	hydrophilic superparamagnetic nanoparticles	magnetic separation	[68]
	magnetic core-hydrophilic shell nanosphere	magnetic separation	[69]
	thermoresponsive Magnetic Nanoparticle	magnetic separation	[42]
	dextran-coated MNPs	magnetic separation	[36]
	hyperbranched polyglycerol coated MNPs	magnetic separation	[70]

In this review article, we critically classified the DS's categories in terms of their energy types applied in recovery process; discuss its potential in recovery process, FO performance, and suitable applications in FO. Specifically, we address the development and the advantages or demerits of existing DSs based on the new DS classification, categorizing them as directly use based, chemical energy based, waste heat based, electric energy based, solar energy based, and magnetic field energy based. Addressing these DS categories and understanding the limits of different DS in FO process will provide vital information to guidance the exploration of successful DS for expanding the range of its application.

2. Classifications of DS Based on the Types of Energy Used in Regeneration Process

The remaining defects of DS, such as high reverse solute flux, low osmotic pressure and high energy cost in recovery seriously hinder FO application. Therefore, the development of a novel DS to solve above problems is still a great challenge. Previously, several reviews have simply classified the known DSs (inorganic and organic compounds) based on the chemical properties. Later, these DSs were further subdivided into inorganic salts, organic salts, polymers, nanoparticles, micelle solutions, gels, and so forth [71]. Recently, the types of DSs based on different regeneration methods were reported, which included directly use, thermal separation, membrane separation (RO, nanofiltration (NF), ultrafiltration (UF), microfiltration (MF), chemical precipitation, stimuli-responsive (e.g., light, electricity, magnetic field). In this review, the classification of DSs based on types of energy employed in regeneration process is summarized (Table 2).

Table 2. Overview of DSs based on the types of energy used in recovery process.

Categories	Recovery Methods	Draw Solutes
Direct use	Without recovery	Saccharides (glucose, fructose) [72,73], fertilizer [20,46], liquid fertilizer [74], sodium lignin sulfonate (NaLS) [75]
Chemical energy	Precipitation	$Al_2(SO_4)_3$ [47,76], $MgSO_4$ [49], $CuSO_4$ [77]
Waste heat	Heating	Sulfur dioxide (SO_2) [78], ammonia and carbon dioxide (NH_3-CO_2) [79]
	heating	*N,N*-dimethylcyclohexylamine (N(Me)$_2$Cy) [56], 1-Cyclohexylpiperidine (CHP) [57], trimethylamine-carbon dioxide (N(CH$_3$)$_2$-CO$_2$) [61], switchable polarity solvents [50], ionic polymer hydrogels with thermal responsive units [80]
	phase separation	Upper critical solution temperature (UCST) ionic liquid [81], ammonium iodide salts [82], lower critical solution temperature (LCST) ionic liquid [83], thermally responsive polyionic liquid hydrogels [84–86], thermosensitive copolymer [54], ionic hydrogels [34], thermo-sensitive polyelectrolyte [87], phase transition materials [88], CO_2 switchable dual responsive polymers [89,90], thermosensitive polymer coated magnetic nanoparticles [38], gas-responsive cationic microgels [91]
	MD	2-Methylimidazole salt [60], Na^+-functionalized carbon quantum dots (Na-CQDs) [92], dendrimer [93], poly (aspartic acid sodium salt) [55], multi-charged oxalic acid complexes [94]
Electric energy	RO	Inorganic salt (NaCl [95,96], $MgCl_2$ [95], KNO_3 [33]), organic ionic salts [97], glucose and sucrose miscible liquids [63]
	NF	divalent metal salt (Na_2SO_4, $MgSO_4$) [98], EDTA sodium salt [21], metal complexes [27,59,99], poly (4-styrenesulfonic acid-co-maleic acid) [69], novel carboxyethyl amine sodium salts [100], organic phosphonate salts [101]
	UF	Thermosensitive polyelectrolyte [102], surface modified MNPs [37,39], phosphatic surfactant [103], micellar solution [58,104], sodium polyacrylate [52], carboxylate polyelectrolyte [105], natural polymer-based cationic polyelectrolyte [106]
	MF	Thermo-responsive copolymers [107]
Solar energy	Irradiating	Bifunctional polymer hydrogel layers [108], graphene gels [109], thermo-responsive nanoparticles [110], composite hydrogel monoliths containing thermoplastic polyurethane microfibers [111], composite hydrogels (carbon particles and sodium acrylate-isopropylacrylamide) (SA-NIPAM) [112,113], composite hydrogels based on graphene and SA-NIPAM [114], dimethyl ether [66]
Magnetic field energy	Magnetic separation	Functionalized MNPs [37], citrate-coated MNPs [115], PAA-Na coated-MNPs [34,116], thermosensitive magnetic nanogels [38,42], dextran coated Fe_3O_4 magnetic nanoparticles [36], triethylene glycol-coated magnetic nanoparticles, polyacrylic acid-coated magnetic nanoparticles [37], poly(oxy-1,2-ethanediyl)-coated magnetic nanoparticles [39], poly(ethylene glycol) diacid-coated (PEG-(COOH)$_2$-coated) MNPs [39], hyperbranched polyglycerol coated MNPs [70], polyacrylic acid-coated MNPs [37,40,41]

2.1. Direct Use without Recovery

Since FO itself does not produce fresh water, a second regeneration process is needed to obtain the pure water. DS regeneration is considered as a potentially energy-intensive process, its energy cost in DS recovery significantly depends on the as-selected DS. For one case, FO without the step of DS regeneration is the lowest energy-saving. These FO processes without DS recovery have been successfully used for drinking, fertilization, irrigation and soil prevention in arid areas [20,46,72–75,78]. Natural sugar DSs (e.g., glucose, fructose and sucrose) [72,73,88] diluted after FO could drink and use in some emergency boats [62]. In addition, other edible sugars (e.g., beverage powders or fructose) were produced to hydration packs for military or emergency purpose. However, the as-obtained product is the sweet water, which is unsuitable for long-term drinking [117]. Another example without DS regeneration is the use of the concentrated fertilizer solution [20,45,118,119]. Phuntsho et al. investigated the conventional inorganic fertilizers as DS. Results showed that 1 kg of fertilizer can take up 11~29 L of fresh water from seawater. In this fertilizer-driven FO process, inorganic fertilizers with high osmotic efficiency can draw water feasibly from wastewater resource or seawater, where the diluted DS can use for fertigation [120]. However, the concentration of the spent DS is commonly too concentrated for the direct fertigation of crops, requiring another dilution with fresh water before it is suitable [20,45]. Recently, Duan et al. [75] also reported an interesting case of using sodium lignin sulfonate (NaLS) as DS in FO, which does not require DS regeneration, the diluted NaLS can use for desert regeneration directly, preventing soil erosion as a soil stabilizer, and providing nutrients for plant growth. In addition, seawater itself can be used as a DS for the reuse of toxic wastewater; it would be safely discharged into the sea after FO, simultaneously retaining the toxic compounds by the membrane [121].

2.2. Chemical Energy

Traditional multivalent inorganic salts, such as $Al_2(SO_4)_3$, $CuSO_4$, $MgSO_4$, have been extensively used as DS for seawater desalination, because of its reasonably high water fluxes. The solubility of these salts may be adjusted with the change of temperature or pH. Once DS transforms into precipitate from the dilutive solution after FO, then the fresh water can be separated by filtration and the clean precipitate may be dissolved with acid into new DS and then return to the next cycle [6]. An early case of using $Al_2(SO_4)_3$ as DS in FO was reported by Frank in 1972 [47]. After FO, the diluted $Al_2(SO_4)_3$ would be reacted with calcium hydroxide ($Ca(OH)_2$) solution to produce aluminum hydroxide precipitation and the product water after filtration, the resultant precipitate was then dissolved with sulfuric acid to recover $Al_2(SO_4)_3$ solution and insoluble $Al(OH)_3$ for reuse. In 2011, a composite of $Al_2(SO_4)_3$ and MNPs as DS was proposed by Liu et al. [76], it can be separated by magnetic field to produce fresh water.

In 2013, Alnaizy et al. [49,77] used $CuSO_4$ and $MgSO_4$ as DS to desalinate brackish water, and the diluted DS may be recovered by adding barium hydroxide $Ba(OH)_2$ to generate precipitation and the product water [18,19]. (Figure 1) However, although this process doesn't consume additional energy except chemical energy itself, the trace of heavy metal salt retained in water will pose a threat to the environment and people's health. The product water requires further purification, which is adverse to reduce the energy cost; the storage of toxic precipitation may arouse the environmental concern. Therefore, the regeneration method by chemical precipitation may not be feasible for large-scale application, although it does not use external energy apart from chemical energy itself.

2.3. Thermal Energy

The use of geothermal energy and low-grade waste heat for DS recovery is deemed to be a more cost-effective solution. In some cases, DS regeneration does not require the heat source strictly; this general low-grade heat (e.g., waste heat or geothermal energy) can be directly used for heating separation process. At present, several kinds of DS regeneration by heating separation were reported. For instance, (1) gas and volatile compounds, (2) phase transition materials (such as lower critical solution temperature (LCST), and upper critical solution temperature (UCST) compounds, and thermo-sensitive gels), and (3) membrane distillation (MD).

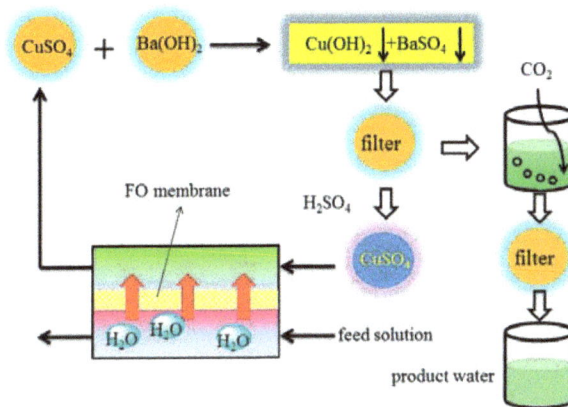

Figure 1. Schematic illustration of the use of precipitate method to regenerate $CuSO_4$ draw solution (DS) after forward osmosis (FO) process.

2.3.1. Gas and Volatile Compounds

Some gases, such as SO_2 and NH_3-CO_2, may easily dissolve in water and their aquo-complexes in water can generate high osmotic pressure. In 1965, Batchelder et al. [78] used volatile SO_2 water solution as DS for seawater desalination. However, the toxic SO_2 gas has strong corrosion to the equipment, which has not been further used. Up to 2005, Elimelech et al. proposed a new DS by the mixing of ammonia and carbon dioxide. McCutcheon et al. [79] found that the solution mixed with a certain proportion of NH_3-CO_2 can generate the extremely high osmotic pressure, mainly due to the generation of ammonium bicarbonate salt (NH_4HCO_3). A high water flux and water recovery can be achieved by using NH_3-CO_2 solution as DS in FO. Moreover, the diluted NH_4HCO_3 solution after FO could be easily decomposed into the gas of NH_3 and CO_2 again when heated to 60 °C with waste heat; meanwhile the product water was also obtained. The collected NH_3 and CO_2 will back for reuse again. Results showed that FO fluxes achieved by this system are about 3.6~36 L/(m^2·h). Since the low-grade waste heat is used to decompose NH_4HCO_3 at low temperature (60 °C), this FO desalination employing NH_4HCO_3 DS can save 85% of energy cost compared with other desalination technologies, showing a great application potential [13]. However, the drawback of NH_4HCO_3 DS is that the high reverse flux caused by its low MW may lead to the reduction water flux and the deterioration the quality of water [33]. Besides, the removal of the trace amounts of ammonia in product water is still a big challenge. Later, to overcome the mentioned defects above, a series of ammonium carbonate salts with similar function of NH_3-CO_2, such as *N,N*-dimethylcyclohexylamine ($N(Me)_2Cy$) [56], 1-Cyclohexylpiperidine (CHP) [57], trimethylamine ($N(CH_3)_2$) [61] and switching polarity solution (SPS) [50] have been reported.

2.3.2. Phase Transition Materials

Phase transition materials [88] may reversibly change the phase between solid and liquid with the various factors (e.g., temperature, pH, CO_2 et al.). According to the different critical dissolution temperatures, they are divided into UCST [81] and LCST materials [83–85,107]. This category of solutes mainly includes the ionic gels [34], the thermo-sensitive copolymer [54], the CO_2 response SPS [89,95], and the thermosensitive polyelectrolyte [87].

A recent study reported the use of a series of LCST materials as DS. Three organic ionic liquids [85] (e.g., tetrabutylphosphonium 2,4-dimethylbenzenesulfonate (P_{4444}-DMBS), tetrabutylphosphonium 2,4,6-trimethylbenzenesulfonate (P_{4444}-TMBS), and tetrabutylphosphonium bromide (P_{4448}-Br)) with different hydrophilicity were investigated. Results showed that the osmolality and molality have a non-linear correlation relationship, due to the occurring of molecules hydrophobic association at high concentration. These solutes whose osmotic pressure is approximately three times higher than that of seawater were applied to dewatering from seawater. DS regeneration after FO was carried out when heated to above LCST (30~50 °C), resulting in the phase separation. Recently, Zhong et al. [81] proposed to the use of a UCST ionic liquid [protonated betaine bis(trifluoromethylsulfonyl)imide (donated as [Hbet] [Tf2N]))] as DS to dewater from high salt water. Results showed that the 3.2 M [Hbet] [Tf2N] solution over the UCST of 56 °C was able to draw water from 3.0 M saline water. Consequently, the diluted solution was cooled to below the UCST to form two phases of ionic liquid and water. The phase of rich ionic liquid could be used directly without additional re-concentration.

The CO_2 response SPS may change their miscibility by injecting of CO_2. Stone et al. [50] first proposed the use of SPS as DS. The phase transformation was easily achieved through the injection of CO_2, because the cloudy solution of amine reacts with CO_2 to form a protonation product, which is readily dissolved in water. The diluted SPS after FO can be recovered to the immiscible state through the removal of CO_2 by heat with a low-grade source. The regeneration process was allowed to mechanically recover the majority of amine solution. However, the residual amine in water still required to be removed using RO. Moreover, the later study found that the cellulose triacetate membrane could be damaged by the DS of $N(Me)_2Cy$, leading to the solute leakage [50]. An improved solution that the polymer with dual temperature and CO_2 response was synthesized as DS for desalination was proposed by Cai et al. [90]. After the injection of CO_2, the ionized polymers can fully dissolve in water and generate high osmotic pressure; the diluted solution can be recovered through the phase separation when heated to 60 °C. The use of dual-response polymer as DS could effectively reduce reverse solute flux. Similarly, 1-cyclohexylpiperidine as a SPS DS showed a comparable performance of that the $N(Me)_2Cy$, but a good compatibility with polyamide membranes [57].

2.3.3. MD

MD is a thermo-driving membrane separation technology, it being not affected by the concentration of feed solution, becomes possible to desalinate high salt wastewater [122]. In MD process, the water vapor on the side of hot feed solution (heated by low-value heat source) diffuses across a hydrophobic porous membrane into the cold liquid water [87,123–127]. In general, a suitable temperature gradient of 10~20 °C on both sides of the hot and cold solutions should be stably maintained to produce distilled water [124]. Since MD has the high selectivity and the 100% theoretical rejection rate for nonvolatile solutes [124,127], it is probably a good candidate for the recovery of diluted DS and can replace traditional separation technologies [55,60,92,93]. In addition, MD can use the low-value waste heat to minimize the capital cost [52,127]. However, there are several limitations about the current MD, including (1) low and unstable permeate flux, (2) membrane fouling, and (3) high heat lost [128–130].

The use of MD to recover 2-methylimidazole compounds was reported by Yen et al. [20]. A water flux of 8 LMH could be achieved. Later, numerous researchers, such as Wang et al. [131], Xie et al. [132], and Zhang et al. [133] investigated MD to recover the diluted solution of NaCl after FO for sustainable water recovery of protein concentration and oil-water separation. Guo et al. [92] used MD to recover Na-CQDs DS for seawater desalination, and a stable water flux of 3.5 L/(m² h) could be maintained after several cycles at 45 °C. Another hybrid FO/MD system for the recovery of polyelectrolytes (PAA-Na) after FO was investigated by Ge et al. [134]. Results showed that the high rejection of PAA-Na was carried out in MD at the temperatures of 50–70 °C. The most efficient performance of this hybrid FO-MD system can be achieved when it is conducted under the condition of 0.48 g/mL solution at 66 °C. Afterwards, the hybrid FO-MD process was also used to treat dye wastewater and toxic wastewater [52,135]. (Figure 2) To minimize the leakage of DS in MD process, the large size of thermoresponsive polymer (PSSS-PNIPAM) is used as DS for seawater desalination [54]. The osmotic pressure of PSSS-PNIPAM decreases with the temperature increases due to the agglomeration of the polymer chains, leading to a higher water vapor pressure. After three recycles, a stable water flux of 2.5 L/(m² h) and product of water could be achieved. In fact, however, MD still needs to overcome lots of defects, including low water flux, membrane scaling and high hot loss, which is adverse to reduce the energy cost.

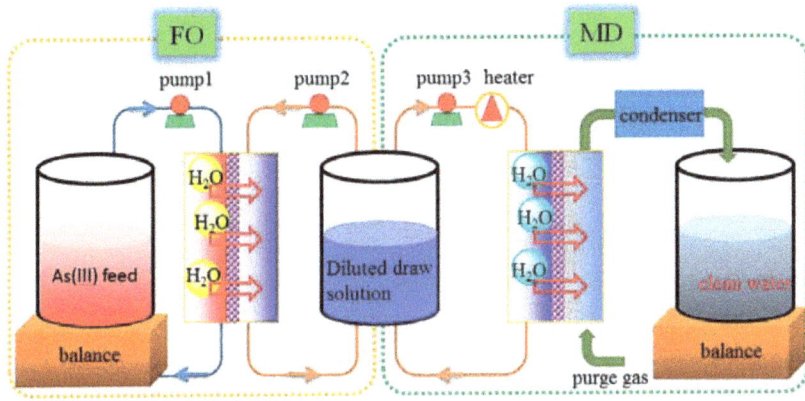

Figure 2. Schematic illustration of using a hybrid FO-MD process for wastewater treatment.

2.4. Electric Energy

At present, membrane separation method is frequently used for DS regeneration, because of its high solute rejection and simple operation. It is well-known that all membrane separation processes consume electricity except MD. Different membrane separation technologies have different energy consumption depending on the osmotic pressure of feed solution and membrane types. Currently, the membrane separation technologies used for DS recovery mainly include RO, NF, and UF.

2.4.1. RO

Considering the low molecular weight cut-offs and high solute rejection, RO is a good option for DS recovery when using monovalent salts (e.g., NaCl, seawater) as DS [136]. In the past few years, most of inorganic salts, including NaCl (seawater), $MgCl_2$, $MgSO_4$ and Na_2SO_4, were frequently used as DS, due to their high osmotic pressure. However, these diluted monovalent salts solution after FO often require extremely high energy costs to be regenerated. It seems impossible to use RO for DS recovery. In spite of this, the hybrid FO-RO for seawater desalination was also proposed (Figure 3) [33], FO and RO can be used as the front end pretreat and post-treatment desalination technologies, respectively [136]. The advantages of this system are greatly reducing membrane fouling after FO treatment [137] and the energy cost in RO units, because of seawater diluted after FO [137].

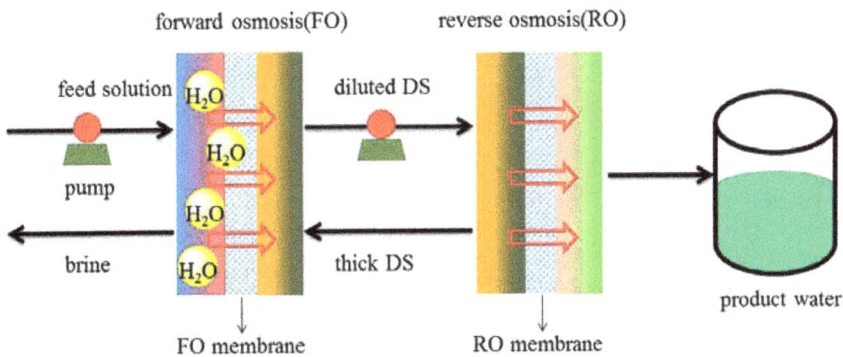

Figure 3. Schematic illustration of using a hybrid FO-RO process for desalination seawater.

Yangali et al. [138] used low-pressure RO for DS recovery employing FO to desalinate seawater. Results showed that this hybrid FO-RO system can save 50% of energy consumption (~1.5 kWh/m^3) compared with independent RO desalination (2.5~4 kWh/m^3). Its cost efficiency would surpass RO only if the water flux above 5.5 L/(m^2 h) can be achieved. In addition, Bowden et al. [97] used RO to recover the organic acid salt. DS regeneration with the average solute rejection rate of near 99% was conducted under the pressure of 28 bar, indicating the feasibility of DS recovery by RO. Another hybrid FO-RO system combination of wastewater as feed using seawater as DS was proposed by Cath et al. [15]. The diluted seawater after FO providing a low saline solution, subsequently, flow into RO units to desalinate. This hybrid system with impaired sources as feed can produce high quality of drinking water and achieve a favorable economic return with the water recovery rate up to 63% [15]. It can be seen that the hybrid FO-RO process may be a competitive choice for desalination of high saline water compared with RO alone. However, considering its energy cost and efficiency, RO recovery for DS regeneration may be discouraged under high operating pressure.

2.4.2. NF

NF membrane with the 1~2 nm of pore size can effectively remove organic compounds with MWs between 200~1000 Da, but its rejection rate of multivalent salts (99%) is higher than that of monovalent salts. A sort of looser membrane, with relative high molecular weight cut-offs, it can produce higher water flux than RO under a given pressure. In recent years, it is intensively used to remove the organic matters and dissolve salts in water treatment, food and medicine industry [21,27,59,98,99]. The use of NF to recover DS has received great attention [71]. Tan and Ng [98] investigated the NF recovery rate of a series of different DSs (e.g., MgCl$_2$, MgSO$_4$, Na$_2$SO$_4$ and ethanol) after FO. The stable water fluxes as high as 10 L/(m^2 h) can be achieved in both FO and NF process, the reachable NF rejection rates of 97.4% can be observed. They found that the total dissolved solid (TDS) is as low as 113.6 mg/L of the product water after two-pass NF treatment process, which meets the drinkable standard (TDS < 500 mg/L) required by world health organization [137]. After that, Zhao et al. [122] used NF to regenerate the diluted bivalent salt solution (MgSO$_4$ and Na$_2$SO$_4$) after FO. According to their findings, the performance of hybrid FO–NF process for desalination of brackish water is superior to that standalone RO process, because of its lower hydraulic pressure, lower membrane fouling, and higher flux recovery after cleaning. Besides, Su et al. [73] proposed the use of NF to regenerate the dilute sucrose after FO, the rejection rates of as high as 99.6% can be achieved, due to the sucrose with large molecule size. Consequently, the concentrated sucrose DS return to FO units may continually draw water from wastewater while producing clean water in NF units. The NF recovery of EDTA sodium salts under external pressure of 5.5 bar was reported by Hau et al. [25]. They found that a rejection rate of 93% for high charged salts was achieved well. Afterward, the regeneration of hydroacid complex solutions using NF with a low-pressure of 10 bar were conducted by Ge et al. [27,59,99]. These compounds had expanded configurations and charged groups, leading to an achievable high rejection rate of 97%. Later, the diluted DS of carboxyethyl amine sodium salts [100], and the organic phosphonate salts [101] were also used NF regeneration after FO. It can be seen that NF is a good choice for DS recovery, due to its low extremal pressure and high rejection rate of valent salts.

2.4.3. UF

UF membrane with a bigger pore size than RO and NF can bring about a higher water flux and lower energy cost [139,140]. UF is suitable for the regeneration of these DSs with big size (such as macromolecules, polyelectrolytes, or particles). Recently, the regeneration of MNPs through magnetic separation was feasible to produce clean water. Ling et al. [38,43] proposed the use of UF to regenerate hydrophilic nanoparticle DS. They found that, unlike magnetic separation, the collected MNPs after UF in concentrated solution didn't agglomerate again, which make it retain the initial osmotic pressure well even after five regeneration cycles. After FO desalination, Ge et al. [52] adopted UF to recover diluted solution of sodium polyacrylate (PAA-Na). For the PAA-Na with the MW of 1800 Da, a rejection rate of 99% can be achieved under the external pressure of 10 bar, because of the MWs and expanded structure of polymer. The water flux in this FO-UF slightly decreased after nine regeneration cycles, which was explained by the loss of PAA-Na during UF [76]. Other DSs, including MNPs modified with PAA (~5 nm) [40], polystyrene sulfonic acid sodium (PAS-Na) [53], phosphate surfactants [103], micellar solution [58,104], thermo-sensitive polyelectrolyte [102], and phase change materials [88], were all used UF to regenerate. As a pressure-dependent membrane process, however, the regeneration performance of UF still relies on the osmotic pressure of DS. In term of the analysis of energy balance, Shaffer et al. [15] reviewed that the object of DS regeneration is to concentrate diluted solution and make it return to the initial osmotic pressure, which the same external pressure will be provided by UF even RO, Therefore, the minimum energy consumption for both processes are the same. The energy cost being equal, RO would be given the priority for its superior rejection rates.

2.5. Magnetic Energy

Magnetic nanoparticles (MNPs) have been intensively investigated as a very promising DS, because it can be recover using magnetic field (Figure 4) [37,38,40,43]. The advantages of magnetic separation are high efficiency and energy conservation compared with other DSs regeneration methods. Recently, to improve the water solubility and surface hydrophilicity, MNPs functionalized with strong hydrophilic groups are considered as one of the feasible solutions to enhance their osmotic pressure, as well as easy recycle utilization [141]. Ling et al. conducted an in-depth study on the first use of hydrophilic functionalized MNPs as DS [37]. The target MNPs coated with 2-pyrrolidine, triethylene glycol, and PAA were synthesized based on the precursor of ferric triacetyl acetonate (Fe(acac)$_3$) through a thermal decomposition method. The sizes of functionalized MNPs were all less than 20 nm and were successfully used in the FO process. At an extreme low concentration of 0.05 mol/L, all functional MNPs showed the improved osmotic pressure and water flux. Water fluxes of three functional MNPs followed the order of MNPs-PPA > MNPs-pyrrolidine > MNPs-triethylene glycol, mainly because of the higher potential energy on the surface of MNPs-PPA. The MNPs separation was conducted with a commercially available magnetic separator. However, these DSs showed a decreased (about 21%) water flux after nine cycles because of MNPs agglomeration.

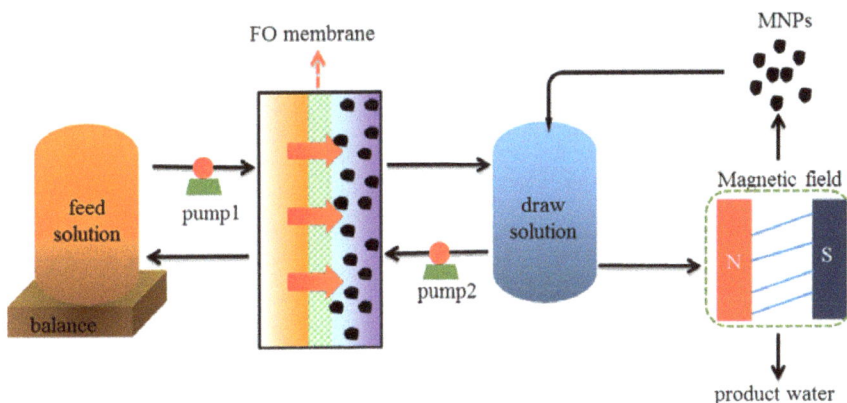

Figure 4. Schematic illustration of using functionalized magnetic nanoparticles as DSs in FO process. (MNPs, magnetic nanoparticles).

Later, Ge et al. [39,68] explored a new MNPs coated with poly (ethylene glycol) diacid (PEG-(COOH)$_2$) with different MWs through a one-pot reaction. Results of transmission electron microscopy showed that these particles had the spherical sizes ranging from 4.2 to 11.5 nm. In addition, higher water flux was found with the use of PEG with lower MWs. The water fluxes of 13.5, 13, and 11.5 L/(m^2·h) belong to those MNPs coated with PEG MWs of 250, 600 and 4000, respectively [43].

The above results indicated that the factors like particle size, hydrophilic group distribution, and the number of hydrophilic groups were greatly related to FO performance. Besides, one type of MNPs coated with dextran [36] and a thermo-responsive PNIPAM-modified MNPs [38] as DS were proposed. Since the reverse solute flux is almost negligible using MNPs as DS, it could be applied to concentrating the proteins and drugs. At present, however, the product water separated by magnetic field is still hard to meet the drink water standards. An additional purification step is needed, which would sacrifice the separation efficiency. Therefore, more studies are dedicated to producing the high quality of water and improve the recovery efficiency.

2.6. Solar Energy

Smart materials may undergo a reversibly change between water absorption and dehydration with the change of surrounding stimuli (e.g., sunlight, pressure, heat). Recently, a series of stimuli-response hydrogels were proposed as DS in FO. This novel type of hydrogels-driven FO process (Figure 5) employing the solar energy for the DS regeneration have been attracting more and more attentions. Especially, the light or thermoresponse hydrogels can release fresh water when it is exposed to sunlight, because of the structure's own shrinkage. For instance, the light or thermoresponse hydrogels can take up enough water under the temperature of maximum volume phase transition and dewater at temperatures above the volume phase transition temperature [34,49,142–145]. The graphene-doped hydrogels as DS were proposed by Li et al. [109]. The efficiency of water release was significantly improved; because of the heat absorption was enhanced by adding an appropriate amount of graphene. In addition, the hydrogels doped with light-absorbing carbon particles (10~25 μm) was also reported [112]. Under the irradiation intensity of 1.0 kW/m^2 for 20 min, a water recovery rate of 50% was achieved, which is superior to than that of p(NIPAM-co-SA) hydrogels (around 17%). However, the doping of the carbon materials, if beyond a certain scope, may scarify water fluxes, for the direct contact area reduced between hydrogels particles and FO membrane. Similarly, other hydrogels, such as dual-functional polymer gels and photosensitive particles [108] and photosensitive particles [37] all show the same effects of taking up and releasing. The eco-friendly dewatering process by using solar energy may obviously reduce the cost. However, the collection of clean water released by the exposure of thermoresponse hydrogels under sunlight remains a big challenge now [146].

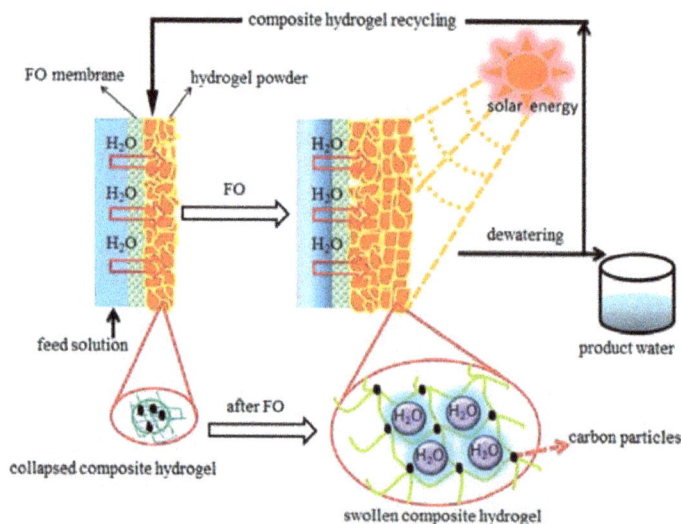

Figure 5. Schematic illustration of using solar energy to recover the thermoresponsive hydrogels DSs in the hydrogels-driven FO process for desalination.

3. Future Challenges or Perspectives

The use of FO for the production of clean water has been proposed since 1960's, it really booms in the improvement of water recovery only within the last decade. FO process with high water flux would be to compare with other existing membrane-based process in seawater desalination and wastewater reuse. Numerous efforts have been dedicated to developing and improving the high performance of FO membranes, but there is also a need for the parallel develop of superior DS with high osmotic pressures to produce high water flux. The lack of powerful DS bearing high water flux, low reverse flux, and easy recovery has become the obstacles to the progress of FO.

FO employed the direct use of DS may be economical and technical sound when using two solutions with a high osmotic pressure difference, due to the lack of DS recovery. However, only few water sources are available to dewater using these specific DSs (e.g., saccharides [72,73], fertilizer [20,46], liquid fertilizer [74], sodium lignin sulfonate [75]. (Table 2) For instance, the drinkable diluted glucose DS after FO can be used for emergency rescue. Besides, the diluted sodium lignin sulfonate solution is utilized in the land as nutrient; whose energy cost in this process only includes the power input of pumping, storage and others. In this case, the total energy cost of FO alone for seawater desalination is approximately 0.59 kWh·m^{-3} [13], which is about one third of that of RO water desalination (1.5 kWh·m^{-3}) [1] and one-seventh of that of other technologies (4 kWh·m^{-3}) [13]. However, most of the expensive DS still requires recovery.

DS regeneration method based on the chemical energy is unlikely to be used on large-scale, due to the environment concern caused by the discharge of by-products. Especially, since the trace amount of heavy-metal ions after precipitation can't be removed completely from the diluted DS, it is often required a secondary treatment using RO to obtain clean water, which leads to the massive energy consumption.

The most commonly used method according to the reported literature is the membrane-based separations (e.g., RO, NF), which uses electricity to recover those small molecules, including inorganic salts, organic salts, and electrolytes. However, it is widely accepted in both academia and industry that a hybrid FO-RO process does not minimize the lower energy consumption for desalination than RO alone. Except the case of using hybrid FO-RO process for the desalination seawater, FO as the front-end treatment and RO as the post-treatment step still attract attentions. Another case is that the use of UF has the potential of higher water fluxes and lower energy costs for regenerating these sufficiently large solute molecules or particles [139,140]. Among all the membrane-based recovery methods, a number of studies are focused on investigating MD regeneration of the diluted DS. In the case of regeneration PSSS-PNIPAM using a hybrid FO-MD system, the calculated energy consumption of 29 kWh m^{-3} is required at 50 °C [54], which is much higher than the use of FO-RO or RO alone. However, it still has the considerable potential for energy cost when the use of low-grade waste heat replaces the electricity. The use of new energy for DS recovery was intensively investigated recently.

Take the magnetic field energy as example, the use of functionalized MNPs as DS after FO may be quickly and cheaply concentrated by a magnetic separator [36,37,39,40,147]. However, the MNPs aggregation during separation process has severely impacted the reused performance, and in some cases, the water quality produced by magnetic separator is not always as good as desired. Moreover, the synthesis of large amount of functionalized nanoparticles are extremely expensive [33,148].

The light or thermo-responsive hydrogels as DS has been demonstrated in the FO process. These smart DSs are of the negligible reverse solute flux and can be recovered with solar energy. However, the water flux in FO is extremely low (most of ranging from 0.5–3 L/(m^2·h)) [113]. The swelling pressure and light or thermo-responsive property of hydrogels need to be further optimized on their physical-chemical properties. More studies need to design continuous smart materials-driven FO process for practical applications. There is no doubt that smart DSs driven FO processes using renewable energy for DS recover may be the first choice in the future.

4. Conclusions

This review has been tried to categorize the current DSs in terms of the types of energy used for DS regeneration process, and also analyzes the potential of DS in the case of seawater desalination and wastewater reuse that it can be capable of reducing energy cost. Publications reported in the past few years are examined apart from some older cases to know how current DS developments have been impacted by past efforts and challenges. Selection of a suitable DS is very important to the energy efficient and cost-effective running in FO. An ideal DS must be cheap and abundant, be high water flux and low reverse solute flux, be nontoxic, and be easy to regenerate. Considering the non-negligible problem with FO is that a second step is needed to recover fresh water from the diluted DS and to simultaneously regenerate the DS. This second step requires having low energy consumption and high efficient output.

All of DS recovery methods have their potentials and limitations in the process of FO application, and that each regeneration method involves at least one or two different types of energy consumption depends on different DS. However, not all DSs are always available for complex water source, which often requires a hybrid technology solution to producing clean water. Considering the selection of DS at a certain application, the main problems of how much energy is consumed and how effective the recovery process (the cost of the DS themselves and the energy cost payed out for solute regeneration) is should be taken into consideration. This review (Table 2) covers most of example matters, which have been employed as potential DS. Several classifications of DS in terms of the different types of energy employed for DS regeneration have been summarized and highlighted, including direct use without recovery, chemical energy, low grade waste heat, electric energy, magnetic field energy, and solar energy. Among them, the "direct use"-based DS is the most save-energy solution in FO, but its application is restricted by the few of available DS. However, DS recovery using chemical energy is unpromising, because it not only requires additional chemicals, but also generates more by-products. Besides, the DS recovered by electric energy (i.e., RO, NF) is still the most popular one in recent years, due to the high salt rejection, but it is incapable of treating that solution of high osmotic pressure. The innovative hydrogels, stimuli responsive polymers, and MNPs coated with hydrophilic groups can be used as DSs, showing the low reverse salt flux and innovative recovery pathways. Those DSs recovered after FO using solar energy, waste heat or magnetic field energy makes DS regeneration process become greener and energy-saving. It is likely that this type of DSs may be the most development promising one in the future, even if there is no any single perfect DS to be discovered for the needs of specific situations. We would like to hope that the main categories of DSs discussed in this review may help the development of further work of DSs innovation.

Author Contributions: All of the authors contributed to publishing this paper. J.L., J.Y., and F.L. collected and summarized the materials; Q.L. analyzed the materials and wrote the manuscript; Y.J., J.Z. and B.Y. revised and polish the manuscript.

Funding: We would like to thank the financial support from the National Natural Science Foundation of China (No. 201705071), the Natural Science Foundation of Guangdong province of China (2016A030310362), Lingnan Normal University (no. 2L1613, 2L1614), Zhangjiang City Science and Technology Program (NO. 2015A02023), and Guandong Province Office of Education Youth Innovation Talent Program (NO. 2016KQNCX096).

Conflicts of Interest: The authors declare no conflict of interest.

References

1. Elimelech, M.; Phillip, W.A. The future of seawater desalination: Energy, technology, and the environment. *Science* **2011**, *333*, 712–717. [CrossRef] [PubMed]
2. Zhao, S.; Zou, L.; Tang, C.Y.; Mulcahy, D. Recent developments in forward osmosis: Opportunities and challenges. *J. Membr. Sci.* **2012**, *396*, 1–21. [CrossRef]
3. Lee, K.P.; Arnot, T.C.; Mattia, D. A review of reverse osmosis membrane materials for desalination—Development to date and future potential. *J. Membr. Sci.* **2011**, *370*, 1–22. [CrossRef]

4.   Montgomery, M.A.; Elimelech, M. Water and sanitation in developing countries: Including health in the equation. *Environ. Sci. Technol.* **2007**, *41*, 17–24. [CrossRef] [PubMed]

5.   Shannon, M.A.; Bohn, P.W.; Elimelech, M.; Georgiadis, J.G.; Mariñas, B.J.; Mayes, A.M. Science and technology for water purification in the coming decades. *Nanosci. Technol.* **2009**, 337–346. [CrossRef]

6.   Ge, Q.; Ling, M.; Chung, T.-S. Draw solutions for forward osmosis processes: Developments, challenges, and prospects for the future. *J. Membr. Sci.* **2013**, *442*, 225–237. [CrossRef]

7.   King, C.W.; Webber, M.E. Water intensity of transportation. *Environ. Sci. Technol.* **2008**, *42*, 7866–7872. [CrossRef] [PubMed]

8.   McGinnis, R.L.; Elimelech, M. Global challenges in energy and water supply: The promise of engineered osmosis. *Environ. Sci. Technol.* **2008**, *42*, 8625–8629. [CrossRef] [PubMed]

9.   Chon, K.; KyongShon, H.; Cho, J. Membrane bioreactor and nanofiltration hybrid system for reclamation of municipal wastewater: Removal of nutrients, organic matter and micropollutants. *Bioresour. Technol.* **2012**, *122*, 181–188. [CrossRef] [PubMed]

10.   Subramani, A.; Badruzzaman, M.; Oppenheimer, J.; Jacangelo, J.G. Energy minimization strategies and renewable energy utilization for desalination: A review. *Water Res.* **2011**, *45*, 1907–1920. [CrossRef] [PubMed]

11.   Greenlee, L.F.; Lawler, D.F.; Freeman, B.D.; Marrot, B.; Moulin, P. Reverse osmosis desalination: Water sources, technology, and today's challenges. *Water Res.* **2009**, *43*, 2317–2348. [CrossRef] [PubMed]

12.   Service, R.F. Desalination freshens up. *Science* **2006**, *313*, 1088–1090. [CrossRef] [PubMed]

13.   McGinnis, R.L.; Elimelech, M. Energy requirements of ammonia–carbon dioxide forward osmosis desalination. *Desalination* **2007**, *207*, 370–382. [CrossRef]

14.   Chung, T.-S.; Li, X.; Ong, R.C.; Ge, Q.; Wang, H.; Han, G. Emerging forward osmosis (fo) technologies and challenges ahead for clean water and clean energy applications. *Curr. Opin. Chem. Eng.* **2012**, *1*, 246–257. [CrossRef]

15.   Shaffer, D.L.; Werber, J.R.; Jaramillo, H.; Lin, S.; Elimelech, M. Forward osmosis: Where are we now? *Desalination* **2015**, *356*, 271–284. [CrossRef]

16.   Mi, B.; Elimelech, M. Organic fouling of forward osmosis membranes: Fouling reversibility and cleaning without chemical reagents. *J. Membr. Sci.* **2010**, *348*, 337–345. [CrossRef]

17.   Zhang, S.; Wang, K.Y.; Chung, T.-S.; Jean, Y.C.; Chen, H. Molecular design of the cellulose ester-based forward osmosis membranes for desalination. *Chem. Eng. Sci.* **2011**, *66*, 2008–2018. [CrossRef]

18.   Qasim, M.; Darwish, N.A.; Sarp, S.; Hilal, N. Water desalination by forward (direct) osmosis phenomenon: A comprehensive review. *Desalination* **2015**, *374*, 47–69. [CrossRef]

19.   Valladares Linares, R.; Li, Z.; Sarp, S.; Bucs, S.S.; Amy, G.; Vrouwenvelder, J.S. Forward osmosis niches in seawater desalination and wastewater reuse. *Water Res.* **2014**, *66*, 122–139. [CrossRef] [PubMed]

20.   Phuntsho, S.; Shon, H.K.; Majeed, T.; El Saliby, I.; Vigneswaran, S.; Kandasamy, J.; Hong, S.; Lee, S. Blended fertilizers as draw solutions for fertilizer-drawn forward osmosis desalination. *Environ. Sci. Technol.* **2012**, *46*, 4567–4575. [CrossRef] [PubMed]

21.   Hau, N.T.; Chen, S.-S.; Nguyen, N.C.; Huang, K.Z.; Ngo, H.H.; Guo, W. Exploration of edta sodium salt as novel draw solution in forward osmosis process for dewatering of high nutrient sludge. *J. Membr. Sci.* **2014**, *455*, 305–311. [CrossRef]

22.   Cornelissen, E.R.; Harmsen, D.; de Korte, K.F.; Ruiken, C.J.; Qin, J.-J.; Oo, H.; Wessels, L.P. Membrane fouling and process performance of forward osmosis membranes on activated sludge. *J. Membr. Sci.* **2008**, *319*, 158–168. [CrossRef]

23.   Klaysom, C.; Cath, T.Y.; Depuydt, T.; Vankelecom, I.F.J. Forward and pressure retarded osmosis: Potential solutions for global challenges in energy and water supply. *Chem. Soc. Rev.* **2013**, *42*, 6959–6989. [CrossRef] [PubMed]

24.   Straub, A.P.; Deshmukh, A.; Elimelech, M. Pressure-retarded osmosis for power generation from salinity gradients: Is it viable? *Energy Environ. Sci.* **2016**, *9*, 31–48. [CrossRef]

25.   Wan, C.F.; Chung, T.-S. Energy recovery by pressure retarded osmosis (pro) in swro–pro integrated processes. *Appl. Energy* **2016**, *162*, 687–698. [CrossRef]

26.   Lutchmiah, K.; Verliefde, A.R.D.; Roest, K.; Rietveld, L.C.; Cornelissen, E.R. Forward osmosis for application in wastewater treatment: A review. *Water Res.* **2014**, *58*, 179–197. [CrossRef] [PubMed]

27.   Ge, Q.; Chung, T.-S. Oxalic acid complexes: Promising draw solutes for forward osmosis (fo) in protein enrichment. *Chem. Commun.* **2015**, *51*, 4854–4857. [CrossRef] [PubMed]

28. Rastogi, N.K. Opportunities and challenges in application of forward osmosis in food processing. *Crit. Rev. Food Sci. Nutr.* **2016**, *56*, 266–291. [CrossRef] [PubMed]

29. LaVan, D.A.; McGuire, T.; Langer, R. Small-scale systems for in vivo drug delivery. *Nat. Biotechnol.* **2003**, *21*, 1184. [CrossRef] [PubMed]

30. Cath, T.Y.; Adams, D.; Childress, A.E. Membrane contactor processes for wastewater reclamation in space: Ii. Combined direct osmosis, osmotic distillation, and membrane distillation for treatment of metabolic wastewater. *J. Membr. Sci.* **2005**, *257*, 111–119. [CrossRef]

31. Sant'Anna, V.; Marczak, L.D.F.; Tessaro, I.C. Membrane concentration of liquid foods by forward osmosis: Process and quality view. *J. Food Eng.* **2012**, *111*, 483–489. [CrossRef]

32. Qin, J.-J.; Lay, W.C.L.; Kekre, K.A. Recent developments and future challenges of forward osmosis for desalination: A review. *Desalin. Water Treat.* **2012**, *39*, 123–136. [CrossRef]

33. Achilli, A.; Cath, T.Y.; Childress, A.E. Selection of inorganic-based draw solutions for forward osmosis applications. *J. Membr. Sci.* **2010**, *364*, 233–241. [CrossRef]

34. Luo, H.; Wang, Q.; Tao, T.; Zhang, T.C.; Zhou, A. Performance of strong ionic hydrogels based on 2-acrylamido-2-methylpropane sulfonate as draw agents for forward osmosis. *J. Environ. Eng.* **2014**, *140*, 04014044. [CrossRef]

35. Cai, Y.; Shen, W.; Loo, S.L.; Krantz, W.B.; Wang, R.; Fane, A.G.; Hu, X. Towards temperature driven forward osmosis desalination using semi-ipn hydrogels as reversible draw agents. *Water Res.* **2013**, *47*, 3773–3781. [CrossRef] [PubMed]

36. Bai, H.; Liu, Z.; Sun, D.D. Highly water soluble and recovered dextran coated fe3o4 magnetic nanoparticles for brackish water desalination. *Sep. Purif. Technol.* **2011**, *81*, 392–399. [CrossRef]

37. Ling, M.M.; Wang, K.Y.; Chung, T.-S. Highly water-soluble magnetic nanoparticles as novel draw solutes in forward osmosis for water reuse. *Ind. Eng. Chem. Res.* **2010**, *49*, 5869–5876. [CrossRef]

38. Ling, M.M.; Chung, T.-S.; Lu, X. Facile synthesis of thermosensitive magnetic nanoparticles as "smart" draw solutes in forward osmosis. *Chem. Commun.* **2011**, *47*, 10788–10790. [CrossRef] [PubMed]

39. Ge, Q.; Su, J.; Chung, T.-S.; Amy, G. Hydrophilic superparamagnetic nanoparticles: Synthesis, characterization, and performance in forward osmosis processes. *Ind. Eng. Chem. Res.* **2011**, *50*, 382–388. [CrossRef]

40. Ling, M.M.; Chung, T.-S. Desalination process using super hydrophilic nanoparticles via forward osmosis integrated with ultrafiltration regeneration. *Desalination* **2011**, *278*, 194–202. [CrossRef]

41. Ling, M.M.; Chung, T.-S. Surface-dissociated nanoparticle draw solutions in forward osmosis and the regeneration in an integrated electric field and nanofiltration system. *Ind. Eng. Chem. Res.* **2012**, *51*, 15463–15471. [CrossRef]

42. Zhao, Q.; Chen, N.; Zhao, D.; Lu, X. Thermoresponsive magnetic nanoparticles for seawater desalination. *ACS Appl. Mater. Interfaces* **2013**, *5*, 11453–11461. [CrossRef] [PubMed]

43. Ling, M.M.; Chung, T.-S. Novel dual-stage fo system for sustainable protein enrichment using nanoparticles as intermediate draw solutes. *J. Membr. Sci.* **2011**, *372*, 201–209. [CrossRef]

44. Phuntsho, S.; Shon, H.K.; Hong, S.; Lee, S.; Vigneswaran, S.; Kandasamy, J. Fertiliser drawn forward osmosis desalination: The concept, performance and limitations for fertigation. *Rev. Environ. Sci. Biol.* **2012**, *11*, 147–168. [CrossRef]

45. Chekli, L.; Kim, Y.; Phuntsho, S.; Li, S.; Ghaffour, N.; Leiknes, T.; Shon, H.K. Evaluation of fertilizer-drawn forward osmosis for sustainable agriculture and water reuse in arid regions. *J. Environ. Manag.* **2017**, *187*, 137–145. [CrossRef] [PubMed]

46. Phuntsho, S.; Shon, H.K.; Hong, S.; Lee, S.; Vigneswaran, S. A novel low energy fertilizer driven forward osmosis desalination for direct fertigation: Evaluating the performance of fertilizer draw solutions. *J. Membr. Sci.* **2011**, *375*, 172–181. [CrossRef]

47. Frank, B.S. Desalination of Sea Water. U.S. Patent US 3670897A, 20 June 1972.

48. Garcia-Castello, E.M.; McCutcheon, J.R.; Elimelech, M. Performance evaluation of sucrose concentration using forward osmosis. *J. Membr. Sci.* **2009**, *338*, 61–66. [CrossRef]

49. Alnaizy, R.; Aidan, A.; Qasim, M. Draw solute recovery by metathesis precipitation in forward osmosis desalination. *Desalin. Water Treat.* **2013**, *51*, 5516–5525. [CrossRef]

50. Stone, M.L.; Rae, C.; Stewart, F.F.; Wilson, A.D. Switchable polarity solvents as draw solutes for forward osmosis. *Desalination* **2013**, *312*, 124–129. [CrossRef]

51. Ge, Q.; Su, J.; Amy, G.L.; Chung, T.-S. Exploration of polyelectrolytes as draw solutes in forward osmosis processes. *Water Res.* **2012**, *46*, 1318–1326. [CrossRef] [PubMed]

52. Ge, Q.; Wang, P.; Wan, C.; Chung, T.-S. Polyelectrolyte-promoted forward osmosis-membrane distillation (fo-md) hybrid process for dye wastewater treatment. *Environ. Sci. Technol.* **2012**, *46*, 6236–6243. [CrossRef] [PubMed]

53. Tian, E.; Hu, C.; Qin, Y.; Ren, Y.; Wang, X.; Wang, X.; Xiao, P.; Yang, X. A study of poly (sodium 4-styrenesulfonate) as draw solute in forward osmosis. *Desalination* **2015**, *360*, 130–137. [CrossRef]

54. Zhao, D.; Wang, P.; Zhao, Q.; Chen, N.; Lu, X. Thermoresponsive copolymer-based draw solution for seawater desalination in a combined process of forward osmosis and membrane distillation. *Desalination* **2014**, *348*, 26–32. [CrossRef]

55. Gwak, G.; Jung, B.; Han, S.; Hong, S. Evaluation of poly (aspartic acid sodium salt) as a draw solute for forward osmosis. *Water Res.* **2015**, *80*, 294–305. [CrossRef] [PubMed]

56. Reimund, K.K.; Coscia, B.J.; Arena, J.T.; Wilson, A.D.; McCutcheon, J.R. Characterization and membrane stability study for the switchable polarity solvent N,N-dimethylcyclohexylamine as a draw solute in forward osmosis. *J. Membr. Sci.* **2016**, *501*, 93–99. [CrossRef]

57. Orme, C.J.; Wilson, A.D. 1-cyclohexylpiperidine as a thermolytic draw solute for osmotically driven membrane processes. *Desalination* **2015**, *371*, 126–133. [CrossRef]

58. Roach, J.D.; Abdulrahman, A.-A.; Alaa, A.-N.; Mohammed, H. Use of micellar solutions as draw agents in forward osmosis. *J. Surfactants Deterg.* **2014**, *17*, 1241–1248. [CrossRef]

59. Ge, Q.; Fu, F.; Chung, T.-S. Ferric and cobaltous hydroacid complexes for forward osmosis (fo) processes. *Water Res.* **2014**, *58*, 230–238. [CrossRef] [PubMed]

60. Yen, S.K.; Mehnas Haja, N.F.; Su, M.; Wang, K.Y.; Chung, T.-S. Study of draw solutes using 2-methylimidazole-based compounds in forward osmosis. *J. Membr. Sci.* **2010**, *364*, 242–252. [CrossRef]

61. Boo, C.; Khalil, Y.F.; Elimelech, M. Performance evaluation of trimethylamine–carbon dioxide thermolytic draw solution for engineered osmosis. *J. Membr. Sci.* **2015**, *473*, 302–309. [CrossRef]

62. Kravath, R.E.; Davis, J.A. Desalination of sea water by direct osmosis. *Desalination* **1975**, *16*, 151–155. [CrossRef]

63. Stache, K. Apparatus for Transforming Sea Water, Brackish Water, Polluted Water or the Like into a Nutrious Drink by Means of Osmosis. U.S. Patent US 4879030A, 7 November 1989.

64. Yaeli, J. Method and Apparatus for Processing Liquid Solutions of Suspensions Particularly Useful in the Desalination of Saline Water. U.S. Patent US 5098575A, 24 March 1992.

65. Ray, S.S.; Chen, S.S.; Nguyen, N.C.; Nguyen, H.T.; Dan, N.P.; Thanh, B.X.; Trang, L.T. Exploration of polyelectrolyte incorporated with triton-x 114 surfactant based osmotic agent for forward osmosis desalination. *J. Environ. Manag.* **2018**, *209*, 346–353. [CrossRef] [PubMed]

66. Monjezi, A.A.; Mahood, H.B.; Campbell, A.N. Regeneration of dimethyl ether as a draw solute in forward osmosis by utilising thermal energy from a solar pond. *Desalination* **2017**, *415*, 104–114. [CrossRef]

67. Huang, J.; Long, Q.; Xiong, S.; Shen, L.; Wang, Y. Application of poly(4-styrenesulfonic acid-co-maleic acid) sodium salt as novel draw solute in forward osmosis for dye-containing wastewater treatment. *Desalination* **2017**, *421*, 40–46. [CrossRef]

68. Li, Z.; Wei, L.; Gao, M.Y.; Lei, H. One-pot reaction to synthesize biocompatible magnetite nanoparticles. *Adv. Mater.* **2005**, *17*, 1001–1005. [CrossRef]

69. Park, S.Y.; Ahn, H.-W.; Chung, J.W.; Kwak, S.-Y. Magnetic core-hydrophilic shell nanosphere as stability-enhanced draw solute for forward osmosis (fo) application. *Desalination* **2016**, *397*, 22–29. [CrossRef]

70. Hee-Man, Y.; Bum-Kyoung, S.E.O.; Kune-Woo, L.E.E.; Jei-Kwon, M. Hyperbranched polyglycerol-coated magnetic nanoparticles as draw solute in forward osmosis. *Asian J. Chem.* **2014**, *26*, 4031–4034.

71. Chekli, L.; Phuntsho, S.; Shon, H.K.; Vigneswaran, S.; Kandasamy, J.; Chanan, A. A review of draw solutes in forward osmosis process and their use in modern applications. *Desalin. Water Treat.* **2012**, *43*, 167–184. [CrossRef]

72. Yong, J.S.; Phillip, W.A.; Elimelech, M. Coupled reverse draw solute permeation and water flux in forward osmosis with neutral draw solutes. *J. Membr. Sci.* **2012**, *392–393*, 9–17. [CrossRef]

73. Su, J.; Chung, T.-S.; Helmer, B.J.; de Wit, J.S. Enhanced double-skinned fo membranes with inner dense layer for wastewater treatment and macromolecule recycle using sucrose as draw solute. *J. Membr. Sci.* **2012**, *396*, 92–100. [CrossRef]

74. Xie, M.; Zheng, M.; Cooper, P.; Price, W.E.; Nghiem, L.D.; Elimelech, M. Osmotic dilution for sustainable greenwall irrigation by liquid fertilizer: Performance and implications. *J. Membr. Sci.* **2015**, *494*, 32–38. [CrossRef]
75. Duan, J.; Litwiller, E.; Choi, S.-H.; Pinnau, I. Evaluation of sodium lignin sulfonate as draw solute in forward osmosis for desert restoration. *J. Membr. Sci.* **2014**, *453*, 463–470. [CrossRef]
76. Liu, Z.; Bai, H.; Lee, J.; Sun, D.D. A low-energy forward osmosis process to produce drinking water. *Energy Environ. Sci.* **2011**, *4*, 2582–2585. [CrossRef]
77. Alnaizy, R.; Aidan, A.; Qasim, M. Copper sulfate as draw solute in forward osmosis desalination. *J. Environ. Chem. Eng.* **2013**, *1*, 424–430. [CrossRef]
78. Batchelder, G.W. Process for the Demineralization of Water. U.S. Patent US 3171799A, 2 March 1965.
79. McCutcheon, J.R.; McGinnis, R.L.; Elimelech, M. Desalination by ammonia–carbon dioxide forward osmosis: Influence of draw and feed solution concentrations on process performance. *J. Membr. Sci.* **2006**, *278*, 114–123. [CrossRef]
80. Li, D.; Zhang, X.; Yao, J.; Simon, G.P.; Wang, H. Stimuli-responsive polymer hydrogels as a new class of draw agent for forward osmosis desalination. *Chem. Commun.* **2011**, *47*, 1710–1712. [CrossRef] [PubMed]
81. Zhong, Y.; Feng, X.; Chen, W.; Wang, X.; Huang, K.-W.; Gnanou, Y.; Lai, Z. Using ucst ionic liquid as a draw solute in forward osmosis to treat high-salinity water. *Environ. Sci. Technol.* **2016**, *50*, 1039–1045. [CrossRef] [PubMed]
82. Park, J.; Joo, H.; Noh, M.; Namkoong, Y.; Lee, S.; Jung, K.H.; Ahn, H.R.; Kim, S.; Lee, J.-C.; Yoon, J.H.; et al. Systematic structure control of ammonium iodide salts as feasible ucst-type forward osmosis draw solutes for the treatment of wastewater. *J. Mater. Chem. A* **2018**, *6*, 1255–1265. [CrossRef]
83. Kim, J.-J.; Kang, H.; Choi, Y.-S.; Yu, Y.A.; Lee, J.-C. Thermo-responsive oligomeric poly(tetrabutylphosphonium styrenesulfonate)s as draw solutes for forward osmosis (fo) applications. *Desalination* **2016**, *381*, 84–94. [CrossRef]
84. Cai, Y.; Wang, R.; Krantz, W.B.; Fane, A.G.; Hu, X.M. Exploration of using thermally responsive polyionic liquid hydrogels as draw agents in forward osmosis. *RSC Adv.* **2015**, *5*, 97143–97150. [CrossRef]
85. Cai, Y.; Shen, W.; Wei, J.; Chong, T.H.; Wang, R.; Krantz, W.B.; Fane, A.G.; Hu, X. Energy-efficient desalination by forward osmosis using responsive ionic liquid draw solutes. *Environ. Sci. Water Res.* **2015**, *1*, 341–347. [CrossRef]
86. Mok, Y.; Nakayama, D.; Noh, M.; Jang, S.; Kim, T.; Lee, Y. Circulatory osmotic desalination driven by a mild temperature gradient based on lower critical solution temperature (lcst) phase transition materials. *Phys. Chem. Chem. Phys.* **2013**, *15*, 19510–19517. [CrossRef] [PubMed]
87. Wang, Y.; Yu, H.; Xie, R.; Zhao, K.; Ju, X.; Wang, W.; Liu, Z.; Chu, L. An easily recoverable thermo-sensitive polyelectrolyte as draw agent for forward osmosis process. *Chin. J. Chem. Eng.* **2016**, *24*, 86–93. [CrossRef]
88. Noh, M.; Mok, Y.; Lee, S.; Kim, H.; Lee, S.H.; Jin, G.-W.; Seo, J.-H.; Koo, H.; Park, T.H.; Lee, Y. Novel lower critical solution temperature phase transition materials effectively control osmosis by mild temperature changes. *Chem. Commun.* **2012**, *48*, 3845–3847. [CrossRef] [PubMed]
89. Wendt, D.S.; Orme, C.J.; Mines, G.L.; Wilson, A.D. Energy requirements of the switchable polarity solvent forward osmosis (sps-fo) water purification process. *Desalination* **2015**, *374*, 81–91. [CrossRef]
90. Cai, Y.; Shen, W.; Wang, R.; Krantz, W.B.; Fane, A.G.; Hu, X. CO_2 switchable dual responsive polymers as draw solutes for forward osmosis desalination. *Chem. Commun.* **2013**, *49*, 8377–8379. [CrossRef] [PubMed]
91. Rabiee, H.; Jin, B.; Yun, S.; Dai, S. Gas-responsive cationic microgels for forward osmosis desalination. *Chem. Eng. J.* **2018**, *347*, 424–431. [CrossRef]
92. Guo, C.X.; Zhao, D.; Zhao, Q.; Wang, P.; Lu, X. Na^{+}-functionalized carbon quantum dots: A new draw solute in forward osmosis for seawater desalination. *Chem. Commun.* **2014**, *50*, 7318–7321. [CrossRef] [PubMed]
93. Zhao, D.; Chen, S.; Wang, P.; Zhao, Q.; Lu, X. A dendrimer-based forward osmosis draw solute for seawater desalination. *Ind. Eng. Chem. Res.* **2014**, *53*, 16170–16175. [CrossRef]
94. Ge, Q.; Amy, G.L.; Chung, T.-S. Forward osmosis for oily wastewater reclamation: Multi-charged oxalic acid complexes as draw solutes. *Water Res.* **2017**, *122*, 580–590. [CrossRef] [PubMed]
95. Hamdan, M.; Sharif, A.O.; Derwish, G.; Al-Aibi, S.; Altaee, A. Draw solutions for forward osmosis process: Osmotic pressure of binary and ternary aqueous solutions of magnesium chloride, sodium chloride, sucrose and maltose. *J. Food Eng.* **2015**, *155*, 10–15. [CrossRef]

96. York, R.J.; Thiel, R.S.; Beaudry, E.G. Full-scale experience of direct osmosis concentration applied to leachate management. In Proceedings of the Seventh International Waste Management and Landfill Symposium (Sardinia'99), S. Margherita di Pula, Cagliari, Sardinia, Italy, 4–8 October 1999.
97. Bowden, K.S.; Achilli, A.; Childress, A.E. Organic ionic salt draw solutions for osmotic membrane bioreactors. *Bioresour. Technol.* **2012**, *122*, 207–216. [CrossRef] [PubMed]
98. Tan, C.H.; Ng, H.Y. A novel hybrid forward osmosis-nanofiltration (fo-nf) process for seawater desalination: Draw solution selection and system configuration. *Desalin. Water Treat.* **2010**, *13*, 356–361. [CrossRef]
99. Ge, Q.; Chung, T.-S. Hydroacid complexes: A new class of draw solutes to promote forward osmosis (fo) processes. *Chem. Commun.* **2013**, *49*, 8471–8473. [CrossRef] [PubMed]
100. Long, Q.; Wang, Y. Novel carboxyethyl amine sodium salts as draw solutes with superior forward osmosis performance. *AIChE J.* **2016**, *62*, 1226–1235. [CrossRef]
101. Long, Q.; Shen, L.; Chen, R.; Huang, J.; Xiong, S.; Wang, Y. Synthesis and application of organic phosphonate salts as draw solutes in forward osmosis for oil-water separation. *Environ. Sci. Technol.* **2016**, *50*, 12022–12029. [CrossRef] [PubMed]
102. Ou, R.; Wang, Y.; Wang, H.; Xu, T. Thermo-sensitive polyelectrolytes as draw solutions in forward osmosis process. *Desalination* **2013**, *318*, 48–55. [CrossRef]
103. Nguyen, H.T.; Chen, S.-S.; Nguyen, N.C.; Ngo, H.H.; Guo, W.; Li, C.-W. Exploring an innovative surfactant and phosphate-based draw solution for forward osmosis desalination. *J. Membr. Sci.* **2015**, *489*, 212–219. [CrossRef]
104. Gadelha, G.; Nawaz, M.S.; Hankins, N.P.; Khan, S.J.; Wang, R.; Tang, C.Y. Assessment of micellar solutions as draw solutions for forward osmosis. *Desalination* **2014**, *354*, 97–106. [CrossRef]
105. Cui, H.; Zhang, H.; Jiang, W.; Yang, F. Preparation and assessment of carboxylate polyelectrolyte as draw solute for forward osmosis. *Environ. Sci. Pollut. Res.* **2018**, *25*, 5752–5761. [CrossRef] [PubMed]
106. Laohaprapanon, S.; Fu, Y.-J.; Hu, C.-C.; You, S.-J.; Tsai, H.-A.; Hung, W.-S.; Lee, K.-R.; Lai, J.-Y. Evaluation of a natural polymer-based cationic polyelectrolyte as a draw solute in forward osmosis. *Desalination* **2017**, *421*, 72–78. [CrossRef]
107. Kim, J.-J.; Chung, J.-S.; Kang, H.; Yu, Y.A.; Choi, W.J.; Kim, H.J.; Lee, J.-C. Thermo-responsive copolymers with ionic group as novel draw solutes for forward osmosis processes. *Macromol. Res.* **2014**, *22*, 963–970. [CrossRef]
108. Razmjou, A.; Liu, Q.; Simon, G.P.; Wang, H. Bifunctional polymer hydrogel layers as forward osmosis draw agents for continuous production of fresh water using solar energy. *Environ. Sci. Technol.* **2013**, *47*, 13160–13166. [CrossRef] [PubMed]
109. Li, D.; Wang, H. Smart draw agents for emerging forward osmosis application. *J. Mater. Chem. A* **2013**, *1*, 14049–14060. [CrossRef]
110. Han, H.; Lee, J.Y.; Lu, X. Thermoresponsive nanoparticles + plasmonic nanoparticles = photoresponsive heterodimers: Facile synthesis and sunlight-induced reversible clustering. *Chem. Commun.* **2013**, *49*, 6122–6124. [CrossRef] [PubMed]
111. Ou, R.; Zhang, H.; Simon, G.P.; Wang, H. Microfiber-polymer hydrogel monolith as forward osmosis draw agent. *J. Membr. Sci.* **2016**, *510*, 426–436. [CrossRef]
112. Li, D.; Zhang, X.; Yao, J.; Zeng, Y.; Simon, G.P.; Wang, H. Composite polymer hydrogels as draw agents in forward osmosis and solar dewatering. *Soft Matter* **2011**, *7*, 10048–10056. [CrossRef]
113. Li, D.; Zhang, X.; Simon, G.P.; Wang, H. Forward osmosis desalination using polymer hydrogels as a draw agent: Influence of draw agent, feed solution and membrane on process performance. *Water Res.* **2013**, *47*, 209–215. [CrossRef] [PubMed]
114. Zeng, Y.; Qiu, L.; Wang, K.; Yao, J.; Li, D.; Simon, G.P.; Wang, R.; Wang, H. Significantly enhanced water flux in forward osmosis desalination with polymer-graphene composite hydrogels as a draw agent. *RSC Adv.* **2013**, *3*, 887–894. [CrossRef]
115. Na, Y.; Yang, S.; Lee, S. Evaluation of citrate-coated magnetic nanoparticles as draw solute for forward osmosis. *Desalination* **2014**, *347*, 34–42. [CrossRef]
116. Dey, P.; Izake, E.L. Magnetic nanoparticles boosting the osmotic efficiency of a polymeric fo draw agent: Effect of polymer conformation. *Desalination* **2015**, *373*, 79–85. [CrossRef]
117. Cath, T.Y.; Childress, A.E.; Elimelech, M. Forward osmosis: Principles, applications, and recent developments. *J. Membr. Sci.* **2006**, *281*, 70–87. [CrossRef]

118. Zou, S.; He, Z. Enhancing wastewater reuse by forward osmosis with self-diluted commercial fertilizers as draw solutes. *Water Res.* **2016**, *99*, 235–243. [CrossRef] [PubMed]
119. Nasr, P.; Sewilam, H. Investigating the performance of ammonium sulphate draw solution in fertilizer drawn forward osmosis process. *Clean Technol. Environ.* **2016**, *18*, 717–727. [CrossRef]
120. Roy, D.; Rahni, M.; Pierre, P.; Yargeau, V. Forward osmosis for the concentration and reuse of process saline wastewater. *Chem. Eng. J.* **2016**, *287*, 277–284. [CrossRef]
121. Ansari, A.J.; Hai, F.I.; Price, W.E.; Nghiem, L.D. Phosphorus recovery from digested sludge centrate using seawater-driven forward osmosis. *Sep. Purif. Technol.* **2016**, *163*, 1–7. [CrossRef]
122. Zhao, S.; Zou, L.; Mulcahy, D. Brackish water desalination by a hybrid forward osmosis–nanofiltration system using divalent draw solute. *Desalination* **2012**, *284*, 175–181. [CrossRef]
123. Wang, K.Y.; Teoh, M.M.; Nugroho, A.; Chung, T.-S. Integrated forward osmosis–membrane distillation (fo–md) hybrid system for the concentration of protein solutions. *Chem. Eng. Sci.* **2011**, *66*, 2421–2430. [CrossRef]
124. Adham, S.; Hussain, A.; Matar, J.M.; Dores, R.; Janson, A. Application of membrane distillation for desalting brines from thermal desalination plants. *Desalination* **2013**, *314*, 101–108. [CrossRef]
125. Wang, P.; Chung, T.-S. A conceptual demonstration of freeze desalination–membrane distillation (fd–md) hybrid desalination process utilizing liquefied natural gas (lng) cold energy. *Water Res.* **2012**, *46*, 4037–4052. [CrossRef] [PubMed]
126. El-Bourawi, M.S.; Ding, Z.; Ma, R.; Khayet, M. A framework for better understanding membrane distillation separation process. *J. Membr. Sci.* **2006**, *285*, 4–29. [CrossRef]
127. Xie, M.; Nghiem, L.D.; Price, W.E.; Elimelech, M. A forward osmosis–membrane distillation hybrid process for direct sewer mining: System performance and limitations. *Environ. Sci. Technol.* **2013**, *47*, 13486–13493. [CrossRef] [PubMed]
128. Cath, T.Y.; Hancock, N.T.; Lundin, C.D.; Hoppe-Jones, C.; Drewes, J.E. A multi-barrier osmotic dilution process for simultaneous desalination and purification of impaired water. *J. Membr. Sci.* **2010**, *362*, 417–426. [CrossRef]
129. Semiat, R. Energy issues in desalination processes. *Environ. Sci. Technol.* **2008**, *42*, 8193–8201. [CrossRef] [PubMed]
130. Mistry, K.H.; McGovern, R.K.; Thiel, G.P.; Summers, E.K.; Zubair, S.M.; Lienhard, J.H. Entropy generation analysis of desalination technologies. *Entropy* **2011**, *13*, 1829–1864. [CrossRef]
131. Zhao, D.; Chen, S.; Guo, C.X.; Zhao, Q.; Lu, X. Multi-functional forward osmosis draw solutes for seawater desalination. *Chin. J. Chem. Eng.* **2016**, *24*, 23–30. [CrossRef]
132. Miller, S.; Shemer, H.; Semiat, R. Energy and environmental issues in desalination. *Desalination* **2015**, *366*, 2–8. [CrossRef]
133. Zhang, S.; Wang, P.; Fu, X.; Chung, T.-S. Sustainable water recovery from oily wastewater via forward osmosis-membrane distillation (fo-md). *Water Res.* **2014**, *52*, 112–121. [CrossRef] [PubMed]
134. Lattemann, S.; Höpner, T. Environmental impact and impact assessment of seawater desalination. *Desalination* **2008**, *220*, 1–15. [CrossRef]
135. Ge, Q.; Lau, C.H.; Liu, M. A novel multi-charged draw solute that removes organic arsenicals from water in a hybrid membrane process. *Environ. Sci. Technol.* **2018**, *52*, 3812–3819. [CrossRef] [PubMed]
136. Altaee, A.; Zaragoza, G.; van Tonningen, H.R. Comparison between forward osmosis-reverse osmosis and reverse osmosis processes for seawater desalination. *Desalination* **2014**, *336*, 50–57. [CrossRef]
137. Hancock, N.T.; Xu, P.; Heil, D.M.; Bellona, C.; Cath, T.Y. Comprehensive bench- and pilot-scale investigation of trace organic compounds rejection by forward osmosis. *Environ. Sci. Technol.* **2011**, *45*, 8483–8490. [CrossRef] [PubMed]
138. Yangali-Quintanilla, V.; Li, Z.; Valladares, R.; Li, Q.; Amy, G. Indirect desalination of red sea water with forward osmosis and low pressure reverse osmosis for water reuse. *Desalination* **2011**, *280*, 160–166. [CrossRef]
139. Fane, A.G.; Fell, C.J.D. A review of fouling and fouling control in ultrafiltration. *Desalination* **1987**, *62*, 117–136. [CrossRef]
140. Gao, W.; Liang, H.; Ma, J.; Han, M.; Chen, Z.-L.; Han, Z.-S.; Li, G.-B. Membrane fouling control in ultrafiltration technology for drinking water production: A review. *Desalination* **2011**, *272*, 1–8. [CrossRef]
141. Chung, T.-S.; Zhang, S.; Wang, K.Y.; Su, J.; Ling, M.M. Forward osmosis processes: Yesterday, today and tomorrow. *Desalination* **2012**, *287*, 78–81. [CrossRef]

142. Fei, R.; George, J.T.; Park, J.; Means, A.K.; Grunlan, M.A. Ultra-strong thermoresponsive double network hydrogels. *Soft Matter* **2013**, *9*, 2912–2919. [CrossRef]

143. Wang, Q.; Zhao, Y.; Yang, Y.; Xu, H.; Yang, X. Thermosensitive phase behavior and drug release of in situ gelable poly(n-isopropylacrylamide-co-acrylamide) microgels. *Colloid Polym. Sci.* **2007**, *285*, 515–521. [CrossRef]

144. Qin, W.; Yanbing, Z.; Huibi, X.; Xiangliang, Y.; Yajiang, Y. Thermosensitive phase transition kinetics of poly(*n*-isopropylacryl amide-*co*-acrylamide) microgel aqueous dispersions. *J. Appl. Polym. Sci.* **2009**, *113*, 321–326.

145. Ward, M.A.; Georgiou, T.K. Thermoresponsive polymers for biomedical applications. *Polymers* **2011**, *3*, 1215–1242. [CrossRef]

146. Mehta, G.D.; Loeb, S. Internal polarization in the porous substructure of a semipermeable membrane under pressure-retarded osmosis. *J. Membr. Sci.* **1978**, *4*, 261–265. [CrossRef]

147. Mishra, T.; Ramola, S.; Shankhwar, A.K.; Srivastava, R.K. Use of synthesized hydrophilic magnetic nanoparticles (hmnps) in forward osmosis for water reuse. *Water Sci. Technol.* **2016**, *16*, 229–236. [CrossRef]

148. Motsa, M.M.; Mamba, B.B.; D'Haese, A.; Hoek, E.M.V.; Verliefde, A.R.D. Organic fouling in forward osmosis membranes: The role of feed solution chemistry and membrane structural properties. *J. Membr. Sci.* **2014**, *460*, 99–109. [CrossRef]

processes

MDPI

Article

Dopamine Incorporated Forward Osmosis Membranes with High Structural Stability and Chlorine Resistance

Yi Wang [1,2], Zhendong Fang [1], Chaoxin Xie [1,*], Shuaifei Zhao [3] , Derrick Ng [2] and Zongli Xie [2,*]

[1]   Water Industry and Environment Engineering Technology Research Centre, Chongqing 401311, China;
     Y.Wang@CSIRO.au (Y.W.); Zhendongfang1962@gmail.com (Z.F.)
[2]   CSIRO Manufacturing, Clayton, VIC 3168, Australia; Derrick.Ng@CSIRO.au
[3]   Department of Environmental Sciences, Macquarie University, Sydney, NSW 2109, Australia;
     shuaifei.zhao@mq.edu.au
*   Correspondence: a86909304@163.com (C.X.); zongli.xie@csiro.au (Z.X.);
     Tel.: +86-023-731001 (C.X.); +61-3-95452938 (Z.X.)

Received: 3 August 2018; Accepted: 22 August 2018; Published: 1 September 2018

Abstract: The degradation and detachment of the polyamide (PA) layer for the conventional thin-film composite (TFC) membranes due to chemical disinfectants cleaning with chlorine and material difference of PA layer and substrate are two major bottlenecks of forward osmosis (FO) technology. In this study, a new type of FO membranes was first prepared by controlling dopamine (DA) as the sole amine in the aqueous phase and the reaction with trimesoyl chloride (TMC) as the acyl chloride during interfacial polymerization (IP) process. The influence of membrane synthesis parameters such as monomer concentration, pH of the aqueous phase, IP reaction time and IP temperature were systematically investigated. The optimized membrane showed both improved structure stability and chlorine resistance, more so than the conventional TFC membrane. In general, novel DA/TMC TFC membranes could be an effective strategy to synthesize high-performance FO membranes with excellent structural stability and chlorine resistance.

Keywords: forward osmosis; thin-film composite; dopamine; interfacial polymerization; structural stability; chlorine resistance

1. Introduction

Recently, FO has attracted increasing interest in seawater desalination [1–3], municipal sewage treatment [4,5], membrane bioreactors [6,7], agriculture fertilizers [8,9] and power generation [10,11], as a potential energy-saving and promising technology [12]. Forward osmosis (FO) is a self-driven process that drives pure water molecules across a semi-permeable membrane from the feed solution (with a low osmotic pressure or high chemical potential) to the draw solution (with a high osmotic pressure or low chemical potential). Thin-film composite polyamide (TFC-PA) membranes, which consist of ultra-thin and dense polyamide (PA) layers and porous and thick substrate layers, have been studied for their applications in FO process due to their high salt rejection capability, wide range pH stability and independently modifiable support layer and active layer ability [13–16].

Conventionally, the thin and dense polyamide (PA) layer is synthesized by a interfacial polymerization process onto the substrate layer by two different monomers with amine and acyl chloride [17]. In order to obtain PA layers with good mechanical properties, at least one selected monomer must have more than two functional groups as well as an aromatic material [18]. However, if the conventional TFC-PA membranes are used in some wastewater containing ethanol, the active PA layer could easily be detached from the top of the substrate layer, because there is no strong linkage between these layers. So the membrane structural stability should be taken into careful consideration for more practical membrane separation applications. Apart from the membrane structural stability, membrane chlorine resistance is also another bottleneck for TFC-PA membrane applications. This is because membrane fouling is inevitable in the FO process, for example, in biological fouling it is the main cause of decreasing membrane performance [5,13,19]. Therefore, membrane cleaning with oxidizing agents to control membrane biofouling is also needed in FO applications. However, the antioxidant ability of the conventional PA layer is poor, and its polymer chain structure would be degraded by the active chlorine in the oxidizing agents easily and consequently bring a significantly reduced membrane separation performance [20,21]. Hence, TFC-PA membranes with both improved structural stability and chorine resistance are highly desirable.

Recently, dopamine (DA) or polydopamine (PDA) inspired chemistry has shown a new route for the synthesis of high-performance membranes. PDA with adhesive proteins has been used to enhance the anti-fouling property of ultra-filtration (UF) [22] and Reverse Osmosis (RO) membranes [23], by forming surface-adherent films onto either the active PA layer or the substrate layer to increase hydrophilic property of membranes. Han et al. first designed time-different PDA coated substrates for FO membrane synthesis and found that FO membrane based on 1 h-PDA-coated substrate showed the best FO performance [16]. Inspired by Han's work, Ping Yu's group [24] prepared three different kinds of UF substrates (PDA coated for 1 h on the topside, on the bottom and on both sides of the substrate before the interfacial polymerization (IP) process, respectively) for TFC-PA membranes fabrication, observing a reduction of reverse solute flux for the PDA topside-coated substrate based TFC-PA membranes. Very recently, Guo et al. [25] proved that only 0.5 h coating was enough for PDA deposition and modification on the PA active layer, which would enhance FO membrane selectivity as well as anti-fouling property.

Inspired by these works, DA was first investigated as a sole amine in aqueous phase in IP process to prepare TFC membranes for FO process. The DA monomers with amine and phenol groups, as the PDA formed by DA self-polymerization with the catechol groups [26,27] can all react with trimesoyl chloride (TMC) with acyl chloride groups to obtain a new type of thin film active layer for separation applications, which is strongly adhered to the top of the substrate. At the same time, the novel active layer, with ester bonds formed by DA/TMC, is much more stable than the amide bonds of the conventional PA layer when exposed to solutions with active chloride. This study sheds light on developing high-performance FO membranes by improving the membrane structure stability as well as the membrane chlorine resistance.

2. Materials and Methods

2.1. Materials

Polysulfone beads (PSf, average molecular weight (MW) ~22 kDa, Sigma-Aldrich, St. Louis, MO, USA), polyvinylpyrrolidone (PVP, average MW ~10 kDa, Sigma Aldrich, St. Louis, MO, USA) and 1-methyl-2-pyrrolidinone (99%, NMP, RCI LABSCAN LIMITED, Samutsakorn, Thailand) were used to fabricate the substrates in this work. Trimesic acid trichloride (98%, TMC, Sigma-Aldrich, St. Louis, MO, USA) monomers were dissolved in n-hexane (Anhydrous, 95%, Sigma-Aldrich, St. Louis, MO, USA) as the organic phase and dopamine hydrochloride (Reagent grade, for research purpose, Sigma-Aldrich, St. Louis, MO, USA) and piperazine (PIP, ReagentPlus®, 99%, Sigma-Aldrich, St. Louis, MO, USA) were dissolved in phosphate buffer saline (PBS) solution as the aqueous phase for IP process. For membrane performance tests, magnesium chloride ($MgCl_2$, 98%, Sigma-Aldrich, St. Louis, MO, USA) was dissolved in deionized water (DI). For membrane structural stability and chlorine resistance tests, ethanol (ACS, ISO, Reag. Ph EMSURE®, Darmstadt, Germany) and sodium hypochlorite (NaClO) solution (Chem Supply Pty Ltd, Adelaide, Australia) with 8–12.5% available chlorine were used, respectively.

2.2. Preparation of Polysulfone Substrates

The porous substrate was synthesized by the classical phase inversion process [28]. To prepare the casting dope for substrate, a mixture containing PSf beads, PVP powders and NMP solvent was kept stirring till the homogenous yellow solution was obtained (this process normally needs 8 h) and then left the solution degassing for at least 12 h. Detail of the procedure can be found in our previous work [28].

2.3. Preparation of Thin Film Composite (TFC) Membranes

The TFC membranes were prepared by interfacial polymerization reaction between the DA monomers and TMC monomers. To be more specific, a series of DA-containing solutions were prepared by mixing PBS and 0.10, 0.30, 0.50 wt% DA, respectively, as shown in Table 1. The PSf substrate was first immersed into DA solution for 0.5 h [29] (0.5 wt% DA saturated PSf substrate, named M-DA, was prepared to verify if DA has become part of this membrane or not). Afterwards, the rest of the DA solution was dried by an air-knife. Then the saturated substrate was followed by immersing into 0.15 wt% TMC solution for 1 min. TMC solution was drain and the substrate was held vertically 120 s to evaporate n-hexane thoroughly. Afterwards, a heating post-treatment process was conducted by putting these as-prepared membranes into an oven at 90 °C for 300 s. After that, the as-prepared TFC membranes were washed 3 times with DI water to remove the unreacted residues and stored in 4 °C DI water in fridge until further testing. The obtained membranes were denoted as M-1, M-2 and M-3, containing 0.10, 0.30 and 0.50 wt% of DA in the aqueous phase in IP process, respectively. For comparison, conventional TFC-PA membranes formed by PIP/TMC (denoted as M-0) were also prepared as the control membranes.

Table 1. Reagents for thin-film composite (TFC) Membranes Preparation.

Membranes	DA (wt%) in Aqueous Phase	PIP (wt%) in Aqueous Phase	TMC (wt%) in Organic Phase
M-0	0	1.0	0.15
M-DA	0.50	0	0
M-1	0.10	0	0.15
M-2	0.30	0	0.15
M-3	0.50	0	0.15

2.4. Membrane Characterizations

The morphology of the obtained membrane surfaces was observed by field emission scanning electron microscopy (FESEM, Merlin ZEISS GEMINI2, Oberkochen, Germany). Membrane surface chemistry of the active layer was determined by X-ray photoelectron spectroscopy (XPS, ESCALab220IXL, Abingdon, UK). Water contact angles (WCA) were tested by a static sessile drop contact angle system (Dataphysics OCA20, Filderstadt, Germany). The functional groups of the membranes were analysed by Attenuated Total Reflection-Fourier Transform Infra-Red, (ATR-FTIR, Thermo Scientific Nicolet 6700, Midland, ON, Canada).

2.5. FO Performance Tests

FO performance tests were conducted by using a commercial PTFE cell (CF042-FO, Sterlitech Company, Kent, WA, USA). The flow velocities for the feed solution side and draw solution side were maintained at 4.9 cm·s^{-1}. The temperatures of both sides were controlled at 20 ± 0.5 °C to reduce the water flux variations caused by temperature changes. 1 M of MgCl$_2$ solution was used as the draw solution and DI water was used as the feed solution. The obtained TFC membranes were tested under only one mode, namely, active layer facing feed solution mode. Each experiment was tested for 1 h in triplicate from two independent batches.

The water permeation flux (J_w) ($L \cdot m^{-2} \cdot h^{-1}$, LMH) and the reverse solute flux, (J_s) ($g \cdot m^{-2} \cdot h^{-1}$) were calculated by the same methods detailed in our previous work [28]. The specific reverse solute flux ($g \cdot L^{-1}$) [30,31], which means the ratio of J_s/J_w, was calculated to measure the FO selectivity of the resultant TFC membranes.

2.6. Evaluation of Membrane Structural Stability and Chlorine Resistance

The structural stability of the as-prepared TFC membranes was studied by immersing them in pure ethanol for a fixed period. And the chlorine resistance of these membranes was evaluated by exposing their surfaces into a 1000 ppm NaClO solution over different periods at ambient temperature. The NaClO solution was kept in dark and replaced every 2 h during the chlorine resistance test to keep the constant concentration. Before the membrane FO testing, these TFC membranes were removed from ethanol or NaClO solution and then washed 3 times with DI water to avoid the residual chlorine attacking and ethanol swelling during performance tests.

3. Results and Discussion

3.1. Membrane Surface Characterization

3.1.1. Membrane Surface Chemistry

In this study, DA/TMC TFC membranes possesses novel dense and thin active layers on top of the substrates, which are critical for high FO water flux and membrane selectivity [32]. ATR-FTIR and XPS tests were utilised to confirm the formation of these layers. The ATR-FTIR spectra for PSf-substrate, M-1, M-2 and M-3 are presented Figure 1a. It is clear that, apart from the bands of the PSf substrates, the DA/TMC TFC series membranes showed enhanced peaks at 1660 cm^{-1} at the same time additional peaks at 1750 cm^{-1}, which are attributed to the C=O breathing in the obtained amide bonds and the C=O stretching in ester bonds, respectively. The increase in the peak intensity of hydroxyl groups at about 1660 cm^{-1} was also found in all the DA/TMC TFC membranes, which are attributed to amide II band stretching vibration in the new-formed active layers. These newly formed bands indicated that the active layers were successfully formed onto the top of the PSf substrates. Moreover, nitrogen was detected by XPS on the DA/TMC membrane surfaces (Figure 1b), which confirmed the existence of the DA or PDA on the top of PSf substrates because the amine groups were the exclusive N source while no N sources can be detected in the PSf substrate [16]. As the concentration of DA increased from 0.10 wt% to 0.50 wt% in the aqueous phase, the nitrogen percent increased from 1.24% to 2.44%, accordingly. High-Resolution XPS spectrum C 1s spectra of the M-2 surface is also shown in Figure 1c, the peaks at bonding energies of 285.2, 286.1, 287.4 and 288.2 eV can be ascribed to the C-C, C-N and C-O, O=C-N and O=C-N groups, respectively. The presence of 287.4 and 288.2 eV peaks further confirmed the successful formation of amide and ester groups in this novel active layer which is synthesised by the reaction between the catechol of PDA (or phenolic hydroxyl and amine groups in DA) and acyl chloride in TMC monomers. Based on ATR-FTIR and XPS analysis results, the formation of additional novel active layers on the top of PSf substrates has been successful. Moreover, these chemical changes indicated that more DA monomers were incorporated in the as-prepared membranes with the monomer concentrations.

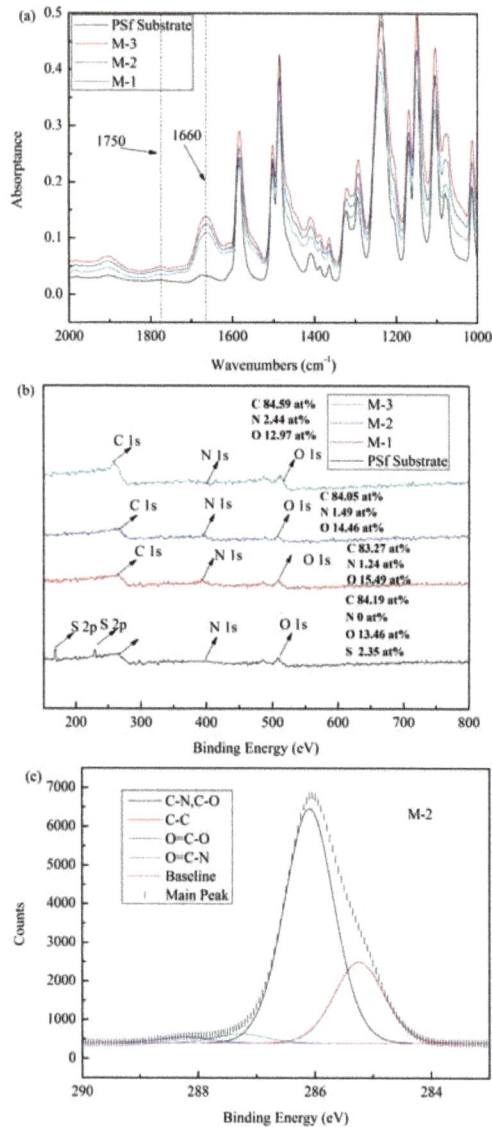

Figure 1. (**a**) Attenuated Total Reflection-Fourier Transform Infra-Red (ATR-FTIR) spectra of PSf substrate, M-1, M-2 and M-3; (**b**) X-ray photoelectron spectroscopy (XPS) spectra of the M-1, M-2 and M-3 membrane surface and (**c**) High-Resolution XPS spectrum C 1s spectra of M-2 surface.

3.1.2. Membrane Surface Morphology

Figure 2 shows the surface SEM images of these as-prepared membranes. Obviously, the PSf substrate surface was smooth, but after DA immersion, the surface of M-DA was rougher due to the DA self-polymerization. As for the PIP/TMC conventional TFC-PA (M-0), a typical "valley and peak" structure was found, confirming the formation of polyamide, which is similar to Dong et al.'s work [33]. As for the DA/TMC membrane surface (M-1, M-2 and M-3), structures with globules on membrane surfaces were found in the SEM images. When the DA concentration was low (0.1 wt%) in the aqueous phase, a relatively smooth surface with small and random globules could be found. As the DA concentration increased to 0.3 wt% (M-2), the appearance of these globules became more uniform and frequent, resulting in a rougher surface could be beneficial for enhancing water molecules transfer while reject salt ions. However, when the DA concentration was increased to 0.5 wt% (M-3), PDA particles agglomeration were clearly observed, which may reduce the membrane rejection due to their limited salt rejection abilities.

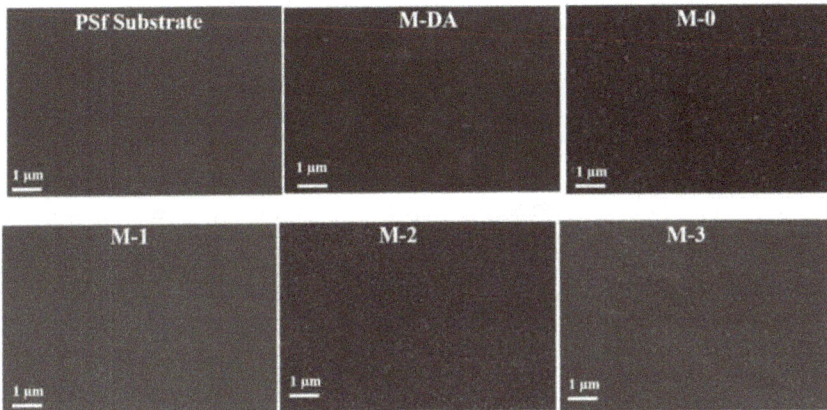

Figure 2. SEM images of PSf substrate, M-DA, M-0, M-1, M-2 and M-3 membrane surface.

3.1.3. Membrane Surface Hydrophilicity

The DA monomers and PDA particles in the aqueous phase with abundant amine and hydroxyl groups would improve the membrane surface hydrophilicity, as shown in Figure 3. WCA is a classical measurement to test the wettability of membrane surfaces: the lower the WCA, the easier for water molecules to wet the membrane surface and the higher the hydrophilicity. The WCA of the PSf substrate was relatively high at 75.3°, after the DA immersion, the WCA of M-DA was improved, indicating DA has been incorporated in the substrate. The WCA of the conventional membrane (M-0) was 51.0°, which is similar to Mohsen et al.'s work [34]. For the DA/TMC membranes with different DA concentrations, the WCAs only decreased from 55.9° to 50.2° when the DA concentration in aqueous phase was increased 5 times (from 0.1 wt% to 0.5 wt%). These results indicate that the incorporation of DA in the aqueous phase would enhance the membrane hydrophilicity while the concentrations of DA had slight influence on the membrane surface hydrophilicity.

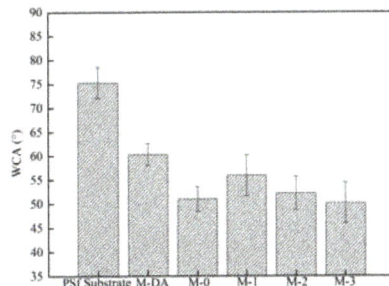

Figure 3. Water contact angles of PSf substrate, M-DA, M-0, M-1, M-2 and M-3.

3.2. Membrane FO Performance and Preparation Parameter Optimization

Figure 4a shows the influence of DA monomers concentration in the aqueous phase on FO performance of the obtained TFC membranes. The FO water flux increased from 3.22 to 6.55 LMH when the DA concentration was increased from 0.1 wt% to 0.3 wt%. However, further increasing the DA concentration showed little influence on the water flux in the meanwhile the specific reverse flux was almost doubled with high error bars. This is because high DA concentration would easily aggregate on the top of the membrane and thus some PDA particles formed globules (see Figure 2) might randomly increase the FO water flux but at the price of FO membrane selectivity. Based on, 0.3 wt% was selected in the following work.

The formation the novel active layer was determined by not only the concentration of DA monomers, but also the pH value of the aqueous phase. This is because, firstly, according to Zhu's study, DA self-polymerization of activity would be highly enhanced in alkaline solutions and reached the most drastic level at pH 8.5 [35]. Apart from this, this IP reaction between amine and phenol groups in DA monomers and acyl chloride groups in TMC monomers would produce hydrochloric acid. Thus, pH value of the aqueous phase would certainly have influence on the processing of the reaction of the chemical equilibrium [36]. Therefore, Figure 4b shows the influence of the pH value of the aqueous phase on the FO performance of the DA/TMC TFC membranes. At the pH of 6.0, the FO water flux was relatively high (9.51 LMH) as well as the specific reverse solute flux (1.62 $g·L^{-1}$), which indicated that acidic aqueous environment had an adverse impact the formation of the active layer. When the pH value was increased to 7, the water flux decreased to 6.55 LMH, and the specific reverse solute flux decreased further to 0.4 $g·L^{-1}$, indicating an enhanced FO membrane selectivity and a dense active layer was formed. When the pH value was further increased to 7.5 and 8.0, the FO water flux continuously decreased while the specific reverse solute flux slightly increased. This could be due to the process of DA self-polymerization was greatly enhanced in alkali solution [37], thus, more PDA particles was formed thus less amine and phenolic hydroxyl groups of DA can react with TMC in IP process. As a result, a low salt selectivity and loose active layer was thus formed. From this, pH 7 was chosen in the following experiments.

Figure 4c shows the influence of IP reaction time on the FO performance of the DA/TMC TFC membranes. When the IP reaction time was 10 s, only a little amount of amine and phenol groups in DA monomers could react with TMC, as a result, the main component of the active layer was loose PDA particles, which is attributed to the high FO water flux and low salt selectivity. When the reaction time increased from 60 s to 120 s, FO water flux decreased from 8.32 to 6.55 LMH and the specific reverse solute fluxes decreased from 1.37 to 0.4 $g·L^{-1}$. However, further increase in the reaction time did not decrease the specific reverse solute flux obviously as well as the FO water flux. Therefore, 120 s was selected in the following experiments to reduce the membrane fabrication time.

Figure 4d shows the influence of IP reaction temperature on the FO performance of the DA/TMC TFC membranes. When the IP reaction temperature was 15 °C, a very high FO water flux (9.08 LMH) and low salt selectivity (0.65 $g·L^{-1}$) was obtained. When the reaction temperature was increased from 25 to 55 °C, FO water flux decreased from 6.55 to 2.54 LMH and the specific reverse solute fluxes gradually decreased from 0.40 to 0.32 $g·L^{-1}$, respectively. This might be attributed to the fact that higher IP reaction temperature increased the reaction between DA and TMC. The obtained relative dense active layer lowered the FO water flux but maintained membrane selectivity. Based on the above experimental results, the synthesis parameters for IP process were fixed at 25 °C and 120 s, 0.3 wt% DA and pH value 7 in the aqueous phase, respectively.

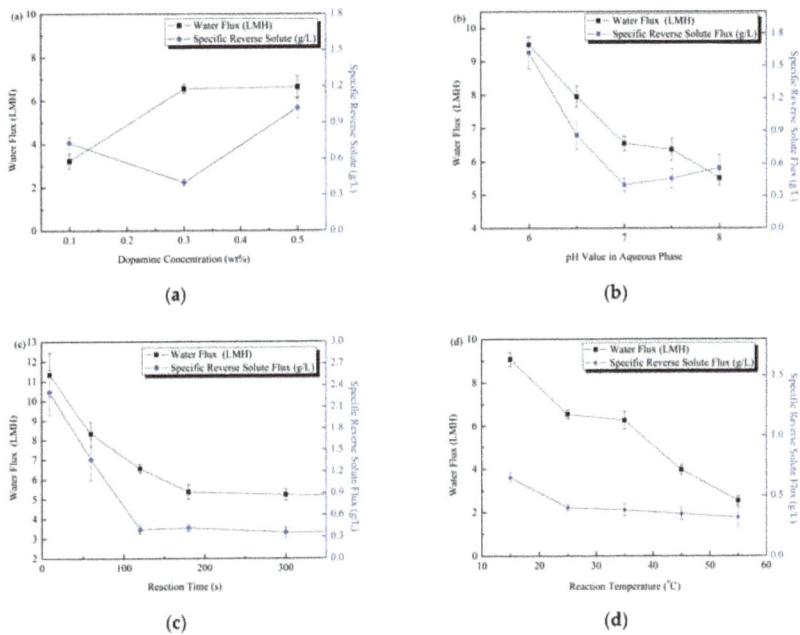

Figure 4. (**a**) The influence of dopamine (DA) concentration (0.1 wt%, 0.3 wt% and 0.5 wt%) in aqueous phase on the forward osmosis (FO) performance of the DA/trimesoyl chloride (TMC) TFC membranes; (**b**) The influence of pH value (6, 6.5, 7, 7.5 and 8) in aqueous phase on the FO performance of the DA/TMC TFC membranes; (**c**) The influence of reaction time (10–300 s) on the FO performance of the DA/TMC TFC membranes and (**d**) The influence of temperature (15, 25, 35, 45 and 55 °C) in aqueous phase on the FO performance of the DA/TMC TFC membranes.

3.3. Membrane Structural Stability and Chloride Resistance

In practical applications, membrane characteristics such as structural stability and chloride resistance are important for the lifetime of FO TFC membranes. In the first scenario, some feed solutions may contain organic solvents like ethanol which could massively swell the substrates and separation layers, which are generally comprised of two different materials, causing the detachment of active layers from the substrates, thus shortening membrane lifetime and increasing the total cost of membrane separation applications. In the second scenario, in order to control the bio-fouling in FO process, some disinfectants with free chloride are widely used, which can inevitably cause the conformational changes of the PA chains, leading to drastic damages for FO membrane separation performance. In order to evaluate the structural stability and chloride resistance of the as-prepared membranes quantitatively, in our study, a doubled increase value in specific reverse solute flux was chosen as an upper limit [38], resembling when the PA layers were detached or degraded too dramatically to be acceptable in practical applications.

3.3.1. Membrane Structural Stability

The optimized DA/TMC membrane (M-2) and the conventional PIP/TMC membrane (M-0) were separately immersed in ethanol to test the structural stability, their FO performances are shown in Figure 5. For M-2, its FO water flux gradually increased from 6.55 to 9.22 LMH after the ethanol immersion, and the specific reverse solute flux also increased to 0.81 g·L^{-1}. The whole process for the initial specific reverse solute flux value to double took as long as 228 h. On the other hand, it only took 32 h for M-0 to double its specific reverse solute flux. This enhancement of structural stability of M-2 (~7.1 times stronger) might be ascribed to the strengthened bond formation between the novel active layer and the substrate under the selected optimal conditions. Namely, when the pH value of the aqueous phase was fixed at 7.0, DA monomers in the aqueous phase were readily able to self-polymerize [35] to form tightly adhesive PDA particles with plenty of π-π and hydrogen-bonding interactions in between the newly-formed active layer and substrate [39].

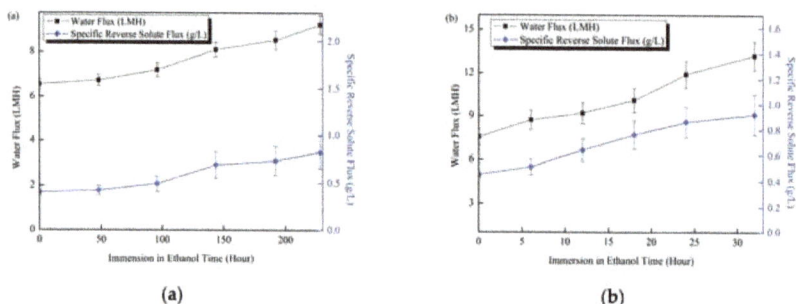

(a)   (b)

Figure 5. (a) The influence of ethanol immersion on the FO performance of M-2 membrane; (b) The influence of ethanol immersion on the FO performance of M-0 membrane. The FO performance was tested in active layer-feed solution (AL-FS) mode with 1 M $MgCl_2$ as draw solution and deionized water (DI) water as the feed solution.

3.3.2. Membrane Chlorine Resistance

Chlorine-resistance studies were conducted for both DA/TMC (M-2) and PIP/TMC membranes (M-0) as shown in Figure 6. For M-2, similar to the test in ethanol immersion, a gradual increase of water flux and specific reverse solute flux were observed during the test of chlorine exposure, which lasted up to as high as 12,000 ppm·h, before the specific reverse solute value was doubled. Both the water flux and specific solute flux increase of M-2 for the whole chlorine resistance test (up to 18,000 ppm·h) were mild and slow. This is because the ester bonds in M-2's active layer were much more stable than amide bonds of conventional PA layers when exposed to active chlorine. Therefore, a gradual increase of water flux due to the slow degradation of the novel active layer was observed. Meanwhile, due to a relatively complete active layer structure, the sudden rise in reverse solute flux on top of the membrane was avoided. As for the M-0 membrane, the water flux increase was firstly observed like the trend for M-2 but eventually decreased to approximately 65% of the initial flux in 500 ppm·h, while the specific reverse solute flux was finally increased from 0.45 to 3.05 $g \cdot L^{-1}$. This is because the amide bonds of the PA layers were vulnerable to free chlorine owing to their high electron density. At the beginning of the degradation of PA layer, the water molecules can easily pass through the membrane, however, because of the drastic degradation of the PA layer a highly increased reverse solute flux was observed, which caused an RO-like concentration polarization on top of the active layer thus decreasing water permeability and FO selectivity. Such significant failure of membrane separation performance may hinder the application of PIP/TMC TFC-PA membranes in wastewater treatment, which usually involves frequently chemical cleaning to avoid high bio-foulants. On the contrary, the M-2 portrayed by the novel dense and stable active layer showed a much slower destruction due to its ester bonds, which are much stronger than amide bonds of conventional PA layers when exposed to chlorine, therefore showing far superior chlorine resistance characteristics (72.3 times enhancement). Additionally, due to its more complete active layer structure, the surging reverse solute flux on the top of membrane was avoided.

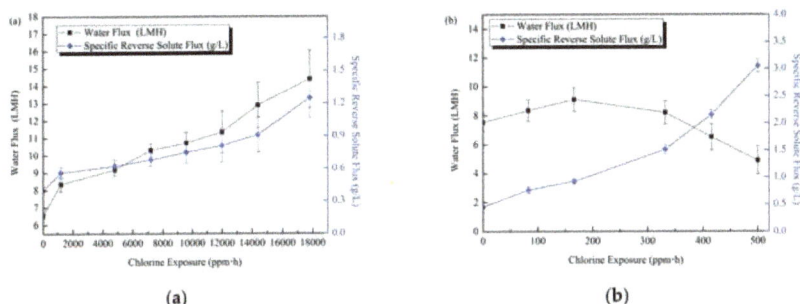

Figure 6. (a) The influence of chlorine immersion on the FO performance of M-2 membrane; (b) The influence of chlorine immersion on the FO performance of PIP/TMC conventional membrane. The FO performance was tested in active layer-feed solution (AL-FS) mode with 1 M MgCl$_2$ as draw solution and DI water as the feed solution.

4. Conclusions

A new type of DA/TMC TFC membranes was prepared via the IP reaction of DA monomers and TMC monomers on PSf substrates. DA as an amine was first explored in aqueous phase in IP process for TFC FO membrane preparation. Uniform hydrophilic structure with globules was generated on the membrane surface of the novel membranes at a moderate (0.3 wt%) DA concentration. The optimized parameters for this novel membrane synthesis process were obtained, including DA concentration (0.3 wt%), IP reaction temperature (25 °C), IP reaction time 120 s and the aqueous phase pH value 7, respectively. The optimized membrane showed enhancement of 7.1 times more structurally stable and 72.3 times more chloride-resistant characteristics than the conventional PIP/TMC membrane due to the poly-dopamine (PDA) strong linkage between the active layer and substrate as well as the polyester groups in the active layer. The excellent structural stability and chloride resistance shown by this new type of membrane could facilitate wider applications of FO membranes in wastewater treatment.

Author Contributions: Conceptualization, Y.W. and Z.X.; Methodology, Y.W.; Investigation, Y.W. and D.N.; Resources, C.X.; Writing-Original Draft Preparation, Y.W.; Writing, Review and Editing, Y.W., D.N., S.Z. and Z.X.; Supervision, Z.X., Z.F. and C.X.; Funding Acquisition, Y.W., C.X. and Z.X.

Funding: This research received no external funding.

Acknowledgments: The authors would like to express their gratitude for the financial support received from CSIRO Manufacturing. The first author acknowledges the scholarship from China Scholarship Council (CSC 201503170336). Special thanks go to Madeline from Swinburne University, Guang Yang from Victoria University and Yi Wang's good friend, Durga, in CSIRO, for the English grammar proof work and Daneshfar in CSIRO for PSf substrates and XPS data analysis.

Conflicts of Interest: The authors declare no conflict of interest.

References

1. Zhao, S.; Zou, L.; Mulcahy, D. Brackish water desalination by a hybrid forward osmosis–nanofiltration system using divalent draw solute. *Desalination* **2012**, *284*, 175–181. [CrossRef]
2. Choi, Y.-J.; Choi, J.-S.; Oh, H.-J.; Lee, S.; Yang, D.R.; Kim, J.H. Toward a combined system of forward osmosis and reverse osmosis for seawater desalination. *Desalination* **2009**, *247*, 239–246. [CrossRef]
3. Zhao, S.; Zou, L. Effects of working temperature on separation performance, membrane scaling and cleaning in forward osmosis desalination. *Desalination* **2011**, *278*, 157–164. [CrossRef]
4. Korenak, J.; Basu, S.; Balakrishnan, M.; Helix-Nielsen, C.; Petrinic, I. Forward osmosis in wastewater treatment processes. *Acta. Chim. Slov.* **2017**, *64*, 83–94. [CrossRef] [PubMed]

5. Xue, W.; Yamamoto, K.; Tobino, T. Membrane fouling and long-term performance of seawater-driven forward osmosis for enrichment of nutrients in treated municipal wastewater. *J. Membr. Sci.* **2016**, *499*, 555–562. [CrossRef]

6. Wang, X.; Zhao, Y.; Yuan, B.; Wang, Z.; Li, X.; Ren, Y. Comparison of biofouling mechanisms between cellulose triacetate (CTA) and thin-film composite (TFC) polyamide forward osmosis membranes in osmotic membrane bioreactors. *Bioresour. Technol.* **2016**, *202*, 50–58. [CrossRef] [PubMed]

7. Wang, X.; Chang, V.W.C.; Tang, C.Y. Osmotic membrane bioreactor (ombr) technology for wastewater treatment and reclamation: Advances, challenges, and prospects for the future. *J. Membr. Sci.* **2016**, *504*, 113–132. [CrossRef]

8. Sahebi, S.; Phuntsho, S.; Eun Kim, J.; Hong, S.; Kyong Shon, H. Pressure assisted fertiliser drawn osmosis process to enhance final dilution of the fertiliser draw solution beyond osmotic equilibrium. *J. Membr. Sci.* **2015**, *481*, 63–72. [CrossRef]

9. Phuntsho, S.; Shon, H.K.; Hong, S.; Lee, S.; Vigneswaran, S.; Kandasamy, J. Fertiliser drawn forward osmosis desalination: The concept, performance and limitations for fertigation. *Rev. Environ. Sci. Bio/Technol.* **2011**, *11*, 147–168. [CrossRef]

10. Altaee, A.; Sharif, A.; Zaragoza, G.; Ismail, A.F. Evaluation of fo-ro and pro-ro designs for power generation and seawater desalination using impaired water feeds. *Desalination* **2015**, *368*, 27–35. [CrossRef]

11. He, W.; Wang, Y.; Shaheed, M.H. Energy and thermodynamic analysis of power generation using a natural salinity gradient based pressure retarded osmosis process. *Desalination* **2014**, *350*, 86–94. [CrossRef]

12. McCutcheon, J.R.; McGinnis, R.L.; Elimelech, M. Desalination by ammonia–carbon dioxide forward osmosis: Influence of draw and feed solution concentrations on process performance. *J. Membr. Sci.* **2006**, *278*, 114–123. [CrossRef]

13. Faria, A.F.; Liu, C.H.; Xie, M.; Perreault, F.; Nghiem, L.D.; Ma, J.; Elimelech, M. Thin-film composite forward osmosis membranes functionalized with graphene oxide-silver nanocomposites for biofouling control. *J. Membr. Sci.* **2017**, *525*, 146–156. [CrossRef]

14. Shen, L.; Xiong, S.; Wang, Y. Graphene oxide incorporated thin-film composite membranes for forward osmosis applications. *Chem. Eng. Sci.* **2016**, *143*, 194–205. [CrossRef]

15. Jia, Q.; Xu, Y.; Shen, J.; Yang, H.; Zhou, L. Effects of hydrophilic solvent and oxidation resistance post surface treatment on molecular structure and forward osmosis performance of polyamide thin-film composite (TFC) membranes. *Appl. Surf. Sci.* **2015**, *356*, 1105–1116. [CrossRef]

16. Han, G.; Zhang, S.; Li, X.; Widjojo, N.; Chung, T.-S. Thin film composite forward osmosis membranes based on polydopamine modified polysulfone substrates with enhancements in both water flux and salt rejection. *Chem. Eng. Sci.* **2012**, *80*, 219–231. [CrossRef]

17. Salter, R.J. Forward osmosis. *Water Cond. Purif.* **2005**, *48*, 36–38.

18. Qin, J.; Chung, T.S. Effect of dope flow rate on the morphology, separation performance, thermal and mechanical properties of ultrafiltration hollow fibre membranes. *J. Membr. Sci.* **1999**, *157*, 35–51. [CrossRef]

19. Mi, B.; Elimelech, M. Organic fouling of forward osmosis membranes: Fouling reversibility and cleaning without chemical reagents. *J. Membr. Sci.* **2010**, *348*, 337–345. [CrossRef]

20. Sun, J.; Zhu, L.-P.; Wang, Z.-H.; Hu, F.; Zhang, P.-B.; Zhu, B.-K. Improved chlorine resistance of polyamide thin-film composite membranes with a terpolymer coating. *Sep. Purif. Technol.* **2016**, *157*, 112–119. [CrossRef]

21. Do, V.T.; Tang, C.Y.; Reinhard, M.; Leckie, J.O. Degradation of polyamide nanofiltration and reverse osmosis membranes by hypochlorite. *Environ. Sci. Technol.* **2012**, *46*, 852–859. [CrossRef] [PubMed]

22. McCloskey, B.D.; Park, H.B.; Ju, H.; Rowe, B.W.; Miller, D.J.; Chun, B.J.; Kin, K.; Freeman, B.D. Influence of polydopamine deposition conditions on pure water flux and foulant adhesion resistance of reverse osmosis, ultrafiltration, and microfiltration membranes. *Polymer* **2010**, *51*, 3472–3485. [CrossRef]

23. Arena, J.T.; McCloskey, B.; Freeman, B.D.; McCutcheon, J.R. Surface modification of thin film composite membrane support layers with polydopamine: Enabling use of reverse osmosis membranes in pressure retarded osmosis. *J. Membr. Sci.* **2011**, *375*, 55–62. [CrossRef]

24. Huang, Y.; Jin, H.; Li, H.; Yu, P.; Luo, Y. Synthesis and characterization of a polyamide thin film composite membrane based on a polydopamine coated support layer for forward osmosis. *RSC Adv.* **2015**, *5*, 106113–106121. [CrossRef]

25. Guo, H.; Yao, Z.; Wang, J.; Yang, Z.; Ma, X.; Tang, C.Y. Polydopamine coating on a thin film composite forward osmosis membrane for enhanced mass transport and antifouling performance. *J. Membr. Sci.* **2018**, *551*, 234–242. [CrossRef]
26. Zangmeister, R.A.; Morris, T.A.; Tarlov, M.J. Characterization of polydopamine thin films deposited at short times by autoxidation of dopamine. *Langmuir* **2013**, *29*, 8619–8628. [CrossRef] [PubMed]
27. Hong, S.; Na, Y.S.; Choi, S.; Song, I.T.; Kim, W.Y.; Lee, H. Non-covalent self-assembly and covalent polymerization co-contribute to polydopamine formation. *Adv. Funct. Mater.* **2012**, *22*, 4711–4717. [CrossRef]
28. Wang, Y.; Fang, Z.; Zhao, S.; Ng, D.; Zhang, J.; Xie, Z. Dopamine incorporating forward osmosis membranes with enhanced selectivity and antifouling properties. *RSC Adv.* **2018**, *8*, 22469–22481. [CrossRef]
29. Zhao, J.; Su, Y.; He, X.; Zhao, X.; Li, Y.; Zhang, R.; Jiang, Z. Dopamine composite nanofiltration membranes prepared by self-polymerization and interfacial polymerization. *J. Membr. Sci.* **2014**, *465*, 41–48. [CrossRef]
30. Zhao, S.; Zou, L.; Tang, C.Y.; Mulcahy, D. Recent developments in forward osmosis: Opportunities and challenges. *J. Membr. Sci.* **2012**, *396*, 1–21. [CrossRef]
31. She, Q.; Wang, R.; Fane, A.G.; Tang, C.Y. Membrane fouling in osmotically driven membrane processes: A review. *J. Membr. Sci.* **2016**, *499*, 201–233. [CrossRef]
32. Freger, V. Nanoscale heterogeneity of polyamide membranes formed by interfacial polymerization. *Langmuir* **2003**, *19*, 4791–4797. [CrossRef]
33. Dong, L.-X.; Huang, X.-C.; Wang, Z.; Yang, Z.; Wang, X.-M.; Tang, C.Y. A thin-film nanocomposite nanofiltration membrane prepared on a support with in situ embedded zeolite nanoparticles. *Sep. Purif. Technol.* **2016**, *166*, 230–239. [CrossRef]
34. Jahanshahi, M.; Rahimpour, A.; Peyravi, M. Developing thin film composite poly(piperazine-amide) and poly(vinyl-alcohol) nanofiltration membranes. *Desalination* **2010**, *257*, 129–136. [CrossRef]
35. Xi, Z.-Y.; Xu, Y.-Y.; Zhu, L.-P.; Wang, Y.; Zhu, B.-K. A facile method of surface modification for hydrophobic polymer membranes based on the adhesive behavior of poly(dopa) and poly(dopamine). *J. Membr. Sci.* **2009**, *327*, 244–253. [CrossRef]
36. Lee, K.P.; Arnot, T.C.; Mattia, D. A review of reverse osmosis membrane materials for desalination—development to date and future potential. *J. Membr. Sci.* **2011**, *370*, 1–22. [CrossRef]
37. Lee, H.; Dellatore, S.M.; Miller, W.M.; Messersmith, P.B. Mussel-inspired surface chemistry for multifunctional coatings. *Science* **2007**, *318*, 426–430. [CrossRef] [PubMed]
38. Lu, P.; Liang, S.; Zhou, T.; Xue, T.; Mei, X.; Wang, Q. Layered double hydroxide nanoparticle modified forward osmosis membranes via polydopamine immobilization with significantly enhanced chlorine and fouling resistance. *Desalination* **2017**, *421*, 99–109. [CrossRef]
39. Li, B.; Liu, W.; Jiang, Z.; Dong, X.; Wang, B.; Zhong, Y. Ultrathin and stable active layer of dense composite membrane enabled by poly(dopamine). *Langmuir* **2009**, *25*, 7368–7374. [CrossRef] [PubMed]

processes

MDPI

Review

Photocatalytic Membranes in Photocatalytic Membrane Reactors

Pietro Argurio [1,*] **, Enrica Fontananova** [2,*] **, Raffaele Molinari** [1] **and Enrico Drioli** [2,3]

1   Department of Environmental and Chemical Engineering, University of Calabria, Via P. Bucci, 44/A, I-87036 Rende (CS), Italy; raffaele.molinari@unical.it
2   Institute on Membrane Technology (ITM), National Research Council of Italy (CNR), Via P. Bucci Cubo 17C, I-87036 Rende (CS), Italy; e.drioli@itm.cnr.it
3   Center of Excellence in Desalination Technology, King Abdulaziz University, P.O. Box 80200, Jeddah 21589, Saudi Arabia
*   Correspondence: pietro.argurio@unical.it (P.A.); e.fontananova@itm.cnr.it (E.F.)

Received: 2 August 2018; Accepted: 5 September 2018; Published: 7 September 2018

Abstract: The present work gives a critical overview of the recent progresses and new perspectives in the field of photocatalytic membranes (PMs) in photocatalytic membrane reactors (PMRs), thus highlighting the main advantages and the still existing limitations for large scale applications in the perspective of a sustainable growth. The classification of the PMRs is mainly based on the location of the photocatalyst with respect to the membranes and distinguished in: (i) PMRs with photocatalyst solubilized or suspended in solution and (ii) PMRs with photocatalyst immobilized in/on a membrane (i.e., a PM). The main factors affecting the two types of PMRs are deeply discussed. A multidisciplinary approach for the progress of research in PMs and PMRs is presented starting from selected case studies. A special attention is dedicated to PMRs employing dispersed TiO_2 confined in the reactor by a membrane for wastewater treatment. Moreover, the design and development of efficient photocatalytic membranes by the heterogenization of polyoxometalates in/on polymeric membranes is discussed for applications in environmental friendly advanced oxidation processes and fine chemical synthesis.

Keywords: photocatalytic membrane reactors; photocatalytic membrane; fine chemistry; wastewater treatment

1. Introduction

Membrane reactors (MRs) are multifunctional reactors that are characterized by the combination of a (catalytic) reaction with a membrane separation processes. The early examples of MRs date back to the 1970s with the use of polymeric membranes in enzymatic reactions and metal membranes for high temperature reactions [1]. Currently, MRs have relevant industrial applications in biotechnological field but also in the environment areas, such as wastewater treatment and gas emissions purification [2,3]. An interesting case is the membrane bioreactors (MBRs) technology recognized as a best available technology (BAT) for wastewater treatments [4]. However, in the last decades, a growing interest towards MRs technology is evident in a wide range of applications, including, not only biotechnological, but also petrochemical, chemical production, environmental remediation, and energy sector. Regarding the environment protection, MRs have been applied in different areas such as wastewater treatment by advanced oxidation processes (AOPs) for organic pollutants degradation and gas emissions purification for CO_2 capture [3]. The main driving force of these academic and industrial research efforts is the necessity to innovate the current resource within an intensive industrial system in the perspective of the sustainable growth.

The Process Intensification (PI) strategy [5,6] is recognized as an efficient tool for the realization of this sustainable growth. The PI strategy comprises an innovative equipment design and process development methods, being able to improve manufacturing and processing, by decreasing production costs, equipment size, energy consumption, waste generation, while improving process efficiency, remote control, information fluxes, and process flexibility [7]. MRs are specific example of reactive separations well responding to the requests of the PI strategy, coupling a reaction with a membrane separation process, not only at equipment level, but by realizing functional synergies between the operations involved [6]. This synergic combination can have several advantages in comparison with traditional reactors depending on the specific functions performed by the membrane [8]. With respect to traditional reactive separations (e.g., reactive distillation, reactive adsorption, reactive crystallization/precipitation), MRs have the advantage to use intrinsically more clean and energy-efficient separation routes [6]. Membrane separations are in fact typically characterized by lower operating temperature, in comparison with thermal separation processes, such as distillation, and they might offer a solution in the case of catalysts or products with a limited thermal stability. Additionally, membrane separation processes are able to separate nonvolatile components by a difference in dimensions, charge, or volatility.

The selective transport of the products and/or the reagents through the membrane can increase the yield and/or the selectivity of some processes. Typical examples are esterification and de-hydrogenation reactions (thermodynamically controlled reactions), in which the removal of water or hydrogen, respectively, increases the reaction yield. The extraction of an instable intermediate through the membrane in a MR might give an improvement of the reaction selectivity; this is the case of partial oxidation reactions and hydrogenation reactions carried out in MRs; also, the selective removal of the reaction products might also improve the downstream processing.

In the case of homogenous catalysts, their immobilization in or on a membrane represents an efficient way to achieve the catalyst recovery, regeneration, and reuse in successive catalytic runs [9].

The membrane can also define the reaction volume providing a contacting zone for two immiscible phases (e.g., in phase transfer catalysis) [10] avoiding the use of polluting auxiliaries, like solvents, in agreement with the green chemistry principles.

The design of the membrane reactor requires a multidisciplinary approach in which different disciplines, like: chemistry, chemical engineering, membrane engineering, and process engineering, give their contribution in order to achieve a synergic combination of the separation and reaction processes that allows optimal performances in terms of productivity and sustainability.

When a membrane is combined with a photocatalytic process the MR is indicated as photocatalytic membrane reactor (PMR) [11] and it can be designed in two main configurations, depending from the catalyst confinement [12–14]: (i) PMRs with solubilized or suspended photocatalyst; and, (ii) PMRs with photocatalyst immobilized in/on a membrane.

Both configurations present specific advantages and limitations depending from the specific application. In a PMR with a solubilized or suspended photocatalyst, a membrane with appropriate molecular weight cut off (MWCO) can be used for the retention of the catalyst in the reactor. Usually globular proteins in aqueous solution are used for MWCO measurements, however in the membrane selection it is necessary to consider that the MWCO depends on the solvent and the solution composition and polarity might change in relevant way during the reaction [10]. The catalyst retention can be optimized by enlargement of the catalyst as: dendrimers, hyperbrached polymers or catalyst bound to a soluble polymer [15]. Moreover, the membrane can have the multiple role of confining, the catalyst, the pollutants, and the degradation intermediates into the reaction environment. In a traditional photoreactor, the degradation intermediates remain in the treated effluent resulting often in a not sufficiently efficient process because these intermediates are frequently more hazardous than the original pollutants. The coupling of photocatalysis and membrane separation can permit to obtain, by using a membrane with an appropriate selectivity, a permeate free of both pollutants and degradation intermediates.

PMRs with solubilized/suspended photocatalyst can be further classified in: (i) integrative-type PMRs and (ii) split type PMRs. In the first one, the photocatalytic reaction and membrane separation processes take place in one apparatus, i.e., an inorganic or polymeric membrane is submerged in the slurry photocatalytic reactor. In split-type PMRs the photocatalytic reaction and membrane separation take place into two separate apparatuses, i.e., the photocatalysis module and the membrane module, which are appropriately coupled [12,16]. Besides, on the basis of a different approach, this kind of PMRs can be also classified while considering the position of the light source, which can be: (i) above/inside the feed tank; (ii) above/inside the membrane unit; and, (iii) above/inside to an additional vessel placed between the feed tank and the membrane unit [13].

In PMRs with immobilized photocatalyst, the photocatalyst is not solubilized/suspended into the reacting environment, but it is immobilized in/on the membrane, giving a photocatalytic membrane (PM). In such a system, the membrane acts both as a selective barrier for the contaminants to be degraded, thus maintaining them into the reaction environment, and as the support for the photocatalyst. PMRs with immobilized photocatalyst are intrinsically integrative-type PMRs, i.e., the photocatalytic and the membrane separation processes take place in the same unit.

In general, in the state of the art literature it has been reported that the PMRs with solubilized/suspended photocatalyst permit to achieve higher efficiency when compared to that of the immobilized system, due to the larger active surface area, which guarantees a good contact between the photocatalyst and the pollutants [12,14]. As a consequence, more numerous are the studies of PMRs with solubilized/suspended photocatalyst, in view of large scale applications [17–20]. However, fouling, which is caused by deposition of the photocatalyst nanoparticles (NPs) on the membrane surface with a consequent flux decline, and light scattering, still limits the performance of this type of membrane photoreactor configuration.

In the immobilized PMRs, the photocatalyst/reagents contact is hindered by the mass transfer limitation over the immobilized photocatalyst. However, in this configuration, it is possible to obtain in general a more easy catalyst recovery, reuse and regeneration in successive catalytic runs than in the previous one.

Besides, PMs generally show better performance with respect to conventional membranes in terms of reduced membrane fouling and improved permeate quality. PMs could degrade the organic pollutants contained into the feed solutions by generating the oxygen-reactive radicals under light irradiation, thereby preventing the formation of a cake layer on the membrane surface, reducing the pollutant concentration in the retentate as well as improving the permeate quality. Moreover, the problem of light scattering by photocatalyst particles, which negatively influence the performance of PMRs with suspended photocatalyst, is avoided by using PMs. So, it can be affirmed that the use of PMs might improve the performance of wastewater treatment by PMRs.

Significant progresses have been also reached in the modelling and simulation of membrane reactors with the aim to improve their performances by the optimization of the reactor design and the operative conditions, as well as, by the understanding and optimization of structural/functional relationships at a molecular level in the systems investigated. New metrics, like the volume index and conversion index, have also been proposed as a simple way to evaluate the efficiency of a membrane reactor in the logic of the process intensification strategy [21].

The present work gives a critical overview of the recent progresses and new perspectives in photocatalytic membranes and photocatalytic membrane reactors. The most important operating parameters affecting the performance of the PMRs are deeply discussed. Then, based on some selected case studies, the criteria for designing and developing a PM are reported. Finally, recent achievements in the use of visible light and in the overcoming of some intrinsic limitations of PMs, like moderate loss of photoactivity and restricted processing capacities, owing to mass transfer limitations and unsatisfactory system lifetime, due to photocatalyst leaching, are overviewed.

2. Operating Parameters and Limits of Photocatalytic Membranes (PMs)

Working under proper operating conditions is important to obtain good PMR performance with both solubilized/suspended and immobilized photocatalyst. The most important factors affecting the PMR performance, influencing the photocatalytic process and/or the membrane process, were reviewed by various authors [12–14]. These factors can be summarized, as follows.

2.1. Operating Mode

PMRs with immobilized photocatalyst can be operated in both dead-end and cross flow modes [22,23]. When the PMR is operated in dead-end mode, the substrates are retained by the membrane and accumulate on its surface thus forming a cake layer. As a result, the membrane permeability and the photocatalytic performance are reduced. Relatively low photocatalytic efficiency is usually observed by operating under dead end mode. This drawback can be explained, as follows: since the feed polluted water was kept into the reactor (no recirculation), a not adequate contact between the pollutants to be degraded, the photocatalyst and the light source was achieved.

On the basis of this, it can be affirmed that the cross flow mode should be the better option in view of industrial applications. By operating in this way, the feed phase flows tangentially to the membrane. This tangential flow tends to remove the deposited particles on the membrane surface, thus resulting in a reduced membrane fouling. The permeate flows perpendicularly across the membrane, while the retentate is usually recirculated in the feed tank.

2.2. Typology of Photocatalyst Immobilization

On the basis of the different procedure of photocatalyst immobilization, PMRs with immobilized photocatalyst can be further classified in three sub-categories.

The first sub-category comprises the PMRs in which the photocatalyst is coated on the PM. The procedure of coating can be performed by using different methods [24] such as dip-coating, electro-spraying of photocatalyst particles, magnetron sputtering of the photocatalyst, and deposition of gas phase photocatalyst NPs. The main drawbacks of using this approach are associated to the diminution of membrane permeability after the coating procedures and to photocatalyst leaching during the photocatalytic runs [25].

The second sub-category includes the PMRs in which the photocatalyst is blended into the membrane matrix. Practically in this kind of PMR, during the membrane preparation the photocatalyst is entrapped in the polymeric matrix. By operating with this type of PMRs, the possibility of photocatalyst leaching was reduced with respect to the PM that was prepared by photocatalyst coating.

The third sub-category of PMRs with immobilized photocatalyst is based on the use of free standing PM. In this case, the PM is manufactured with a pure photocatalyst, so that the immobilization of the photocatalyst in/on the membrane support is unnecessary. Thus, a further reduction of the possibility of photocatalyst leaching was obtained.

2.3. Photocatalyst Type and Its Characteristics

Photocatalyst is the key component of a photocatalytic process. The properties of photocatalyst and its concentration in the reacting environment represent key parameters, playing an important role on photocatalytic performance.

Key properties of photocatalyst, having significant effects on photocatalytic efficiency, are the band gap energy, the specific surface area and particle size distribution, the crystallographic structure and composition, etc.

A photocatalyst is a semiconductor material able to convert the energy of irradiated photons in the chemical necessary to excite electrons from its valence band (VB) to its conduction band (CB). Then, it is clear that band gap energy plays the most important role in its selection. A photocatalyst characterized by lower band gap requires less photon energy. In particular, a photocatalyst can achieve a satisfactory visible light response when its band gap is sufficiently low, which is an important characteristic in view of large scale applications. Other important characteristics to be considered to choose a photocatalyst are good chemical and physical stability, non-toxic nature, availability, and low cost.

TiO$_2$-based photocatalyst is the most utilized photocatalyst in PMR (used in suspended form) because it is characterized by strong catalytic activity, relatively long lifetime of electron-hole pairs, high (photo)chemical stability in aqueous media and in a wide range of pH (0–14), limited cost, and low toxicity. So, it can be affirmed that TiO$_2$ is the archetypical photocatalyst, virtual synonym for photocatalysis. However, this material is not active under visible light irradiation, because of its wide band gap. Thus, TiO$_2$ is capable to use only less than about 5% of the solar energy, which is in the UV range. On these bases, the development of photocatalysts that are able to utilize visible light represents a challenge in view of large-scale application of PMR systems, permitting the use of a greener light source (the sun) [26,27]. When considering that the quantum efficiency of the photocatalytic processes usually decreases by increasing the intensity of the radiation, the preparation of photocatalysts active under visible light is convenient only when the quantum efficiency remains relatively high, resulting in a wide use of photons from visible light.

Other metal oxide semiconductors that are used in PMRs are ZnO [28,29], WO$_3$ [30], and CuO/ZnO [31]. These photocatalysts are characterized by a different band gap in comparison with TiO$_2$ [14].

Polyoxometalates (POMs) are instead soluble metal oxide analogs with tunable solubility depending from the counterion used [32]. The heterogenization of POMs in/on polymeric membranes for PMRs application is reviewed in Section 4.2.

2.4. Ligth Source

The starting step in a photocatalytic process is the irradiation of the photocatalyst surface with photons having energy (hv) equal to or higher than the band gap, so that the electrons are promoted from VB to CB. Several authors studied the influence of wavelength and intensity of the light source on the rate of the photocatalytic reaction [33–35].

The photocatalytic performance depends strongly on the value of the light intensity [36]. In general, it can be observed that the reaction rate increases by increasing the light intensity till to the mass transfer limit is achieved. In general, the reaction rate increases with the light intensity when the photocatalytic process is operated at low light intensity, which means a light intensity in the range 0–20 mW·cm^{-2}. This trend can be explained when considering that, by operating with this light intensity, the reaction giving the formation of electron-hole couples is predominant with respect to the electron-hole recombination. By operating with a light source emitting at medium light intensity (approximately 25 mW·cm^{-2}), the reaction rate is function of the square root of light intensity. This trend can be ascribed to the competition effect between the electron-hole formation and recombination reactions. The reaction rate is not influenced by this parameter when the photocatalytic process is performed with a high light intensity (>25 mW·cm^{-2}) [12,13].

The UV spectrum is usually divided into UV-A (315–400 nm), UV-B (280–315 nm), and UV-C (100–280 nm), depending on the emitting wavelength. Artificial UV lamps are the most commonly used light sources, because of their higher photon flux with respect to solar irradiation. These lamps can be used in immersed or external configuration, and emit light in the UV-A range (max intensity in the range 355–365 nm) or in the UV-C range (germicidal lamps, max intensity at 254 nm).

Depending on the emission range of interest, different typologies of light sources are available, like xenon, mercury, and deuterium lamps [37]. Besides, regarding the different radiation sources, it can be distinguished between pulsed and continuous light sources [38], which can affect the photocatalytic performance.

Several works on laser induced photocatalysis have been also recently carried out [38–40]. Yamani [38], compared the photocatalytic performance of 355 nm pulsed laser and continuous Hg arc lamp equipped with visible light band pass filter. The results evidenced that the photocatalytic degradation efficiencies that were obtained with the different radiation sources were comparable and the pulse energy has an important influence on the photocatalytic performances.

Among the various types of lamps, the led ones, emitting in the UV or UV-vis range, are gaining interest for their efficiency and they can be also powered by photovoltaic panels [41].

2.5. Feed Characteristics

The properties of feed that influence the performance of a PMR with entrapped photocatalyst comprise the initial concentration of the pollutants to be degraded, the pH of the feed, the temperature, and the presence and concentration of inorganic ions.

It is generally accepted that the reaction rate increases with substrate concentration till to a certain level, over which the rate starts to decrease. This trend can be explained considering that a further increase of substrate concentration results in a reduction of the generation rate of •OH radicals due to light scattering by the dissolved substrate and to the decrease of the active sites available for •OH radicals generation (photocatalyst saturation).

In their study on photocatalytic degradation of the dye tartrazine, Aoudjit et al. [42] evidenced that the dye concentration in the feed phase to the PMR system strongly affects system performance. Indeed, the photodegradation percentage decreased from ca. 78 to 46% by increasing the initial concentration of the dye from 10 to 30 mg·L^{-1}. This trend was explained by considering that, maintaining the amount of the photocatalyst constant, the ratio hydroxyl radical/tartrazine molecules decreases by increasing dye concentration, resulting in the lower photodegradation efficiency observed. Moreover, the higher tartrazine concentrations also increase the light scattering, thus affecting the UV light absorbance by TiO$_2$ nanoparticles surface and decreasing the amount of hydroxyl radicals formed.

The increase of the initial substrate concentration influences not only the photocatalytic performance of the system, but also the fouling of the membrane, and thus the permeate flux. So, also when the increase of pollutants concentration has a positive effect on the photodegradation rate, it has to be taken into account that the mass transfer resistance of the membrane also increases, resulting in the deterioration of membrane performance [43].

The operating pH strongly influences the performance of a PM. First of all, the adsorption of the substrate on the active sites of the photocatalyst and desorption of the products are two pH-influenced processes. In the case of the TiO$_2$-Degussa photocatalyst, considering that its isoelectric point is 6.8, it can be affirmed that a TiO$_2$ based PM is negatively charged when the pH of the feed solution is higher than 6.8. In this condition, the adsorption of positively charged molecules is favored. On the contrary, the TiO$_2$ based PM is positively charged when the pH of the feed phase is lower than 6.8, and the adsorption of negatively charged substrates is favored. The operating pH also influences the permeation of the different solutes contained into the feed solution across the membrane. Indeed, the molecules dissolved in the reacting environment could change their state from positively charged to neutral, and then from neutral to negatively charged by changing the pH, having an important effect on their permeation across the membrane.

It is accepted that a temperature in the range 20–80 °C is optimal for conducting photocatalytic processes [33]. By operating at temperature lower than 0 °C, the desorption of the photodegradation products from the surface of the photocatalyst becomes the limiting step, and the photocatalytic activity decreases. On the other hand, at a temperature higher than 80 °C, the electron-hole recombination on the photocatalyst is favored, which has a negative effect on photocatalytic efficiency. Besides, at those higher temperatures the adsorption of the substrate (exothermic process) is inhibited.

The presence of inorganic ions in the feed solution can give positive or negative effects on the rate of the photocatalytic process depending on the reaction mechanism and its nature. In particular, the presence of inorganic ions as Cl$^-$, NO$_3^-$, SO$_4^{2-}$, CO$_3^{2-}$ and HCO$_3^-$, which usually exist in waters, decreases photocatalytic activity by scavenging holes (h$_{CB}$) and hydroxyl radical (•OH) [44–46]. On the other hand, the addition of oxyanion oxidants, such as S$_2$O$_8^{2-}$, BrO$_3^-$, IO$_4^-$, ClO^{2-}, and ClO^{3-}, has a positive effect on the photoreactivity. Indeed, these ions could act as scavenger for the CB electrons and reduce the electron-hole recombination, as follows:

$$S_2O_8^{2-} + e^- \rightarrow SO_4^{2-} + SO_4^{\bullet-} \tag{1}$$

$$SO_4^{\bullet-} + H_2O \rightarrow SO_4^{2-} + \bullet OH + H^+ \tag{2}$$

H_2O_2 is another oxidant species that increases the photoreactivity of the system, thanks to the formation of •OH radicals. Nevertheless, as reported by some authors [47] an excessive amount of hydrogen peroxide can have a negative effect on system reactivity. This behavior can be explained considering that H_2O_2 acts as scavenger of VB holes and •OH, producing hydroperoxyl radicals $(HO_2\bullet)$ as follows:

$$H_2O_2 + 2h^+ \rightarrow O_2 + 2H^+ \tag{3}$$

$$H_2O_2 + \bullet OH \rightarrow H_2O + HO_2\bullet \tag{4}$$

The $HO_2\bullet$ radical has a less oxidizing power than •OH.

The presence of inorganic ions into the feed solution, affecting the photocatalytic performance as described, can also influence the performance of the membrane separation process. The influence of inorganic salts and humic acids on fouling and stability of a polyethersulfone ultrafiltration (UF) membrane operated in a PMR was studied by Dorowna et al. [48]. A more significant influence of inorganic ions on membrane performance was observed by increasing the concentration of the salts.

2.6. Flow Rate over and across the Membrane

In the case of PMRs that operated under cross-flow filtration mode, where the feed solution is continuously recirculated tangentially to the membrane, the recirculation flow rate is an important operating parameter affecting the photocatalytic performance of the system. Usually, the photocatalytic efficiency increases by increasing the recirculation flow rate. This trend can be explained by considering the larger turbulence in the solution, which promotes the mass transfer from the bulk of the feed solution to the surface of the PM, while reducing the membrane fouling.

Flow rate across the membrane depend on the applied driving force (e.g., transmembrane pressure in a pressure driven membrane separation), membrane structure, and composition. It is a key parameter for the PMR because it determines the contact times between the photocatalyst and the reagents/products. Mass transfer of the reagents to the catalytic sites, and of the product away from them, should be fast enough in order to avoid reaction limitation meanwhile the contact time catalyst/reagent should be also appropriate to control reaction selectivity.

3. Preparation and Choice of Materials to Manufacture PMs

An "ideal" (photo)catalyst should conjugate the following properties: low costs, high selectivity, elevated stability under reaction conditions, non-toxicity and "green properties", like "recoverability", i.e., the possibility to recover and reuse the catalyst [49,50]. In this perspective, the heterogenization of homogeneous catalysts in/on a support has interesting implications because it permits to recover and reuse the same catalytic system several times. Among the different heterogenization techniques, the entrapping of a catalyst in or on a membrane offers new possibility for the design of new efficient catalytic processes.

In the design of a catalytic membrane major issues in the membrane material selection are the mechanical, thermal and chemical stability under reaction conditions. In the case of photocatalytic membranes also the photostability and transparency of the membrane are critical concerns. Moreover, the membrane transport properties, like permeability and selectivity, need to be considered and can be tailored by an appropriate selection of the membrane structure and composition. In the case of a (photo)catalytic membrane, the incorporation of the (photo)catalyst introduces additional complexity to the fabrication process, because the catalyst should be firmly entrapped in the membrane and the catalyst structure and activity has to be preserved during the membrane preparation procedure. Moreover, the properties (including transport properties) of a catalytic membrane are usually different from those of a pristine polymeric membrane that is prepared in the same experimental conditions.

A variety of materials have been used for the preparation of PMs: organic, inorganic, and metallic. In particular, polymeric membranes represent the most popular organic materials used to prepare PMs. Typical polymers that have been considered as membrane material include polyamide [51], polyvinylidene fluoride (PVDF) [52], polyethersulfone (PES) [53], polyurethane (PU) [54], polyethylene terephthalate (PET) [55], polyacrylonitrile (PAN) [56], and polytetrafluoroethylene (PTFE) [57].

Since the PM is in direct contact with the reacting environment, the possibility of its damage by the irradiation and oxidizing species (e.g., •OH radical) has to be taken into account. More often, the abrasion of membrane surface because of the mechanical action of the particles suspended in the reacting environments causes more damage to the membrane than the action of the oxidant environment [58]. On the basis of this, inorganic ceramic membranes, thanks to their excellent thermal, chemical, and mechanical stability, may represent a good option to polymeric membranes in PM preparation. Actually, the main limitation to ceramic membrane utilization is represented by the higher manufacturing cost when compared to polymeric membrane.

It is important to realize a stable catalyst immobilization in order to avoid catalyst leaching out from the membrane and to have a good adhesion between the membrane material and the catalyst, with an optimal dispersion of the second one. Different immobilization strategies can be used in order to achieve these goals. Covalent bindings in general guarantee a stable immobilization. However, also electrostatic interactions, weak interactions (Van der Waals or hydrogen bonds), or catalyst encapsulation are exploited for an efficient catalyst immobilization [59].

In several cases, the catalyst is immobilized during membrane formation processes by dispersing the catalyst in the polymer dope solution. The polymer/catalyst affinity can be tailored by an appropriate chemical functionalization of one or both components. Ideally, the solvent used for the reaction needs to be a non-solvent for the catalyst, and it is necessary that the membrane does not swell excessively.

For a porous membrane, the choice of the membrane material is of less importance for transport properties in comparison with a dense membrane, because permeation occurs through the membrane pores [10]. However, the membrane material is relevant for the membrane stability, as well as for wettability and fouling tendency.

The membrane preparation technique and conditions depend from the membrane material and the desired structure and morphology. Different techniques for membrane preparation can be opportunely employed to prepare catalytic membranes: phase separation, coating, interfacial polymerization, sintering, stretching, and track-etching, are some examples [60]. The most used and versatile for polymeric membranes is the phase separation technique.

3.1. Dip Coating with Photocatalyst Particles

A widely employed surface immobilization technique consists in dip coating. Wang et al. [61] studied the synthesis of high quality TiO_2 membranes on alumina supports and tested the so prepared PMs in the degradation of the dye Methyl Orange (MO). In particular, the PMs were synthesized by the spin-coating technique using the sol-gel method with water-soluble chitosan (WSC) as an additive. Obtained results evidenced that the proposed method of preparation permitted obtaining TiO_2 membranes on alumina supports that were characterized by enhanced structural and catalytic properties. High surface area (116–164 $m^2 \cdot g^{-1}$) and porosity (47.3%–52.2%) were obtained, as well as homogeneity without cracks and high degradation of MO (61.2%–49.2%).

Chakraborty et al. [62] immobilized TiO_2 NPs on polymeric hollow fibers (HFs) UF membranes while using three immobilization methods: (i) spray coating, (ii) vacuum coating, and (iii) sol-gel dip coating. PES and polyvinylchloride (PVC)-PAN HF membranes were used as catalyst support.

In the case of spray and vacuum coating, a significant leaching/detachment of NPs has been observed. This result was explained by the authors while considering that the NPs were immobilized physically and not chemically. Thus, a not stable coating on the membrane surface was obtained. Besides, spray coated and vacuum coated PMs showed negligible permeability due to the clustering of TiO_2 adding extra resistance layers to water permeation across the membrane and to the blocking of the membrane pores by TiO_2 agglomerates, respectively. This was also confirmed by Scanning Electron Microscopy (SEM) images.

The sol-gel coating was tested as the third method of photocatalyst immobilization on the membrane surface. In this case the coating solution was a diluted sol, coated on the membrane by using a simple lab-scale dip coater. The coating method was optimized in order to obtain a better immobilization of TiO_2 NPs on the membrane while limiting the permeability loss in the coated membrane. SEM images and Raman spectra confirmed TiO_2 immobilization on the polymeric support and evidenced that sol-gel coating permitted obtaining a stable TiO_2 layer. The pure water permeability of the coated membranes was about 65–80% below those of uncoated membranes for both PES and PVC-PAN.

3.2. Electrosprying of Photocatalyst Particles

Daels et al. [51] prepared polyamide nanofiber membranes containing TiO_2 NPs by electrospinning. The photocatalytic activity of the prepared PMs was demonstrated by considering the photodegradation of the dye MB. The results of photodegradation (up to 84% after 2 h of UV irradiation) evidenced the excellent photocatalytic activity of TiO_2 NPs immobilized on the nanofibrous structure. After 6 h of UV irradiation the photocatalytic degradation increased up to 99%, thus obtaining a quantitative removal of MB. The stability of the PMs was confirmed, demonstrating that the NPs of the photocatalyst were successfully incorporated onto the electrospun fibers.

By using a similar approach, Nor et al. [63] prepared electrospun PVDF/titanium dioxide nanofibers (PVDF/TNF) by hot pressing TNF onto PVDF flat sheet membrane. The membranes were prepared by three consecutive steps. First of all, the polymeric support, i.e., the flat sheet PVDF membrane was prepared by a slight modified diffusion phase inversion technique. Then, the TiO_2-NFs were synthesized by the electrospinning technique at room temperature. The spun TiO_2-NFs were collected on an aluminum foil. Finally, the TiO_2-NFs were placed and hot pressed for 30 min onto the PVDF membrane. The hot press technique was carried out at constant pressure (80 bar) by varying the operating temperatures (60 °C, 100 °C, 160 °C, and 180 °C). The membranes that were hot pressed at 60 °C exhibited unfavorable structure, since the TiO_2-NFs were not adhered completely.

3.3. Sputtering of Photocatalyst Particles

Another possible technique for manufacturing PMs with the photocatalyst coated on the solid support is represented by sputtering a titanium film on top of the membrane [64]. By using this approach, Fischer et al. [65] easily synthesized TiO_2 nanotubes by sputtering a titanium film on top of a porous polymer material, namely a PES membrane. The sputtering step was followed by the anodization of titanium to TiO_2 nanotubes and the subsequent crystallization to photocatalytically active anatase under mild conditions (temperatures below 120 °C). The so-prepared PMs were characterized by a photocatalytic activity increased up to six times when compared to a TiO_2 film on the membrane. This result was ascribed to the enhanced surface area and light-harvesting capability of the anatase nanotubular structures on the membrane.

3.4. Deposition of Gas Phase Photocatalyst Nanoparticles

PMs coated with nanostructured thin films of TiO_2 and Pt/TiO_2 have been also produced by one-step deposition of gas phase NPs on glass fiber filters [66]. The obtained PMs, coated with Pt/TiO_2 nanostructured films, were tested in the photocatalytic hydrogen production by ethanol photo-steam reforming. An increased reaction rate with respect to that obtained with the same photoactive Pt/TiO$_2$ films deposited by using more conventional methods on a quartz substrate was observed. This enhanced H$_2$ production rate was ascribed to the increased contact time of the reactant mixture with the photoactive films, thanks to the diffusion through the coated glass fiber filters.

The described methods permit obtaining coated PMs, where the photocatalyst is coated tightly on the surface of the support membrane. The main limitations of these photocatalytic systems consist in photocatalyst leaching and in the diminution of membrane permeability because of the increased membrane resistance.

3.5. Blended and Free-Standing PMs

In the case of blended PMs, the photocatalyst is blended into the membrane matrix. By operating in this way, the possibility of photocatalyst leaching is reduced when compared to coated PMs. The most used membranes for the photocatalyst entrapment were PVDF polymeric membranes [67,68]. Other polymeric membranes were also used [69–71]. Aoudjit et al. [42] prepared poly(vinylidenefluoride–trifluoroethylene) (P(VDF–TrFE)) PMs containing titanium dioxide P25 by solvent evaporation (see Figure 1). The analysis of the PM morphological properties evidenced an interconnected porous microstructure and the homogeneous distribution of the TiO$_2$ NPs within the membrane pores.

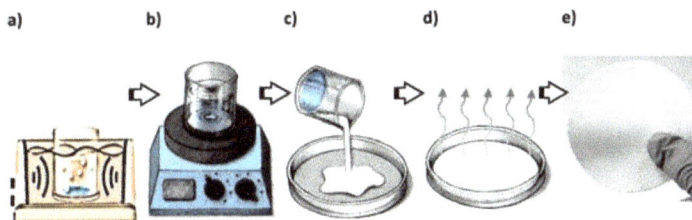

Figure 1. Schematic representation of the preparation of TiO$_2$/poly(vinylidenefluoride –trifluoroethylene) (TiO$_2$/PVDF-TrFE) nanocomposite photocatalytic membranes (PMs) by solvent casting; (**a**) ultrasonic bath of *N,N*-dimethylformamide (DMF) and TiO$_2$ nanoparticles (NPs); (**b**) magnetic stirring of DMF, TiO$_2$ NPs, and PVDF-TrFE; (**c**) pouring of the solution on a glass support; (**d**) solvent evaporation at room temperature; and, (**e**) membrane after complete evaporation of the solvent (Aoudjit et al. [42]).

The production of free-standing PMs, in which the PM is manufactured with a pure photocatalyst, is often performed via the electrochemical anodization of a metallic substrate of the photocatalyst, followed by the separation of the photocatalyst nanotube film and different annealing treatments. Compared to coated or blended PMs, in the case of free-standing PMs the immobilization step is unnecessary, thus further reducing the possibility of photocatalyst leaching. On this topic, Zhao et al. [72] designed and prepared crystallized doubly open-ended Ag/TiNT (NT: Nano Tube) array membranes. The PMs were prepared by a lift-off process. It was based on an anodization-annealing -anodization-etching sequence, followed by uniform Ag NPs decoration. The prepared PMs were tested in the photocatalytic degradation of gaseous *N,N*-dimethylformamide (DMF), which is a typical volatile organic compound contained in the gaseous effluents from manufacturing factories, obtaining very interesting results.

As reported, in view of efficient utilization of PMs, it is fundamental to prepare systems with appropriate stability, porosity, and durable immobilization of the catalyst particles. When considering these features, the implementation and use of fiber based membranes represent an interesting and promising approach in view of enhancing the photocatalytic activity of PMs [73].

4. Selected Case Studies

4.1. Ti- and Ag-Based Photocatalytic Membrane Reactors

The pioneering studies on the use of TiO_2-based membranes in photocatalytic membrane reactors were performed by Anderson and coworkers [74,75] about 30 years ago.

Molinari et al. [76] in 2002 entrapped TiO_2 photocatalyst in a polymeric membrane and compared the performance of the PM with that one obtained by using the same photocatalyst in suspension. The used light source was a medium pressure Hg lamp emitting in the range between 300 and 400 nm with a maximum emission peak of 366 nm. The photocatalytic efficiency of the entrapped catalyst was lower with respect to that of the suspended configuration. The presence of the polymeric chain around the particles of catalyst was the main cause of this trend, since it reduces the surface area of the photocatalyst that is available for light irradiation and substrate adsorption. The attack of •OH radicals to the PM represented another disadvantage of this system in which the photocatalyst was entrapped in polymeric membranes. The comparison of the results that were obtained by degrading different pollutants evidenced that the photodegradation rate was influenced by the presence of the reaction intermediate into the reacting environment. These molecules competed with the photocatalyst for light adsorption and with the substrate for catalyst adsorption.

Some important features of inorganic membranes are their higher chemical, thermal, and mechanical stability, with respect to the polymeric materials. Starting from this consideration, Zhang et al. [77] prepared, by using the sol-gel technique, TiO_2/Al_2O_3 PMs and tested them in the degradation of the azo-dye Direct Black 168 (DB). A high-pressure mercury lamp, with a maximum emission at 365 nm, was used as radiation source. The photolytic degradation of the dye achieved 70% and dye retention of the membrane was 65%, after 300 min by operating in continuous. The removal efficiency of the dye was improved markedly by coupling the UV irradiation with membrane filtration, and reached 82% after 300 min of UV irradiation, evidencing a synergistic effect. As emphasized by the authors, these TiO_2/Al_2O_3 composite PMs permitted to perform at the same time the photocatalytic degradation and the separation, obtaining a high permeate flux (82 L·m^{-2}·h^{-1}) under low pressure (0.5 bar), while preventing membrane fouling, thanks to the high porosity and pore size of the PM. Despite these important achievements, the permeate quality was not excellent, evidencing that the membrane possessed an inadequate rejection to the pollutant.

The possibility to use a green light source, like the sun, to activate the photocatalyst represents a fundamental step in view of applying PMR systems at large scale. Thus, it is of great importance to develop immobilized photocatalytic systems with high quantum efficiency under visible light. In this direction it has to be taken into consideration that the increase of the radiation intensity usually causes a decrease of the quantum efficiency of the photocatalytic processes, which means a narrow use of photon from visible light.

On this topic, Athanasekou et al. [26] prepared and tested three kinds of active photocatalytic ceramic UF membranes. These PMs were prepared by dip-coating on the internal and external surface of UF mono-channel monoliths three different TiO_2 bases photocatalyst: nitrogen doped TiO_2 (N-TiO_2), graphene oxide doped TiO_2 (GO-TiO_2) and organic shell layered TiO_2. The photocatalytic activity of these PMs in the degradation of two azo dyes, MB and MO, was evaluated in a photocatalytic device operating in continuous dead-end conditions. Both the internal and the external side of the PMs were irradiated. The external side (shell) of the PMs was irradiated with two different radiation sources: (i) near-UV radiation in the range 315–380 nm, with a maximum emission at 365 nm; and, (ii) visible light. 10 UVA miniature LEDs emitting in the range 360–420 nm or six visible LEDs emitting at 460 nm were used to irradiate the internal side (lumen) of the PMs. The results evidenced that all the PMs were effective against common problems encountered in dye removal by conventional membrane filtration, like fouling tendency, increased energy consumption and the formation of retentate effluents containing the pollutant at high concentration. Best dye degradations (57% and 27% for MB and MO) were obtained by using the N-TiO_2 PM under UV irradiation. These unsatisfactory degradations of the dyes gave evidence that the used PMs are not adequate for rejecting the dyes, despite the photocatalyst deposition. Lower dye degradations (29% and 15% for MB and MO, respectively) were obtained by using the same PM under visible light, despite the higher light irradiation density (7.2 vs. 2.1 mW·cm^{-2}). This result evidenced that the N-TiO$_2$ PM did not achieve efficient quantum efficiency under visible light. By considering the cost of the process in terms of energetic consumption, the GO-TiO$_2$ membrane was the best one. Indeed, the energy consumption of GO-TiO$_2$ PM was only 28% when compared with that of consumed by the N-TiO$_2$ PM while providing 63% rejection of MB of N-TiO$_2$ PM. On this basis, it can be affirmed that the GO-TiO$_2$ PM can permit to achieve a higher dye rejection with a lower power energy consumption by recycling two or more times the permeate.

Considering the inadequateness in terms of pollutant rejection of the PMs prepared by Athanasekou, Wang et al. [78] prepared and tested in an experimental set-up operated under dead-end mode N-TiO$_2$ ceramic PMs characterized by higher photocatalytic efficiency. These PMs were synthesized by dip coating, similarly to Athanasekou et al., with a substantial difference: the dip-coating of the photocatalyst NPs on membrane support was repeated seven times. Thus, composite PMs characterized by an average pore size of ca. 2 nm, adequate for retaining the dye, were synthesized. SEM and X-Ray Diffraction (XRD) analyses evidenced that the doping of the TiO$_2$ photocatalyst with N permitted obtaining small nanocrystals with high surface and interfacial area. This feature, due to the inhibition of the growth of TiO$_2$ crystal grains by N doping, permitted obtaining improved photocatalytic performance. In particular, a permeation flux equal to ca. 96% of pure water flux, a close to 99% dye rejection and good membrane stability were obtained. Despite these interesting achievements, a rapid diminution of dye rejection, probably caused by the formation of cake layer and concentration polarization, was observed. This trend was ascribed to the used filtration mode. Besides, the N-TiO$_2$ PMs showed an unsatisfactory photoactivity under visible light.

On the same topic, Aoudjit et al. [42] prepared some PMs by immobilizing TiO$_2$-P25 NPs into a poly(vinylidenefluoride–trifluoroethylene) (P(VDF–TrFE)) membrane. The prepared NPs were tested in the photocatalytic degradation of tartrazine in the solar photoreactor schematized in Figure 2.

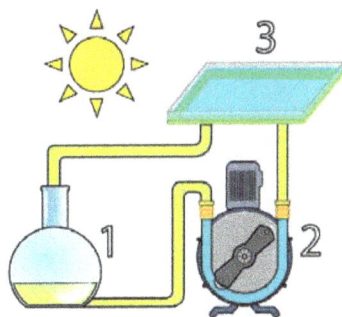

Figure 2. Schematization of the proposed solar photoreactor: 1—flask containing tartrazine solution; 2—peristaltic pump; and, 3—photoreactor containing the PM (Aoudjit et al. [42]).

The method used to prepare the P(VDF-TrFE) PMs was solvent evaporation, as described in Section 3. The photocatalytic degradation tests were performed in a solar photoreactor sited in northern Algeria, during August and September, under solar irradiation. The effective volume of the photoreactor was 1 L. The tartrazine solution was continuously recirculated at 28 mL·s^{-1} flow rate. Obtained results evidenced that the PM containing 8 wt.% TiO$_2$ was the best one, exhibiting a remarkable sunlight photocatalytic activity, with 78% of the pollutant degraded over five hours. As reported in Section 2.5, the initial dye concentration strongly affected system performance. In particular, the photodegradation efficiency decreased from 78 to 46% by increasing the initial dye concentration from 10 to 30 mg·L^{-1} because of: (i) the decreased ratio hydroxyl radical/tartrazine molecules; and, (ii) the higher light scattering, which affects the adsorption of UV light by the surface of

TiO$_2$ NPs, thus decreasing the amount of •OH formed. The degradation percentage of the considered dye after 5 h of irradiation was 37% and 77% for 9.78 and 28 mL·s^{-1}, respectively. This trend confirms what is reported in Section 2.6, i.e., the recirculation flow rate strongly influences the photocatalytic performance of the PM. This trend was due to the presence of more turbulence in the solution, which promoted the mass transfer from the feed solution to the surface of the PM. The reusability of the produced PM was also cheeked. Obtained results evidenced a 10% of efficiency loss by passing from the first use of the PM (ca. 78% of dye degradation) to the second use (ca. 67% of dye degradation). The observed efficiency loss was ascribed to photocatalyst leaching both during the photocatalytic tests and during the system cleaning with distilled water. However, no further reuses were possible because of the relevant leaching out of the NPs. The obtained dye degradations, unsatisfactory by an environmental point of view, were interesting by considering that they were obtained by using natural solar light.

Multi-walled carbon nanotubes (MWCNTs) are characterized by interesting properties, such as huge surface area, great electronic mobility, and stability. These characteristics make this material a good additive for preparing photocatalytic systems that are characterized by satisfactory photocatalytic activity and stability. On the basis of this and always with the aim of using solar radiation Wu et al. [79] prepared, by combining the electrospinning technique with an in situ reaction to obtain Ag$_3$PO$_4$, ternary composite fiber membranes (TCFMs), which possessed good photocatalytic performance. These TCFMs were composed by MWCNTs and Ag$_3$PO$_4$ supported on PAN. The results evidenced that the addition of MWCNTs into the Ag$_3$PO$_4$/PAN binary composite fiber membranes (BCFMs) is fundamental for obtaining a PM that is active under visible light. Indeed, the band gap of Ag$_3$PO$_4$/PAN BCFMs decreased by adding MWCNTs, making the ternary system able to use light characterized by higher wavelengths. The photocatalytic activity and stability of the prepared PMs were evaluated in a batch processing system by using a Xenon arc lamp. The emitted radiation was filtered with a 420-nm cut-off filter. Rhodamine B (RhB) was considered as model contaminant. By operating with the MWCNTs/Ag$_3$PO$_4$/PAN TCFMs, the RhB solution gradually discolored and the dye concentration drastically decreased and became practically zero after 80 min. When compared with the binary system, the ternary system was characterized by a higher and more stable photocatalytic activity for degrading RhB. In particular, the kinetic constant of dye degradation, calculated for the ternary system after observing that the degradation follows pseudo-first-order kinetics model, was 1.8-fold higher than that of the binary system. This result was ascribed to the fast electron transfer from Ag$_3$PO$_4$ to MWCNTs, resulting in a reduced electron-hole recombination and less photocorrosion of Ag$_3$PO$_4$.

Regarding system stability, the photocatalytic efficiency of TCFMs quantified in terms of dye removal after three recycles decreased by 20%, instead of the 45% decrease that was obtained without the MWCNTs. The slight decline of the activity was ascribed by the authors to the leaching of some Ag$_3$PO$_4$ NPs. When considering the photodegradation mechanism, it was observed that holes (h$^+$) and superoxide radicals (•O$_2$$^-$) played fundamental functions for degrading RhB.

The possibility of continuously degrading RhB by using flexible TCFMs under visible-light irradiation was checked in a PMR operated under dead end mode. On the basis of the results that are summarized in Table 1, it can be affirmed that the most important operating parameter affecting the removal rates was the initial RhB concentration. While considering the influence of initial RhB concentration in the feed solution on the removal percentage, it is confirmed that the reaction rate increases with substrate concentration till to a certain level, over which the rate starts to decrease. This trend, as reported in Section 2.4, can be ascribed to both diminution of •OH/pollutant molar ratio and light scattering, which reduced the production rate of •OH. The increase of the solution flow rate was also able to improve the removal percentage (Table 1).

Table 1. Influence of feed flow rate and initial RhB concentration on the photocatalytic degradation of RhB by using the MWCNTs-1%/Ag$_3$PO$_4$/PAN TCFMs (data from Wu et al. [79]).

Flow Rate (mL·min^{-1})	Feed RhB Concentration (mg·L^{-1})	Removal Percentage (%)
10	2	88.3
30	2	89.4
50	2	96.9
30	1	89.9
30	4	96.4
30	8	54.4

As emphasized by the authors, the overall results showed a high potentiality in continuous wastewater treatment when using a sustainable PMR. Modified methods to further improve the intensity of the interaction between Ag$_3$PO$_4$ NPs and fibers have to be employed for reducing leaching problems and increasing the stability of the photocatalytic performance.

The influence of salts and dissolved organic matter on the performance of PMs is another crucial point to be considered in the view of developing photocatalytic processes that are applicable at a large scale. Indeed, these substances, which are frequently present in real industrial effluents, can affect system performance by decreasing photocatalytic performance, due to scavenging effects, and/or by increasing membrane fouling.

On these aspects, Pastrana-Martínez et al. [27] determined the influence of dissolved NaCl on the performance of three PMs, indicated as M-P25, M-TiO$_2$, and M-GOT. These membranes were obtained by immobilizing on a plane cellulosic membrane TiO$_2$ P25, a lab-made TiO$_2$ and TiO$_2$ modified with graphene oxide (GO-TiO$_2$). These PMs were tested for photodegrading MO. A medium pressure mercury vapor lamp was used to operate under a near UV irradiation. A 430 nm cut-off filter was used for performing the experiments under visible light irradiations. The tests were performed in continuous under dead-end filtration. Three different aqueous matrices, distilled water (DW), simulated brackish water (SBW, 0.5 g·L^{-1} NaCl), and seawater (35 g·L^{-1} NaCl) were considered.

In Figure 3, the results that were obtained by using DW as aqueous matrix are summarized. The photocatalytic activity of M-GOT PM for MO abatement was higher than the other. This trend was also confirmed in terms of total organic carbon (TOC) removal. By using DW as aqueous matrix, MO abatement and permeate flux for two successive use were practically constants, regardless of the used PM. This result evidenced the maintaining of the photocatalytic activity of the tested PMs (membrane stability) without fouling problems. Practically, by using DW as the solvent, the photodegradation intermediates did not accumulate on the membrane surface. Thus, high permeate fluxes can be maintained by the PMs for long operation periods. The activity of M-P25 and M-TiO$_2$ membranes under visible light were negligible, while the M-GOT PM possessed a moderate visible light activity, due to the decrease of the band gap. The lower degradation that was obtained by operating under visible light was due to the lower light intensity entering into the photoreactor (2.8 vs. 33 mW·cm^{-2}).

The presence of NaCl in SBW (0.5 g·L^{-1}) lead to a little decrease of photodegradation performance with respect to the use of DW, regardless of the PM employed. This trend was ascribed to Cl$^-$ ion, having a scavenging effect on holes and hydroxyl radicals. This trend was also confirmed by TOC removal. However, the better performance (52% and 13% MO degradation for near UV–vis and visible light irradiation, respectively) was obtained by using the M-GOT PM. Moreover, the photocatalytic performance decreased from the first to the second use, confirming that chlorine anions act as scavengers of holes and hydroxyl radicals.

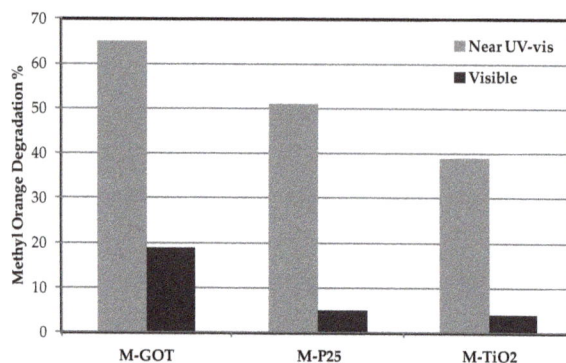

Figure 3. Photocatalytic degradation of Methyl Orange (MO) in distilled water (DW) in continuous mode under near UV-vis and visible light irradiation (data from Pastrana-Martínez et al. [27]).

The authors observed that, despite the encouraging results, these PMs showed lower photocatalytic efficiency with respect to PMRs with photocatalyst in suspension, as also evidenced in other works [80,81].

Fiber based membranes are characterized by good porosity and efficient dispersion and stabilization of catalyst NPs into the solid matrix. On the basis of this, Papageorgiou et al. [73] prepared and tested for MO degradation Ca alginate polymer fibers, which represent good candidates for photocatalytic applications thanks to some of their properties as the possibility to achieve good porosity together with the high transparency of their matrixes.

The results that were obtained during preliminary batch tests evidenced that Ca alginate/TiO$_2$ fibers possessed a photodegradation efficiency (measured in terms of MO degradation) higher than that of TiO$_2$ powder. This result was attributed to the good adsorption capacity and the high surface area of the fibers coupled with the excellent dispersion and stability of the TiO$_2$ NPs into the solid matrix. Despite these encouraging results, Ca alginate/TiO$_2$ polymer fibers gradually degraded, because of the •OH attack on the polymeric material, as previously observed by Molinari et al. [76].

Continuous flow experiments were performed in the hybrid photocatalytic/UF experimental set-up schematized in Figure 4, where a photocatalytic UF membrane (UF-PM) was coupled with the composite Ca alginate/TiO$_2$ porous fibers. The UF-PM was obtained by chemical vapor deposition of TiO$_2$ NPs on both the surfaces of a γ-alumina UF support. As evidenced in Figure 4, the permeate effluent, which represents the treated water, was collected inside the UF-PM, indicated as tubular membrane (Figure 4). 15 UVA LEDs emitting in the UV range 360–420 nm with a maximum emission peak at 383–392 nm irradiated the internal surface of this membrane. The polluted water has been fed in the lumen of Ca alginate/TiO$_2$ polymer fibers in a flow channel that is defined by the intermediate and the outer Plexiglas tubes (Figure 4). Four UV lamps emitting in the near-UV range 315–380 nm with a maximum emission peak at 365 nm, were used to irradiate the alginate fibers.

The TiO$_2$/Ca alginate fibers, working as pretreatment stage of the UF-PM, permitted to obtain an increase of the photocatalytic efficiency in comparison with the use of the UF-PM alone. Moreover, these photocatalytic active fibers in the process permitted obtaining an increased permeate flux across the UF-PM. The degraded Ca alginate/TiO$_2$ polymer fibers were retained by the UF-PM, thus preventing permeate quality deterioration. Despite these advantages that were achieved thanks to the use of these fibers as pretreatment stage, an unsatisfactory 40% dye removal was obtained.

Summarizing, it can be affirmed that particle agglomeration, which is an important drawback of TiO$_2$ promoted photocatalysis, can be limited/eliminated by developing TiO$_2$-nanofiber (TiO$_2$-NF). Despite this achievement some fundamental aspect have to be addressed in view of industrial application: (i) the degradation of the TiO$_2$/Ca alginate fibers, which affects both permeate quality and system lifetime, and (ii) the poor dye removal meaning unsatisfactory permeate quality.

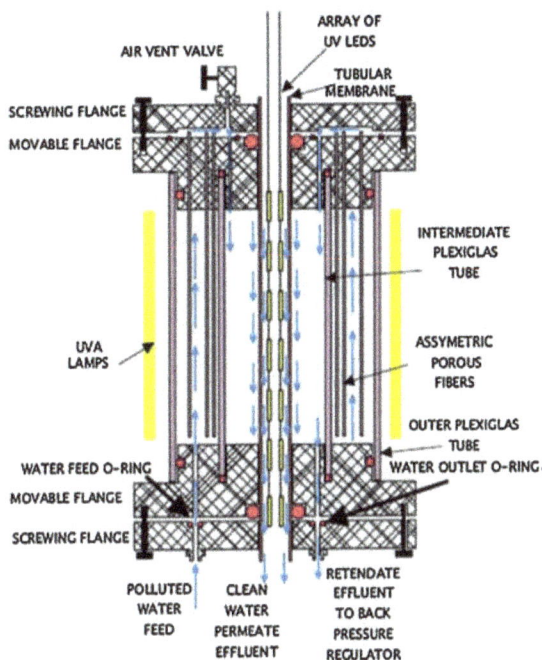

Figure 4. Schematization of the hybrid experimental set-up used by Papageorgiou et al. [73].

On the same topic, Nor et al. [63] prepared nanocomposite PMs consisting of TiO_2 nanofibers (TiO_2-NFs) electrospun on PVDF supports. The photocatalytic efficiency of these PVDF/TiO_2 PMs was evaluated in the photocatalytic degradation of bisphenol A (BPA). The catalytic test were performed by irradiating a polluted solution with a UV lamp having a maximum emission peak $\lambda = 312$ nm. The results that were obtained during the photodegradation tests evidenced that the PVDF/TiO_2 PM hot-pressed at 100 °C (PVDF/TiO_2-100), permitted to obtain a higher UV absorbance and pure water flux in comparison with the PMs PVDF/TiO_2-160 and PVDF/TiO_2-180 hot pressed at higher temperatures. This is because the morphological structure of the nanocomposite PMs was strongly influenced by the hot pressing temperature. In particular, the structure density of the PMs increased with the hot pressing temperature. Good BPA photodegradation (84.53%) was obtained by using the PVDF/TiO_2-100 under UV for 300 min, followed by PVDF/TiO_2-160 (77.61% BPA degradation) and PVDF/TiO_2-180 (62.54% BPA degradation). These results evidenced that the introduction of TiO_2-NFs in on PVDF supports via hot pressing is a preparation method, which permits obtaining an enhanced photodegradation of organic pollutants, such as BPA.

Ramasundaram et al. [80] prepared TiO_2-NFs integrated stainless steel filter (SSF) by some sequential steps (Figure 5), which also included a hot-pressing step. First of all, a free-standing TiO_2-NFs layer was electrospun. The so prepared layer was then bounded to SSF surface by a hot pressing process while using PVDF nanofibers (PVDF-NFs) interlayer as a binder. Five different thicknesses of PVDF-NFs layer were considered in order to evaluate the influence of this parameter on the stability and the activity of the prepared SSF PMs. The stability of the five different SSF PMs was evaluated by submitting to sonication for 30 min five solutions containing them in water. The obtained results evidenced that 42 nm was the optimal PVDF-NFs layer for firmly bounding the TiO_2-NFs to the surface of SSF.

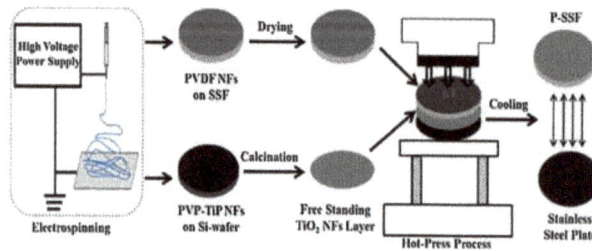

Figure 5. Schematization of the preparation process used for preparing the stainless steel filter PMs (SSF PMs) [80].

The photocatalytic performance of the so prepared and characterized PMs was tested in the degradation of cimetidine (CMD), a molecule of pharmaceutical interest, in an experimental set-up operated in dead-end mode. The results evidenced that the CMD photodegradation increased from 42% to 90% when the thickness of TiO_2-NFs increased from 10 to 29 mm. A further increase of the thickness did not have significant effect on drug degradation. This trend was probably due to the decreased light penetration. Moreover, CMD photodegradation was influenced by water flux across the membrane. In particular, it decreased from 89% to 64% and 47% by increasing the permeation flux from 10 to 20 and then 50 $L \cdot h^{-1} \cdot m^{-2}$. This trend can be explained by considering that, when the permeation flux increases, the contact time between the photocatalytic TiO_2 layer and drug molecules is decreased, resulting in the observed lower photodegradation.

Metal doping of TiO_2 photocatalyst, thanks to the passage of the promoted electron from the CB of the photocatalyst to that one of the metal, enhances: (i) the photocatalytic activity under visible light by reducing the band gap and (ii) the photocatalytic efficiency by minimizing the recombination of electron–hole couples. On these bases, Liu et al. [82] prepared an Ag/TiO_2-NF PM and tested it in the photocatalytic degradation of MB using a solar simulator (Xenon arc lamp) as the light source. These Ag/TiO_2-NF PMs permitted to obtain advanced performance in terms of both membrane fouling and loss of photocatalytic efficiency. Those enhancements were ascribed to the uniform dispersion of Ag NPs on TiO_2-NFs, while maintaining sufficient active sites. The latter achievement was evidenced by Brunauer-Emmet-Teller (BET) measurements (see Table 2): an increased BET specific surface area was obtained by doping with Ag the TiO_2-NFs (102.3 $m^2 \cdot g^{-1}$ vs. 85.6 $m^2 \cdot g^{-1}$). According to these data, the photocatalytic activity under solar irradiation of the Ag/TiO_2-NF membrane (expressed in terms of MB degradation rate constant (min^{-1})) was significantly enhanced. As a consequence, 30 min of solar irradiation were enough to obtain a MB removal of nearly 80%. Complete mineralization was obtained after 80 min.

Table 2. MB degradation rate constant under solar irradiation by using different TiO_2-based photocatalytic membranes (Data from Liu et al. [82]).

	BET Specific Surface Area $(m^2 \cdot g^{-1})$	MB Degradation Rate Constant (min^{-1})
Ag/TiO_2-nanofiber membrane	102.3	0.0211
TiO_2-nanofiber membrane	85.6	0.0137
P25 membrane	/	0.0076

The results that were obtained by Liu et al. [82] show that the intrinsic diminution of the photocatalytic activity of a PMR using immobilized instead of suspended photocatalyst, caused by presence of the polymeric chain around the particles of catalyst, can be minimized by using NF based PMs, thanks to their high surface area.

Concerning the other classical drawbacks of photocatalyst immobilization, i.e., photocatalyst leaching, Liu et al. [82] found an excellent reusability of the prepared PMs. Indeed, the photocatalytic activity did not decrease during five successive photocatalytic runs. Moreover, the intrinsic antibacterial capability of the Ag/TiO$_2$-NF PMs could have a benefic effect with respect to control of the membrane biofouling, which is an important achievement in view of large scale applications.

A different approach to effectively disperse the photocatalyst particle into a PM was developed by Fischer et al. [83]. By hydrolysis of titanium tetraisopropoxide, TiO$_2$ NPs were formed directly onto the surface of two PES and PVDF hydrophilic membranes, and one PVDF hydrophobic membrane. By means of this method, a thin layer of disaggregated TiO$_2$ NPs was directly synthesized on membrane surface, thus achieving a strong binding between the photocatalyst and the membrane supports. The photocatalytic performance of the prepared PMs was studied by considering the photodegradation of three different pollutants, the dye MB and two drugs (ibuprofen (IBU) and diclofenac (DCF)). On the basis of the results it can be affirmed that the properties of pollutant adsorption gave a greater contribution with respect to TiO$_2$ content onto the PM. Indeed, better photocatalytic performances (70% MB photodegradation) were obtained with the hydrophilic TiO$_2$/PVDF PM, despite its lower TiO$_2$ content with respect to the TiO$_2$/PES PM (0.092% vs. 0.809%). The worst performance was obtained by using the hydrophobic PVDF-modified membrane. This trend was ascribed to the lower contact with MB, which was dissolved in water. Lower photodegradation efficiencies were obtained for DCF and IBU (68% and 44%, respectively), because of their higher concentration in water. An acceptable permeate quality was obtained after two or more permeate recycles. The stability of the photocatalytic performance of the prepared PMs was also tested. The results evidenced a complete stability of PMs after five successive runs obtaining the complete photocatalytic degradation of MB without damaging the PMs. This important achievement was due to the complete covering of the support by the TiO$_2$ layer, thus avoiding the direct irradiation of the polymeric support with UV irradiation.

Another possible way to overcome the intrinsic limitations of PMs is represented by the use of TiO$_2$ nanotubes (TiO$_2$ NTs). Fundamental characteristics in view of achieving excellent photocatalytic performance are their high surface area/volume ratio, short distance for charge carrier diffusion, and high photon-collection efficiency. Fischer et al. [84] prepared nanotubular TiO$_2$-PES membranes. The prepared PMs were tested for degrading DCF under UV light. The photocatalytic tests were carried by using two different operating modes: a static mode, i.e., with the membrane operating only as a photocatalytic support, and dead-end mode. A significant lower degradation rate constant was obtained by using the dead end mode with respect to the static mode (0.085×10^{-3} vs. 9.96×10^{-3} min^{-1}). This difference was ascribed to consideration that, during the dead-end flow tests, only 20% of the solution was irradiated, while the remaining 80%, contained in the feed tank and tubes, was not irradiated. This consideration permit to introduce an important parameter to be appropriately considered in view of yielding satisfactory degradation rates: the ratio between the irradiated and the not irradiated volume.

4.2. Polyoxometalates-Based Photocatalytic Membranes

An interesting example of photocatalytic membrane preparation is the heterogenization in/on polymeric membranes of the polyoxometalate decatungstate (W$_{10}$O$_{32}$$^{4-}$). POM are polyanionic metal oxide clusters of early transition metals [85], having promising properties for application in oxidation reactions for fine chemistry and wastewater treatments.

Decatungstate shows remarkable properties for the photocatalytic treatment of wastewater since its absorption spectrum, characterized by a maximum absorption at 324 nm, partially overlaps the solar emission spectrum opening the potential route for an environmentally friendly solar-assisted application [86]. The decatungstate-promoted photocatalysis is a multi-step process that can occur by substrate activation, or, when the reaction is carried out in water, by solvent activation (Figure 6) [86,87]. The dioxygen intercepts the organic radicals giving rise to an autooxidation chain, and provides the re-oxidation of the photocatalyst closing the cycle.

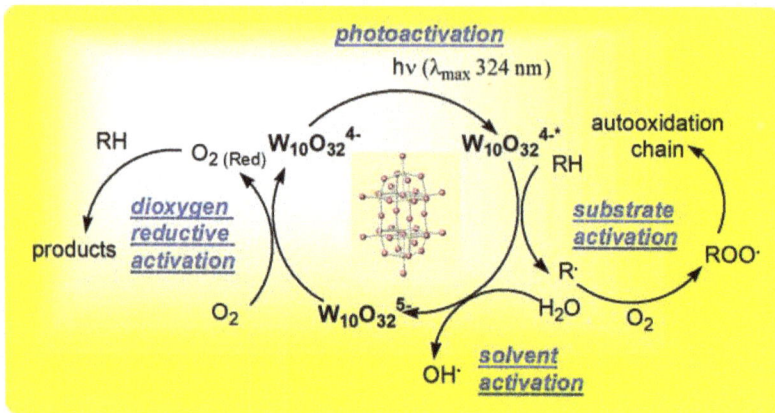

Figure 6. Schematization of the decatungstate-promoted photocatalysis in water [87].

However, decatungstate has also some relevant limitations, like: low quantum yields, small surface area, poor selectivity, and limited stability at pH higher than 2.5 [88,89]. Membrane technology can contribute to overcome these limitations by: the multi-turnover recycling of the heterogeneous photocatalyst, the possibility to tune reaction selectivity as a function of the substrate-membrane affinity, moreover, the structured polymeric micro-environment that is offered by the membrane can influence catalyst stability and activity.

In this perspective, innovative heterogeneous photo-oxygenation systems able to employ visible light, oxygen, mild temperatures, and solvent with a low environmental impact (like water or neat reactions), were designed and developed by the immobilization of decatungstate and other POMs in polymeric membranes made of polydimethylsiloxane (PDMS), PVDF, and Hyflon (Figure 7) [87,90–93].

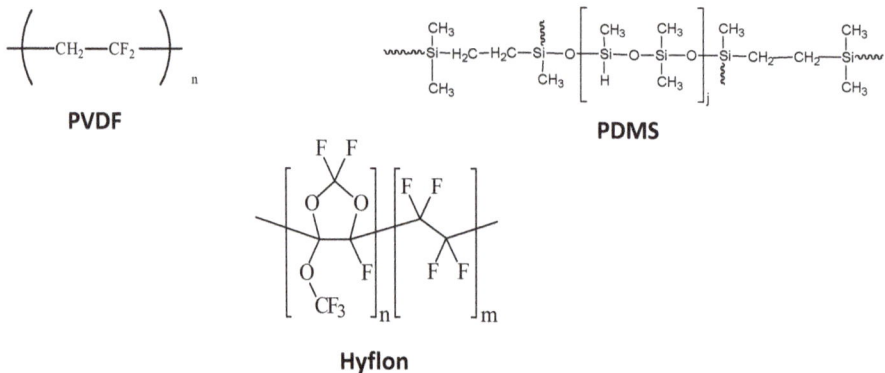

PVDF

PDMS

Hyflon

Figure 7. Chemical formula of the polymers used as membrane materials for the heterogenization of polyoxometalates.

These polymers were selected because they are transparent in the region of interest of the catalyst and are characterized by a high chemical, thermal and photostability. Moreover, in the case of PVDF and Hyflon, the use of a fluorinated media to carry out aerobic oxidation reactions is particularly useful because the high solubility of oxygen in fluorinated environment.

With the aim to improve the catalyst/polymer interactions, and to avoid catalyst leaching out from the membrane, liphophilic (insoluble in water) derivatives of the decatungstate were employed:

the tetrabutilamonium salt ((n-$C_4H_9N)_4W_{10}O_{32}$ indicated as TBAW10) and a fluorous-tagged decatungstate, ($[CF_3(CF_2)_7(CH_2)_3]_3CH_3N)_4W_{10}O_{32}$, indicated as ($R_fN)_4$ $W_{10}O_{32}$).

The photocatalytic membranes prepared were characterized by different and tailored properties depending on the nature and structure of the polymeric micro-environment in which the catalyst was immobilized.

Solid state characterization techniques, like FT-IR and UV-vis spectroscopy, confirmed that the structure and spectroscopic properties of the catalyst were preserved in the heterogenized form. The results evidenced as the appropriate catalyst/polymer design allowed for realizing new heterogeneous photocatalysts for the oxidation of organic substrates performed under oxygen atmosphere at room temperature, with improved performance concerning catalyst's stability and selectivity than the analogues homogeneous reactions.

A membrane induced discrimination of the substrate was observed in the oxidations of a series of alcohols at different molecular weight and polarity carried with PDMS- and PVDF-based catalytic membranes containing TBAW10 operating in a batch setup while using a Hg-Xe arc lamp as light source [87]. The alcohol oxidation occurred following a degradation pathway in which the aldehyde was formed as an intermediate [87].

PVDF membranes containing TBAW10 were also successfully used in the aerobic mineralization of phenol in water carried out in a continuous flow-through photocatalytic membrane reactors [92] operating with a mercury vapor lamp, emitting from 310 nm to visible light. The results indicate that the photocatalytic membrane were stable and recyclable in successive runs [92].

The catalyst heterogenization was carried out by solubilizing the catalyst into the polymer dope solution using a common solvent (dimethylacetamide) to prepare nano-hybrid membrane (PVDF-W10) by the phase separation technique. The catalytic membranes were characterized by a homogeneous distribution of the catalyst in membrane, as evident from SEM in back scattered electrons mode (BSE) and linear RX maps on the cross sections.

The catalyst stability was positively influenced by the polymeric environment, in which it was confined. Moreover, the selective separation function of the membrane resulted in an enhancement of the phenol mineralization in comparison with an analogous homogeneous reaction carried out with sodium salt of the decatungstate [92]. The dependence of the phenol degradation rate by the catalyst loading in membrane and the transmembrane pressure was investigated, allowing to identify the catalytic membrane with catalyst loading 25.0 wt.% and operating at 1 bar (contact time 22 s), as the more efficient system [92].

The rates of phenol degradation catalysed by homogenous $Na_4W_{10}O_{32}$ and heterogeneous PVDF-W10 (25.0 wt.%) were compared in similar operative conditions. The percentages of phenol degraded in the homogeneous and heterogeneous reaction were similar. In both cases, after 5 h, about 50% of the phenol was converted. However, in the case of the homogenous reaction various persistent intermediates (e.g., benzoquinone, hydroquinone, and catechol) were observed and only the 34.0% of mineralization to CO_2 and water was achieved. On the contrary, during photodegradation performed by PVDF catalytic membrane, the phenol converted was completely mineralized to CO_2 and H_2O.

The higher catalytic activity of the PVDF-W10 membranes, in comparison to the homogeneous catalyst, was ascribed to the selective absorption of the organic substrate from water on the hydrophobic PVDF polymer membrane that increased the effective phenol concentration around the catalytic sites, optimizing the catalyst-substrate contact in the flow-through PMR.

Moreover, the polymeric hydrophobic environment protected the decatungstate from the conversion to an isomer, which absorbs only light under 280 nm, which instead occurs in homogeneous solution at pH > 2.5 [89,92].

Polymeric catalytic membranes were also prepared by immobilizing sodium decatungstate ($Na_4W_{10}O_{32}$; W10) [94,95] and phosphotungstic acid ($H_3PW_{12}O_{40}$; W12) [91], on the surface of PVDF membranes functionalized by Ar/NH_3 plasma discharges (PVDF-NH_2).

Polar chemical groups (mainly NH$_2$, but also OH, CN, NH, and CO) were grafted by a NH$_3$ plasma discharge on the upper surface of PVDF membranes [91,94,95], pre-treated with Ar in order to control hydrophobic recovery phenomena [96].

The groups grafted on the membrane surface were able to interact with the POMs solubilized in the aqueous solution, forming charge transfer complexes.

Surface-diagnostic techniques, such as X-ray photoelectron spectroscopy, contact angle measurements (CA), and RX maps, were used to support the successfully surface modification.

The catalytic membranes obtained showed superior performances (higher reaction rates) when compared to the corresponding homogeneous catalysts, in the aerobic phenol degradation reaction [91,94,95].

Decatungstate was also heterogenized in membrane made of Hyflon, an amorphous perfluoro co-polymer. It is important to note that the TBAW10 formed irregular catalyst aggregates in Hyflon membranes because of the low affinity with the polymeric matrix. However, the affinity between the polymer and the catalyst was improved by an appropriate functionalization of the second one. The fluorous-tagged $(R_fN)_4$ $W_{10}O_{32}$, was well dispersed in the Hyflon membranes as spherical clusters with uniform size.

The cationic amphiphilic R_fN^+ groups induced the self-assembly of the surfactant-encapsulated clusters (i.e., RfN^+ groups capped on $W_{10}O_{32}{}^{4-}$) which, during membrane formation process, gave supramolecular assemblies of the catalyst, stabilized by the polymeric matrix.

This self-assembling process was tuned by a proper choice of the membrane preparation conditions, like polymer concentration, catalyst loading, cast film thickness, and temperature [9].

The Hyflon-based catalytic membranes were applied in the solvent-free photo-oxygenation of benzylic C-H bonds with up to 6100 turnover number (moles of products for moles of catalyst) in 4 h and remarkable alcohol selectivity, thus providing a convenient alternative to other radical centered oxygenation systems [94].

5. Conclusions

In the last years, significant progresses were reached in the design and development photocatalytic membranes and photocatalytic membrane reactors.

PMRs are reactive separations that realize functional synergies between a membrane-based separation process and a photocatalytic conversion in agreement with the fundamentals of the process intensification strategy. However, significant improvements in membranes performance and durability, as well as plant design optimization, are still required in order to reach a mature technological stage and offer a real challenge to conventional photocatalytic systems in the perspective of the realization of a sustainable growth.

TiO$_2$ is the most widely investigated photocatalyst in PMRs. This photocatalyst is used in two main configurations: (i) dispersed in solution and compartmentalized in the reactors by the membrane; and, (ii) immobilized in or on the photocatalytic membrane.

However, interesting results were also achieved with photocatalytic membranes functionalized with polyoxometalates applied into oxidation reactions for wastewater treatment and fine chemistry.

The design and realization of photocatalytic membranes by the immobilization of a photocatalyst in/on a membrane, is characterized by relevant technical complexities but these systems can lead specific advantages in terms of productivity and sustainability in comparison to traditional heterogeneous photocatalysts.

Main issues to be addressed are the development of tailored photocatalytic membranes and membrane modules with acceptable costs, stable in a wide range of operative conditions, resistant to fouling, and showing high and reproducible performance over long terms.

Author Contributions: E.F. and P.A. elaborated the literature overview, wrote and revised the paper; E.D. and R.M. participated in valuable discussion and gave insightful comments on the work.

Funding: This research received no external funding.

Conflicts of Interest: The authors declare no conflict of interest.

Abbreviation

AOPs	advanced oxidation processes
BAT	best available technology
BCFM	binary composite fiber membrane
BPA	bisphenol A
BET	Brunauer-Emmet-Teller
CHD	chlorhexidine digluconate
CMD	cimetidine
CB	conduction band
DCF	diclofenac
DB	direct black 168
DW	distilled water
DMF	N,N-dimethylformamide
GO-TiO$_2$	graphene oxide doped TiO$_2$
HF	hollow fiber
IBU	ibuprofen
MBR	membrane bioreactor
MR	membrane reactor
MB	methylene blue
MO	methyl orange
MWCO	molecular weight cut off
MWCNT	multi-walled carbon nanotubes
NP	nanoparticle
NT	nanotube
N-TiO$_2$	nitrogen doped TiO$_2$
PM	photocatalytic membrane
PMR	photocatalytic membrane reactor
PAN	polyacrylonitrile
PDMS	polydimethylsiloxane
PES	polyethersulfone
PET	polyethylene terephthalate
POMs	polyoxometalates
PTFE	polytetrafluoroethylene
PU	polyurethane
PVC	polyvinylchloride
PVDF	polyvinylidene fluoride
P(VDF-TrFE)	poly(vinylidenefluoride–trifluoroethylene)
PI	process intensification
(R$_f$N)$_4$ W$_{10}$O$_{32}$	[CF$_3$(CF$_2$)$_7$(CH$_2$)$_3$]$_3$CH$_3$N)$_4$W$_{10}$O$_{32}$
RhB	rhodamine B
SEM	scanning electron microscopy
SBW	simulated brackish water
SSF	stainless steel filter
TBAW10	(n-C$_4$H$_9$N)$_4$W$_{10}$O$_{32}$
TCFMs	ternary composite fiber membranes
TiO$_2$NTs	TiO$_2$ nanotubes

TNF	titanium dioxide nanofiber
TOC	total organic carbon
UF	ultrafiltration
UF-PM	photocatalytic UF membrane
UV	ultraviolet
VB	valence band
XRD	X-ray diffraction
WSC	water-soluble chitosan

References

1. Sanchez Marcano, J.G.; Tsotsis, T.T. Membrane Reactors. In *Ullmann's Encyclopedia of Industrial Chemistry*; Wiley-VCH: Weinheim, Germany, 2005; ISBN 9783527303854.
2. Wöltinger, J.; Karau, A.; Leuchtenberger, W.; Drauz, K. Membrane Reactors at Degussa. In *Technology Transfer in Biotechnology. Advances in Biochemical Engineering*; Kragl, U., Ed.; Springer: Berlin, Germany, 2005; Volume 92, ISBN 978-3-540-22412-9. [CrossRef]
3. Ibrahim, M.H.; El-Naas, M.H.; Zhang, Z.; Van der Bruggen, B. CO_2 Capture Using Hollow Fiber Membranes: A Review of Membrane Wetting. *Energy Fuels* **2018**, *32*, 963–978. [CrossRef]
4. Le-Clech, P.; Chen, V.; Fane, T.A.G. Fouling in membrane bioreactors used in wastewater treatment. *J. Membr. Sci.* **2006**, *284*, 17–53. [CrossRef]
5. Charpentier, J.-C. Modern Chemical Engineering in the Framework of Globalization, Sustainability, and Technical Innovation. *Ind. Eng. Chem. Res.* **2007**, *46*, 3465–3485. [CrossRef]
6. Stankiewicz, A. Reactive separations for process intensification: An industrial perspective. *Chem. Eng. Process.* **2003**, *42*, 137–144. [CrossRef]
7. Drioli, E.; Stankiewicz, A.; Macedonio, F. Membrane engineering in process intensification—An overview. *J. Membr. Sci.* **2011**, *380*, 1–8. [CrossRef]
8. Sirkar, K.K.; Shanbhag, P.V.; Kovvali, A.S. Membrane in a Reactor: A Functional Perspective. *Ind. Eng. Chem. Res.* **1999**, *38*, 3715–3737. [CrossRef]
9. Fontananova, E.; Drioli, E. Catalytic Membranes and Membrane Reactors. In *Comprehensive Membrane Science and Engineering*; Drioli, E., Giorno, L., Eds.; Elsevier: Oxford, UK, 2010; Volume 3, pp. 109–133, ISBN 978-0-08-093250-7.
10. Vankelecom, I.F.J. Polymeric Membranes in Catalytic Reactors. *Chem. Rev.* **2002**, *102*, 3779–3810. [CrossRef] [PubMed]
11. Molinari, R.; Marino, T.; Argurio, P. Photocatalytic membrane reactors for hydrogen production from water. *Int. J. Hydrogen Energy* **2014**, *39*, 7247–7261. [CrossRef]
12. Zheng, X.; Shen, Z.-P.; Shi, L.; Cheng, R.; Yuan, D.-Y. Photocatalytic Membrane Reactors (PMRs) in Water Treatment: Configurations and Influencing Factors. *Catalysts* **2017**, *7*, 224. [CrossRef]
13. Mozia, S. Photocatalytic membrane reactors (PMRs) in water and wastewater treatment. A review. *Sep. Purif. Technol.* **2010**, *73*, 71–91. [CrossRef]
14. Molinari, R.; Lavorato, C.; Argurio, P. Recent progress of photocatalytic membrane reactors in water treatment and in synthesis of organic compounds. A review. *Catal. Today* **2017**, *281*, 144–164. [CrossRef]
15. Dioos, B.M.L.; Vankelecom, I.F.J.; Jacobs, P.A. Aspects of Immobilisation of Catalysts on Polymeric Supports. *Adv. Synth. Catal.* **2006**, *348*, 1413–1446. [CrossRef]
16. Molinari, R.; Lavorato, C.; Argurio, P. Photocatalytic reduction of acetophenone in membrane reactors under UV and visible light using TiO_2 and Pd/TiO_2 catalysts. *Chem. Eng. J.* **2015**, *274*, 307–316. [CrossRef]
17. Sclafani, A.; Palmisano, L.; Schiavello, M. Phenol and nitrophenol photodegradation using aqueous TiO_2 dispersions. In *Aquatic and Surface Photochemistry*; Helz, G.R., Zepp, R.G., Crosby, D.G., Eds.; Lewis Publishers: London, UK, 1994; p. 419. [CrossRef]
18. Hairom, N.H.H.; Mohammad, A.W.; Kadhum, A.A.H. Effect of various zinc oxide nanoparticles in membrane photocatalytic reactor for Congo red dye treatment. *Sep. Purif. Technol.* **2014**, *137*, 74–81. [CrossRef]
19. Molinari, R.; Caruso, A.; Argurio, P.; Poerio, T. Degradation of the drugs Gemfibrozil and Tamoxifen in pressurized and de-pressurized membrane photoreactors using suspended polycrystalline TiO_2 as catalyst. *J. Membr. Sci.* **2008**, *319*, 54–63. [CrossRef]

20. Lavorato, C.; Argurio, P.; Mastropietro, T.F.; Pirri, G.; Poerio, T.; Molinari, R. Pd/TiO$_2$ doped faujasite photocatalysts for acetophenone transfer hydrogenation in a photocatalytic membrane reactor. *J. Catal.* **2017**, *353*, 152–161. [CrossRef]

21. Brunetti, A.; Barbieri, G.; Drioli, E. Upgrading of a syngas mixture for pure hydrogen production in a Pd–Ag membrane reactor. *Chem. Eng. Sci.* **2009**, *64*, 3448–3454. [CrossRef]

22. Song, H.; Shao, J.; He, Y.; Liu, B.; Zhong, X. Natural organic matter removal and flux decline with PEG–TiO$_2$-doped PVDF membranes by integration of ultrafiltration with photocatalysis. *J. Membr. Sci.* **2012**, *405–406*, 48–56. [CrossRef]

23. Wang, W.Y.; Irawan, A.; Ku, Y. Photocatalytic degradation of Acid Red 4 using a titanium dioxide membrane supported on a porous ceramic tube. *Water Res.* **2008**, *42*, 4725–4732. [CrossRef] [PubMed]

24. Iglesias, O.; Rivero, M.J.; Urtiaga, A.M.; Ortiz, I. Membrane-based photocatalytic systems for process intensification. *Chem. Eng. J.* **2016**, *305*, 136–148. [CrossRef]

25. Horovitz, I.; Avisar, D.; Baker, M.A.; Grilli, R.; Lozzi, L.; Di Camillo, D.; Mamane, H. Carbamazepine degradation using a N-doped TiO$_2$ coated photocatalytic membrane reactor: Influence of physical parameters. *J. Hazard. Mater.* **2016**, *310*, 98–107. [CrossRef] [PubMed]

26. Athanasekou, C.P.; Moustakas, N.G.; Morales-Torres, S.; Pastrana-Martínez, L.M.; Figueiredo, J.L.; Faria, J.L.; Silva, A.M.T.; Dona-Rodriguez, J.M.; Romanos, G.E.; Falaras, P. Ceramic Photocatalytic Membranes for Water Filtration Under UV and Visible Light. *Appl. Catal. B Environ.* **2015**, *178*, 12–19. [CrossRef]

27. Pastrana-Martínez, L.M.; Morales-Torres, S.; Figueiredo, J.L.; Faria, J.L.; Silva, A.M.T. Graphene Oxide Based Ultrafiltration Membranes for Photocatalytic Degradation of Organic Pollutants in Salty Water. *Water Res.* **2015**, *77*, 179–190. [CrossRef] [PubMed]

28. Misra, A.J.; Das, S.; Habeeb Rahman, A.P.; Das, B.; Jayabalan, R.; Behera, S.K.; Suar, M.; Tamhankar, A.J.; Mishra, A.; Lundborg, C.S.; et al. Doped ZnO nanoparticles impregnated on Kaolinite (Clay): A reusable nanocomposite for photocatalytic disinfection of multidrug resistant Enterobacter sp. under visible light. *J. Colloid Interface Sci.* **2018**, *530*, 610–623. [CrossRef] [PubMed]

29. Najma, B.; Kasi, A.K.; Khan Kasi, J.; Akbar, A.; Bokhari, S.M.A.; Stroe, I.R. ZnO/AAO photocatalytic membranes for efficient water disinfection: Synthesis, characterization and antibacterial assay. *Appl. Surf. Sci.* **2018**, *448*, 104–114. [CrossRef]

30. Peyravi, M.; Jahanshahi, M.; Khalili, S. Fouling of WO$_3$ nanoparticle-incorporated PSf membranes in ultrafiltration of landfill leachate and dairy a combined wastewaters: An investigation using model. *Chin. J. Chem. Eng.* **2017**, *25*, 741–751. [CrossRef]

31. Alzahrani, E. Chitosan membrane embedded with ZnO/CuO nanocomposites for the photodegradation of fast green dye under artificial and solar irradiation. *Anal. Chem. Insights* **2018**, *13*. [CrossRef] [PubMed]

32. Katsoulis, D.E. A Survey of Applications of Polyoxometalates. *Chem. Rev.* **1998**, *98*, 359–388. [CrossRef] [PubMed]

33. Herrmann, J.-M. Heterogeneous Photocatalysis: State of the Art and Present Applications. *Top. Catal.* **2005**, *34*, 49–65. [CrossRef]

34. Brosillon, S.; Lhomme, L.; Vallet, C.; Bouzaza, A.; Wolbert, D. Gas Phase Photocatalysis and Liquid Phase Photocatalysis: Interdependence and Influence of Substrate Concentration and Photon Flow on Degradation Reaction Kinetics. *Appl. Catal. B Environ.* **2008**, *78*, 232–241. [CrossRef]

35. Sabaghi, V.; Davar, F.; Fereshteh, Z. ZnS nanoparticles prepared via simple reflux and hydrothermal method: Optical and photocatalytic properties. *Ceram. Int.* **2018**, *44*, 7545–7556. [CrossRef]

36. Zhang, W.; Ding, L.; Luo, J.; Jaffrin, M.Y.; Tang, B. Membrane fouling in photocatalytic membrane reactors (PMRs) for water and wastewater treatment: A critical review. *Chem. Eng. J.* **2016**, *302*, 446–458. [CrossRef]

37. Reddy, N.L.; Rao, V.N.; Kumari, M.M.; Kakarla, R.R.; Ravi, P.; Sathish, M.; Karthik, M.; Venkatakrishnan, S.M.; Inamuddin. Nanostructured semiconducting materials for efficient hydrogen generation. *Environ. Chem. Lett.* **2018**, *16*, 765–796. [CrossRef]

38. Yamani, Z.H. Comparative Study on Photocatalytic Degradation of Methylene Blue by Degussa P25 Titania: Pulsed Laser Light Versus Continuous Broad Spectrum Lamp Irradiation. *Arab. J. Sci. Eng.* **2018**, *43*, 423–432. [CrossRef]

39. Hayat, K.; Gondal, M.A.; Khaleda, M.M.; Yamani, Z.H.; Ahmed, S. Laser induced photocatalytic degradation of hazardous dye (Safranin-O) using self synthesized nanocrystalline WO$_3$. *J. Hazard. Mater.* **2011**, *186*, 1226–1233. [CrossRef] [PubMed]

40. Li, S.; Lin, Q.; Liu, X.; Yang, L.; Ding, J.; Dong, F.; Li, Y.; Irfana, M.; Zhang, P. Fast photocatalytic degradation of dyes using low power laser-fabricated Cu_2O–Cu nanocomposites. *RSC Adv.* **2018**, *8*, 20277–20286. [CrossRef]

41. Dominguez, S.; Laso, J.; Margallo, M.; Aldaco, R.; Rivero, M.J.; Irabien, Á.; Ortiz, I. LCA of greywater management within a water circular economy restorative thinking framework. *Sci. Total Environ.* **2018**, *621*, 1047–1056. [CrossRef] [PubMed]

42. Aoudjit, L.; Martins, P.M.; Madjene, F.; Petrovykh, D.Y.; Lanceros-Mendez, S. Photocatalytic reusable membranes for the effective degradation of tartrazine with a solar photoreactor. *J. Hazard. Mat.* **2018**, *344*, 408–416. [CrossRef] [PubMed]

43. Halim, R.; Utama, R.; Cox, S.; Le-Clech, P. Performances of submerged membrane photocatalysis reactor during treatment of humic substances. *Membr. Water Treat.* **2010**, *1*, 283–296. [CrossRef]

44. Song, L.; Zhu, B.; Jegatheesan, V.; Gray, S.; Duke, M.; Muthukumaran, S. Treatment of secondary effluent by sequential combination of photocatalytic oxidation with ceramic membrane filtration. *Environ. Sci. Pollut. Res.* **2018**, *25*, 5191–5202. [CrossRef] [PubMed]

45. Tang, T.; Lu, G.; Wang, W.; Wang, R.; Huang, K.; Qiu, Z.; Tao, X.; Dang, Z. Photocatalytic removal of organic phosphate esters by TiO_2: Effect of inorganic ions and humic acid. *Chemosphere* **2018**, *206*, 26–32. [CrossRef] [PubMed]

46. Borthakur, P.; Das, M.R. Hydrothermal assisted decoration of NiS_2 and CoS nanoparticles on the reduced graphene oxide nanosheets for sunlight driven photocatalytic degradation of azo dye: Effect of background electrolyte and surface charge. *J. Colloid Interface Sci.* **2018**, *516*, 342–354. [CrossRef] [PubMed]

47. Augugliaro, V.; Litter, M.; Palmisano, L.; Soria, J. The Combination of Heterogeneous Photocatalysis With Chemical and Physical Operations: A Tool for Improving the Photoprocess Performance. *J. Photochem. Photobiol. C Photochem. Rev.* **2006**, *7*, 127–144. [CrossRef]

48. Darowna, D.; Wróbel, R.; Morawski, A.W.; Mozia, S. The influence of feed composition on fouling and stability of a polyethersulfone ultrafiltration membrane in a photocatalytic membrane reactor. *Chem. Eng. J.* **2017**, *310*, 360–367. [CrossRef]

49. Gladysz, J.A. Recoverable catalysts. Ultimate goals, criteria of evaluation, and the green chemistry interface. *Pure Appl. Chem.* **2001**, *73*, 1319–1324. [CrossRef]

50. Hill, C.L. Homogeneous catalysis. Controlled green oxidation. *Nature* **1999**, *401*, 436–437. [CrossRef] [PubMed]

51. Daels, N.; Radoicic, M.; Radetic, M.; Van Hulle, S.W.; De Clerck, K. Functionalisation of electrospun polymer nanofibre membranes with TiO_2 nanoparticles in view of dissolved organic matter photodegradation. *Sep. Purif. Technol.* **2014**, *133*, 282–290. [CrossRef]

52. Wu, G.; Cui, L.; Xu, Y.; Lu, X. Photocatalytic membrane reactor for degradation of phenol in aqueous solution. *Fresenius Environ. Bull.* **2007**, *16*, 812–816.

53. Fischer, K.; Gawel, A.; Rosen, D.; Krause, M.; Latif, A.A.; Griebel, J.; Prager, A.; Schulze, A. Low-temperature synthesis of anatase/rutile/brookite TiO_2 nanoparticles on a polymer membrane for photocatalysis. *Catalysts* **2017**, *7*, 209. [CrossRef]

54. Kim, J.H.; Joshi, M.K.; Lee, J.; Park, C.H.; Kim, C.S. Polydopamine-assisted immobilization of hierarchical zinc oxide nanostructures on electrospun nanofibrous membrane for photocatalysis and antimicrobial activity. *J. Colloid Interface Sci.* **2018**, *513*, 566–574. [CrossRef] [PubMed]

55. Artoshina, O.V.; Rossouw, A.; Semina, V.K.; Nechaev, A.N.; Apel, P.Y. Structural and physicochemical properties of titanium dioxide thin films obtained by reactive magnetron sputtering, on the surface of track-etched membranes. *Pet. Chem.* **2015**, *55*, 759–768. [CrossRef]

56. Shi, Y.; Yang, D.; Li, Y.; Qu, J.; Yu, Z.-Z. Fabrication of PAN@TiO_2/Ag nanofibrous membrane with high visible light response and satisfactory recyclability for dye photocatalytic degradation. *Appl. Surf. Sci.* **2017**, *426*, 622–629. [CrossRef]

57. Li, N.; Tian, Y.; Zhang, J.; Sun, Z.; Zhao, J.; Zhang, J.; Zuo, W. Precisely-controlled modification of PVDF membranes with 3D TiO_2/ZnO nanolayer: Enhanced anti-fouling performance by changing hydrophilicity and photocatalysis under visible light irradiation. *J. Membr. Sci.* **2017**, *528*, 359–368. [CrossRef]

58. Mozia, S.; Darowna, D.; Wróbel, R.; Morawski, A.W. A study on the stability of polyethersulfone ultrafiltration membranes in a photocatalytic membrane reactor. *J. Membr. Sci.* **2015**, *495*, 176–186. [CrossRef]

59. Drioli, E.; Fontananova, E. Catalytic membranes embedding selective catalysts: Preparation and applications. In *Heterogenized Homogeneous Catalysts for Fine Chemicals Production: Materials and Processes*; Barbaro, P., Liguori, F., Eds.; Springer: Berlin, Germany, 2010; Chapter 6; ISBN 978-90-481-3696-4.

60. Strathmann, H.; Giorno, L.; Drioli, E. *An Introduction to Membrane Science and Technology*; CNR: Rome, Italy, 2006; ISBN 88-8080-063-9.

61. Wang, X.; Shi, F.; Huang, W.; Fan, C. Synthesis of high quality TiO_2 membranes on alumina supports and their photocatalytic activity. *Thin Solid Films* **2012**, *520*, 2488–2492. [CrossRef]

62. Chakraborty, S.; Loutatidou, S.; Palmisano, G.; Kujawa, J.; Mavukkandy, M.A.; Al-Gharabli, S.; Curcio, E.; Arafat, H.A. Photocatalytic hollow fiber membranes for the degradation of pharmaceutical compounds in wastewater. *J. Environ. Chem. Eng.* **2017**, *5*, 5014–5024. [CrossRef]

63. Nor, N.A.M.; Jaafar, J.; Ismail, A.F.; Mohamed, M.A.; Rahman, M.A.; Othman, M.H.D.; Lau, W.J.; Yusof, N. Preparation and performance of PVDF-based nanocomposite membrane consisting of TiO_2 nanofibers for organic pollutant decomposition in wastewater under UV irradiation. *Desalination* **2016**, *391*, 89–97. [CrossRef]

64. Kaijun, Z.; Qingshan, L.; Yu, W. Preparation and performance of PMMA/R-TiO_2 and PMMA/A-TiO_2 electrospun fibrous films. *Integr. Ferroelectr.* **2018**, *188*, 31–43. [CrossRef]

65. Fischer, K.; Gläser, R.; Schulze, A. Nanoneedle and nanotubular titanium dioxide–PES mixed matrix membrane for photocatalysis. *Appl. Catal. B Environ.* **2014**, *160*, 456–464. [CrossRef]

66. Della Foglia, F.; Chiarello, G.L.; Dozzi, M.V.; Piseri, P.; Bettini, L.G.; Vinati, S.; Ducati, C.; Milani, P.; Selli, E. Hydrogen production by photocatalytic membranes fabricated by supersonic cluster beam deposition on glass fiber filters. *Int. J. Hydrogen Energy* **2014**, *39*, 13098–13104. [CrossRef]

67. Zhang, E.; Wang, L.; Zhang, B.; Xie, Y.; Sun, C.; Jiang, C.; Zhang, Y.; Wang, G. Modification of polyvinylidene fluoride membrane with different shaped α-Fe_2O_3 nanocrystals for enhanced photocatalytic oxidation performance. *Mater. Chem. Phys.* **2018**, *214*, 41–47. [CrossRef]

68. Chen, Q.; Yu, Z.; Pan, Y.; Zeng, G.; Shi, H.; Yang, X.; Li, F.; Yang, S.; He, Y. Enhancing the photocatalytic and antibacterial property of polyvinylidene fluoride membrane by blending Ag–TiO_2 nanocomposites. *J. Mater. Sci.-Mater. Electron.* **2017**, *28*, 3865–3874. [CrossRef]

69. Hoseini, S.N.; Pirzaman, A.K.; Aroon, M.A.; Pirbazari, A.E. Photocatalytic degradation of 2,4-dichlorophenol by Co-doped TiO_2(Co/TiO_2) nanoparticles and Co/TiO_2 containing mixed matrix membranes. *J. Water Proc. Eng.* **2017**, *17*, 124–134. [CrossRef]

70. Rajeswari, A.; Vismaiya, S.; Pius, A. Preparation, characterization of nano ZnO-blended cellulose acetate-polyurethane membrane for photocatalytic degradation of dyes from water. *Chem. Eng. J.* **2017**, *313*, 928–937. [CrossRef]

71. Yin, J.; Deng, B. Polymer-matrix nanocomposite membranes for water treatment. *J. Membr. Sci.* **2015**, *479*, 256–275. [CrossRef]

72. Zhao, Y.; Ma, L.; Chang, W.; Huang, Z.; Feng, X.; Qi, X.; Li, Z. Efficient photocatalytic degradation of gaseous N,N-dimethylformamide in tannery waste gas using doubly open-ended Ag/TiO_2 nanotube array membranes. *Appl. Surf. Sci.* **2018**, *444*, 610–620. [CrossRef]

73. Papageorgiou, S.K.; Katsaros, F.K.; Favvas, E.P.; Romanos, G.E.; Athanasekou, C.P.; Beltsios, K.G.; Tzialla, O.I.; Falaras, P. Alginate Fibers as Photocatalyst Immobilizing Agents Applied in Hybrid Photocatalytic/Ultrafiltration Water Treatment Processes. *Water Res.* **2012**, *46*, 1858–1872. [CrossRef] [PubMed]

74. Anderson, M.A.; Gieselmann, M.J.; Xu, Q.J. Titania and Alumina Ceramic Membranes. *J. Membr. Sci.* **1988**, *39*, 243–258. [CrossRef]

75. Moosemiller, M.D.; Hill, C.G., Jr.; Anderson, M.A. Physicochemical Properties of Supported γ-Al_2O_3 and TiO_2 Ceramic Membranes. *Sep. Sci. Technol.* **1989**, *24*, 641–657. [CrossRef]

76. Molinari, R.; Palmisano, L.; Drioli, E.; Schiavello, M. Studies on Various Reactor Configurations for Coupling Photocatalysis and Membrane Processes in Water Purification. *J. Membr. Sci.* **2002**, *206*, 399–415. [CrossRef]

77. Zhang, H.; Quan, X.; Chen, S.; Zhao, H.; Zhao, Y. Fabrication of Photocatalytic Membrane and Evaluation Its Efficiency in Removal of Organic Pollutants From Water. *Sep. Purif. Technol.* **2006**, *50*, 147–155. [CrossRef]

78. Wang, Z.-B.; Guan, Y.-J.; Chen, B.; Bai, S.-L. Retention and Separation of 4BS Dye from Wastewater by the N-TiO_2 Ceramic Membrane. *Desalin. Water Treat.* **2016**, *57*, 16963–16969. [CrossRef]

79. Wu, X.-Q.; Shen, J.-S.; Zhao, F.; Shao, Z.-D.; Zhong, L.-B.; Zheng, Y.-M. Flexible electrospun MWCNTs/Ag$_3$PO$_4$/PAN ternary composite fiber membranes with enhanced photocatalytic activity and stability under visible-light irradiation. *J. Mater. Sci.* **2018**, *53*, 10147–10159. [CrossRef]

80. Ramasundaram, S.; Yoo, H.N.; Song, K.G.; Lee, J.; Choi, K.J.; Hong, S.W. Titanium Dioxide Nanofibers Integrated Stainless Steel Filter for Photocatalytic Degradation of Pharmaceutical Compounds. *J. Hazard. Mater.* **2013**, *258–259*, 124–132. [CrossRef] [PubMed]

81. Georgi, A.; Kopinke, F.-D. Interaction of Adsorption and Catalytic Reactions in Water Decontamination Processes: Part I. Oxidation of Organic Contaminants with Hydrogen Peroxide Catalyzed by Activated Carbon. *Appl. Catal. B Environ.* **2005**, *58*, 9–18. [CrossRef]

82. Liu, L.; Liu, Z.; Bai, H.; Sun, D.D. Concurrent Filtration and Solar Photocatalytic Disinfection/Degradation Using High-Performance Ag/TiO$_2$ Nanofiber Membrane. *Water Res.* **2012**, *46*, 1101–1112. [CrossRef] [PubMed]

83. Fischer, K.; Grimm, M.; Meyers, J.; Dietrich, C.; Gläser, R.; Schulze, A. Photoactive Microfiltration Membranes via Directed Synthesis of TiO$_2$ Nanoparticles on the Polymer Surface for Removal of Drugs From Water. *J. Membr. Sci.* **2015**, *478*, 49–57. [CrossRef]

84. Fischer, K.; Kuhnert, M.; Glaser, R.; Schulze, A. Photocatalytic Degradation and Toxicity Evaluation of Diclofenac by Nanotubular Titanium Dioxide–PES Membrane in a Static and Continuous Setup. *RSC Adv.* **2015**, *5*, 16340–16348. [CrossRef]

85. Pope, M. *Heteropoly and Isopoly Oxometalates*; Springer: New York, NY, USA, 1983; ISBN 978-3-662-12006-4.

86. Maldotti, A.; Molinari, A.; Amadelli, R. Photocatalysis with Organized Systems for the Oxofunctionalization of Hydrocarbons by O$_2$. *Chem. Rev.* **2002**, *102*, 3811–3836. [CrossRef] [PubMed]

87. Bonchio, M.; Carraro, M.; Scorrano, G.; Fontananova, E.; Drioli, E. Heterogeneous Photooxidation of Alchols in Water by Photocatalytic Membranes Incorporating Decatungstate. *Adv. Synth. Catal.* **2003**, *345*, 1119–1126. [CrossRef]

88. Mylonas, A.; Papaconstantinou, E. Photocatalytic degradation of phenol and p-cresol by polyoxotungstates. Mechanistic implications. *Polyhedron* **1996**, *15*, 3211–3217. [CrossRef]

89. Texier, I.; Giannotti, C.; Malato, S.; Richter, C.; Delaire, J. Solar photodegradation of pesticides in water by sodium decatungstate. *Catal. Today* **1999**, *54*, 297–307. [CrossRef]

90. Bonchio, M.; Carraro, M.; Gardan, M.; Scorrano, G.; Drioli, E.; Fontananova, E. Hybrid photocatalytic membranes embedding decatungstate for heterogeneous photooxygenation. *Top. Catal.* **2006**, *40*, 133–140. [CrossRef]

91. Fontananova, E.; Donato, L.; Drioli, E.; Lopez, L.; Favia, P.; d'Agostino, R. Heterogenization of Polyoxometalates on the Surface of Plasma-Modified Polymeric Membranes. *Chem. Mater.* **2006**, *18*, 1561–1568. [CrossRef]

92. Drioli, E.; Fontananova, E.; Bonchio, M.; Carraro, M.; Gardan, M.; Scorrano, G. Catalytic Membranes and Membrane Reactors: An Integrated Approach to Catalytic Process with a High Efficiency and a Low Environmental Impact. *Chin. J. Catal.* **2008**, *29*, 1152–1158. [CrossRef]

93. Carraro, M.; Gardan, M.; Scorrano, G.; Drioli, E.; Fontananova, E.; Bonchio, M. Solvent-free, heterogeneous photooxygenation of hydrocarbons by Hyflon® membranes embedding a fluorous-tagged decatungstate: The importance of being fluorous. *Chem. Commun.* **2006**, *43*, 4533–4535. [CrossRef]

94. Lopez, L.C.; Buonomenna, M.G.; Fontananova, E.; Iacoviello, G.; Drioli, E.; d'Agostino, R.; Favia, P. A New Generation of Catalytic Poly(vinylidene fluoride) Membranes: Coupling Plasma Treatment with Chemical Immobilization of Tungsten-Based Catalysts. *Adv. Funct. Mater.* **2006**, *16*, 1417–1424. [CrossRef]

95. Lopez, L.C.; Buonomenna, M.G.; Fontananova, E.; Drioli, E.; Favia, P.; d'Agostino, R. Immobilization of Tungsten Catalysts on Plasma Modified Membranes. *Plasma Processes Polym.* **2007**, *4*, 326–333. [CrossRef]

96. Favia, P.; Sardella, E.; Gristina, R.; d'Agostino, R. Novel plasma processes for biomaterials: Micro-scale patterning of biomedical polymers. *Surf. Coat. Technol.* **2003**, *169–170*, 707–711. [CrossRef]

Article

Increasing Salt Rejection of Polybenzimidazole Nanofiltration Membranes via the Addition of Immobilized and Aligned Aquaporins

Priyesh Wagh [1] , **Xinyi Zhang** [2] , **Ryan Blood** [1], **Peter M. Kekenes-Huskey** [1,2] ,
Prasangi Rajapaksha [2], **Yinan Wei** [2] **and Isabel C. Escobar** [1,*]

[1]   Chemical and Materials Engineering Department, University of Kentucky, Lexington, KY 40506, USA;
      priyesh.wagh@uky.edu (P.W.); stuart.blood@uky.edu (R.B.); pkekeneshuskey@uky.edu (P.M.K.-H.)
[2]   Department of Chemistry, University of Kentucky, Lexington, KY 40506, USA; xinyi.zhang@uky.edu (X.Z.);
      prasangi_iro@uky.edu (P.R.); Yinan.Wei@uky.edu (Y.W.)
*   Correspondence: isabel.escobar@uky.edu; Tel.: +859-257-7990

Received: 21 December 2018; Accepted: 29 January 2019; Published: 3 February 2019

Abstract: Aquaporins are water channel proteins in cell membrane, highly specific for water molecules while restricting the passage of contaminants and small molecules, such as urea and boric acid. Cysteine functional groups were installed on aquaporin Z for covalent attachment to the polymer membrane matrix so that the proteins could be immobilized to the membranes and aligned in the direction of the flow. Depth profiling using x-ray photoelectron spectrometer (XPS) analysis showed the presence of functional groups corresponding to aquaporin Z modified with cysteine (Aqp-SH). Aqp-SH modified membranes showed a higher salt rejection as compared to unmodified membranes. For 2 M NaCl and CaCl$_2$ solutions, the rejection obtained from Aqp-SH membranes was $49.3 \pm 7.5\%$ and $59.1 \pm 5.1\%$. On the other hand, the rejections obtained for 2 M NaCl and CaCl$_2$ solutions from unmodified membranes were $0.8 \pm 0.4\%$ and $1.3 \pm 0.2\%$ respectively. Furthermore, Aqp-SH membranes did not show a significant decrease in salt rejection with increasing feed concentrations, as was observed with other membranes. Through simulation studies, it was determined that there was approximately 24% capping of membrane pores by dispersed aquaporins.

Keywords: aquaporins; nanofiltration; immobilization; biomimetic

1. Introduction

A growing research area in water purification is the incorporation of transmembrane water channel proteins, known as aquaporins in the synthetic membranes owing to the excellent permeability and selectivity of aquaporins (aqp) towards water molecules [1–6]. These membranes are called biomimetic membranes because they mimic the function of aquaporins present in lipid bilayer within cell membranes. In recent years, a number of approaches have been adapted from biological concepts and principles to develop biomimetic membranes [2,6–27]. However, there are still many challenges associated with aquaporins based membranes. Generally, aquaporin-based biomimetic membranes developed to date consist of three building blocks: aquaporins, amphiphilic molecules in which the aquaporins are embedded in order to simulate the environment of the lipid bilayer in the cell membranes, and a polymer support structure [22]. These amphiphilic molecules in which the aquaporins are incorporated can be either lipids or polymers. Due to the superior mechanical and chemical properties, block copolymers and amphiphilic polymers have been predominantly investigated for the development of aquaporin-based membranes [2,6,19,22,28–37]. Studies using lipids as the amphiphilic molecules to support aquaporins have shown that these systems were able to maintain membrane integrity [7,12–14,18,23,38,39].

The widespread application of aquaporin-based membranes faces several challenges with respect to synthesis, stability and function of the membrane assembly. One of the challenges is to design and prepare biomimetic membranes with embedded and aligned aquaporin proteins without losing their integrity and performance, while providing an additional solid support that is sufficiently porous [23]. Toward this goal, aquaporin constructs were modified to bear affinity tags or unique amino acids at the N-terminus of the aquaporin molecule, which was used to facilitate directional immobilization. Each aquaporin monomer was modified with a unique amino acid Cys group at the N-terminus right after the first Met (Figure 1A), and due to the aquaporin tetrameric nature, these Cys groups became four anchors for attachment. There are two intrinsic Cys groups in the sequence of aquaporin. Studies have shown that they are not chemically reactive toward modifications due to their limited accessibility [4]. Therefore, the engineered Cys would be the only site in aquaporin reactive toward the thiol-specific modification. The presence of these four Cys anchors per aquaporin tetramer was used to ensure that all tetramers were attached on the membrane surface in alignment with the feed direction.

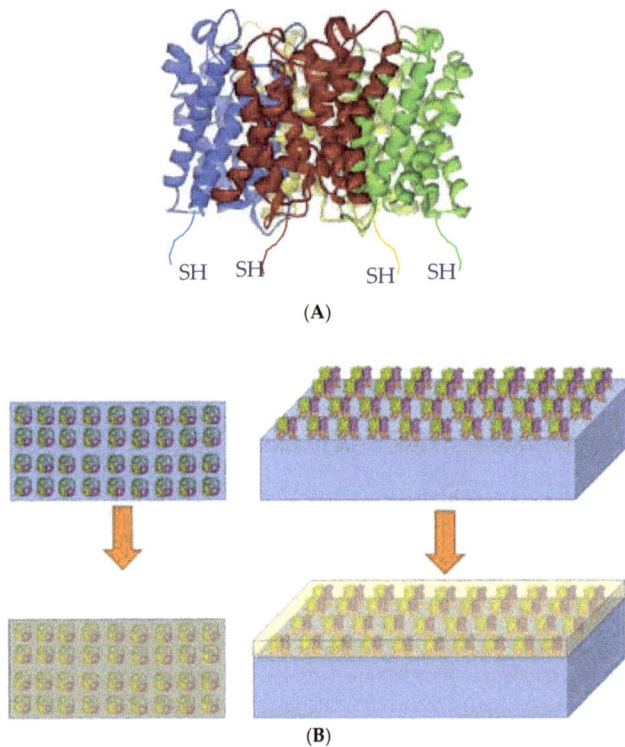

(A)

(B)

Figure 1. (**A**) Cys modified aquaporin molecule (Aqp-SH). (**B**) Chemical attachment of Aqp-SH on –COOH modified Polybenzimidazole (PBI-COOH) membrane and in-situ addition of a layer of Polyvinyl alcohol carrying long alkyl chains (PVA-alkyl).

The objective of this study was to covalently attach Cys modified aquaporins (Aqp-SH) to a polymeric membrane backbone in order to align them in the direction of flow. Polyvinyl alcohol carrying long alkyl chains (PVA-alkyl) was used to bind the remaining sites present on the backbone and to seal the gaps in between attached aquaporin molecules. PVA-alkyl has been previously used to enhance the mechanical strength of the membrane assembly and simulate the natural environment for attached aligned aquaporins [40]. Figure 1B provides a schematic of the attachment of Cys modified aquaporins to the membrane backbone.

2. Experimental

2.1. Materials

2.1.1. Polybenzimidazole (PBI)

Polybenzimidazole (PBI, PBI Performance Products, Inc, Charlotte, NC, USA) was used as the backbone of membrane assembly. PBI has found applications in ion-exchange membranes for fuel cells because of its excellent mechanical, thermal and chemical stability over a wide range of pH [41–43]. The specific polymer composition used in these membranes is poly [2,2'-(1,3-phenylene)-5,5'-bibenzimidazole]. The absence of aliphatic groups and stability of benzimidazole group in PBI are responsible for its applications in a wide range of pH [44]. Hydrogen bonds can be formed intramolecularly or intermolecularly due to the heterocycle imidazole ring presented in the repeating unit of PBI molecules [45]. PBI was dissolved in N,N-Dimethylacetamide (Sigma Aldrich, St. Louis, MO, USA) to prepare dope solution.

2.1.2. PVA-Alkyl

PVA-alkyl is amphiphilic in nature with the high hydrophilicity of PVA (Sigma Aldrich, St. Louis, MO, USA) and hydrophobicity of the long alkyl side chains. Being amphiphilic in nature, it can be an excellent synthetic alternative for the lipid bilayer in cell membrane where aquaporins are constituted naturally [22]. PVA-alkyl spontaneously attaches to cell surface, anchoring through hydrophobic interactions between the alkyl chains and the lipid bilayer of the cell membrane without reducing cell viability [46]. It carries 28 alkyl side chains per molecule, and interacts strongly with the lipid bilayer in cell membrane because the alkyl chains anchor to the cell surface at multiple points [47]. Because of these hydrophobic interactions and the tendency of the polymer to protect the cells, PVA-alkyl is proposed to be an excellent material to support aquaporins. Materials required in order to synthesize PVA-alkyl, and the polymer PBI were of the same source and grade that were used in previous studies [40].

2.1.3. AquaporinZ Modification with Single Cysteine at the N-Terminus

Cysteine contains a thiol group in its side chain, which can be used for immobilization. To prevent cysteine from being buried in the structure of AqpZ with limited accessibility for binding, a cysteine was added before the his-tag which was used to facilitate protein purification via the conventional metal-affinity chromatograph as shown in Figure 2. In this study, cysteine group was added using QuikChange site-directed mutagenesis following manufacturer's instruction (Agilent). Primers are: 5'-GAGATATACCATGGGTTGCTCTGGTCTGAACGAC-3', and 5'- GTCGTTCAGACCAGAGCAACCCATGGTATATCTC-3', using pET28a-ApqZ as template [48]. The modification was verified by DNA sequencing.

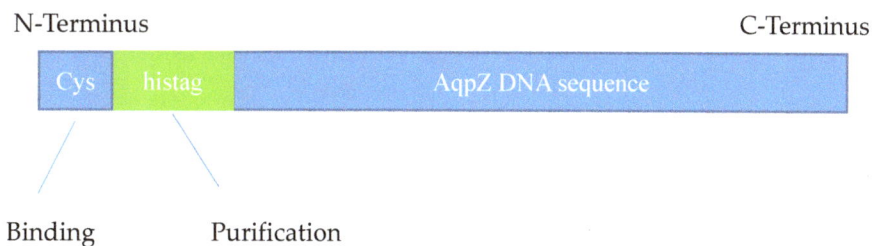

Figure 2. Schematic of Cysteine attachment at the N-terminus of aquaporins using site-directed mutagenesis. Cysteine groups were added to act as anchors in order to attach Aqp on PBI membrane surface.

2.1.4. AquaporinZ Expression and Purification

The constructed plasmid was transformed into *Escherichia coli* strain C43 (DE3). Single colony was cultured overnight at 37 °C in 5 mL Luria Broth (LB) medium containing 50 μg/mL kanamycin (Thermo Fisher Scientific, Waltham, MA, USA). The overnight culture was then inoculated into 300 mL fresh LB medium with 50 μg/mL kanamycin and shaking at 250 rpm at 37 °C. The cells were induced with 1 mM Isopropyl β-D-1-thiogalactopyranoside (IPTG) (Sigma Aldrich, St. Louis, MO, USA) when the absorbance at 600 nm reached 0.8. After 4 h incubation, the cells were collected by centrifugation at $8000\times g$ for 10 min.

To purify the protein, cell pellet was re-suspended with 30 mL Phosphate buffered saline (PBS) buffer (Thermo Fisher Scientific, Waltham, MA, USA) (20 mM $NaPO_4$, 0.3 M NaCl and pH 7.9) supplied with 0.5 mM protease inhibitor phenylmethylsulfonyl fluoride (PMSF, Sigma Aldrich, St. Louis, MO, USA) and sonicated for 20 min on an ice-water bath. The cell lysate was clarified by centrifugation at $15{,}317\times g$, 4 °C for 20 min. Then cell debris was dissolved using 2% Triton in PBS buffer and incubated with shaking for 2 h at 4 °C to extract membrane protein. The re-suspension was clarified with centrifugation at $15{,}317\times g$, 4 °C for 20 min and the supernatant was collected. Ni-NTA agarose beads (Qiagen, Germantown, MD, USA) was mixed with the supernatant for 40 min at 4 °C with shaking. The resin was then loaded into an empty column, drained, and washed with PBS buffer supplemented with 0.03% DDM (n-Dodecyl β-D-maltoside, Sigma Aldrich, St. Louis, MO, USA) and 40 mM imidazole (Sigma Aldrich, St. Louis, MO, USA). Protein was eluted with 500 mM imidazole and 0.03% DDM in PBS buffer. Imidazole was removed by dialysis against PBS buffer supplemented with 0.03% DDM overnight.

An inactive mutant Aqp-SH R189A was also expressed, according to previously published protocol [11], to be used as a negative control to the –Cys modified Aqp. In order to express the inactive mutant, the arginine (R) residue at position 189 in AqpZ was replaced with alanine (A) using site-directed mutagenesis. The Arginine constriction region in aquaporins not only provides the selectivity filter but it also ensures that proton transport is blocked. The replacement of Arginine with Alanine causes a proton transport through aquaporins [49]. This inactive mutant of aquaporin shows no selectivity towards water [50]. Hence, it was used to prepare a negative control of the functional Aqp-SH modified membrane. The protocol used to incorporate Aqp-SH R189A was the same as that used to incorporate functional Aqp-SH into PBI membranes.

2.2. Methodology

2.2.1. PBI Membranes Casting

PBI membranes were prepared according to previously published studies [40,43]. The dope solution was diluted to 21% PBI by adding solvent. The non-solvent phase that was used in this process was water. Flat sheet membranes were prepared using casting knife, or doctor's blade (Paul N Gardner Co, U.S. Pat 4869200, Pompano Beach, FL, USA) The membranes were stored in a 50/50 glycerol-DI water solution in order to prevent their drying and collapse of pore structure. The membranes were kept in the solution at least one day before they were analyzed.

2.2.2. Surface Activation of Membranes

In order to attach Aqp-SH and PVA-alkyl to PBI membranes, membrane surfaces were activated following previous techniques [40,43], in which 4-chloromethyl benzoic acid (CMBA) purchased from Sigma-Aldrich (St. Louis, MO, USA) was used in order for functionalization. CMBA added a carboxylic acid group to the surface of PBI membrane, which could be used as a platform for subsequent functionalization of the membrane [41,42].

2.2.3. Preparation of PVA–Alkyl

PVA-alkyl was prepared in a two-step process, according to literature protocols [40,51–53]. Briefly, a reaction between PVA and sodium monochloroacetate (Sigma Aldrich, St. Louis, MO, USA) yielded carboxy-methyl PVA (PVA-COOH), and PVA-alkyl was prepared by reacting PVA-COOH with hexadecanal [40,52].

2.2.4. Chemical Attachment of –Cys Modified Aqp to PBI Backbone

Immobilization of aquaporins into polymer matrix was done in order to align their channels with the direction of water flux and to optimize their performance. Aquaporins were covalently attached to the modified PBI backbone with carbodiimide chemistry. For this task, flat sheet PBI membranes were prepared and modified with CMBA. In the next step, Cys modified Aquaporin (Aqp-SH) were covalently attached in a reducing environment to the –COOH modified PBI membrane using carbodiimide chemistry, as shown in Figure 3. In this mechanism, Aqp-SH acted as a nucleophile to get covalently attached to the –COOH group present on the surface of PBI membrane. Cys groups present after the N-terminus acted as anchors the Aquaporin molecules to prevent the swaying and to help with the alignment of the aquaporin molecules in the direction of the flow. PVA-alkyl was used in order to bind to the remaining –COOH groups present in the membrane and to seal the gaps in between the attached Aqp-SH molecules following the EDCH chemistry previously used.

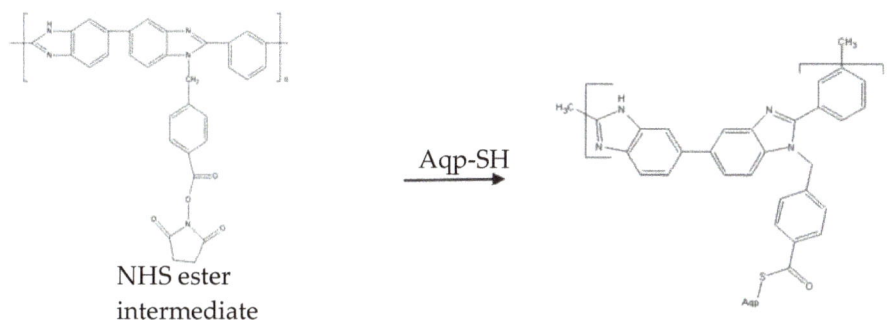

Figure 3. Chemical attachment of Aqp-SH to –COOH modified PBI membranes using carbodiimide chemistry.

2.2.5. Surface Modification of PBI Membrane Using PVA-Alkyl

PVA-alkyl was attached to the membrane using carbodiimide chemistry as reported in previous studies [40].

2.2.6. Membrane Characterization

Dynamic Light Scattering

Since aquaporins form the functional element of biomimetic membranes, producing high quality proteins is critical. Before immobilizing proteins on membrane surface, it is important to evaluate the proteins for their concentration, purity and aggregation state. For this purpose, analysis of protein solution with dynamic light scattering to determine the presence and extent of aggregation was carried out in Litesizer 500 particle analyzer by Anton Paar (Ashland, VA, USA). An aquaporin solution was taken in a glass cuvette and a plot of particle size vs relative frequency and polydispersity index (PDI) of the solution was obtained. Good quality protein samples would have PDI of 0.08, acceptable quality protein would have PDI of 0.1 to 0.4, while the precipitated protein would have PDI of 0.4 to 0.9 [19].

Molecular Weight Cut Off (MWCO)

The MWCO analysis of unmodified PBI, CMBA modified PBI (PBI-CMBA), and PVA-alkyl modified PBI membranes was conducted using 100 ppm solutions of various molecular weights of polyethylene glycol (PEG) and sucrose solutions. The total organic carbon (TOC) of both feed and permeate solutions were measured using Teledyne Tekmar Fusion TOC analyzer (Mason, OH, USA). The various samples that were used in this study along with their Stokes-Einstein radii are shown in Table 1. The rejection values of all solutes were used to determine the MWCO of both unmodified and modified PBI membranes. The molecular weight of solute in feed solution for which the membranes showed more than 90% rejection was considered the MWCO of the membranes. The apparent solute rejection R (%) was calculated using Equation (1).

$$R = \left(1 - \frac{C_p}{C_f}\right) \times 100\% \tag{1}$$

where C_p and C_f are solute concentrations in permeate and feed solutions respectively.

Table 1. Neutral solutes used for Molecular Weight Cut Off (MWCO) analysis and their Stokes-Einstein radii in nm [54–58]. PEG: polyethylene glycol.

Solute	Mol. Wt. (gm/mol)	Stokes-Einstein Radii (nm)
PEG 200	200	0.41
Sucrose	342.3	0.47
PEG 400	400	0.57
PEG 600	600	0.68
PEG 1000	1000	0.94

Contact Angle Measurements

Contact angle was used as a measure to determine the hydrophilicity of the membrane surface. A drop shape analyzer—DSA 100 (KRUSS USA, Matthews, NC, USA) was used for contact angle measurements using sessile drop technique.

Zeta Potential and Surface Charge Analysis

Zeta potential is used to analyze the surface charge of membranes at different pH environments. It is particularly important to analyze the separation efficiency of membranes based on charge and also a confirmation test for surface modification [59]. Surface charge was analyzed by measuring the zeta potential using an Anton Paar SurPASS electrokinetic analyzer (Anton Paar, Ashland, VA, USA) in surface analysis mode. Before analysis, membranes were rinsed with copious amounts of DI water to remove any residual solvent or glycerol from the storage solution in the case of PBI membranes. The KCl electrolyte solution (Sigma Aldrich, St. Louis, MO) used in these measurements had an ionic strength of 1.0 mM. The pH values for the various readings were adjusted using 0.5 M HCl (Sigma Aldrich, St. Louis, MO, USA) and 0.5 M NaOH (Sigma Aldrich, St. Louis, MO, USA) solutions for acid and base titrations.

Elemental Analysis

Membranes modified with Aqp-SH were analyzed for changes in the concentration of sulfur since unmodified PBI, –COOH modified PBI, and PVA-alkyl modified PBI membranes do not contain any sulfur present in their structures. Hence, Aqp-SH modified membranes were analyzed for the sulfur concentration in them as a confirmation for attachment of aquaporins to the membranes. K-Alpha x-ray photoelectron spectrometer (XPS, Thermo Fisher Scientific, Waltham, MA, USA) was used in order to analyze the elemental composition along the cross section of both unmodified and Aqp-SH modified membranes. Depth profiling was performed using an ion beam to etch layers of membrane surfaces and elemental composition was measured after each etching cycle. An ion beam of 200 eV was used to etch the surface. Three etching cycles were performed for 120 s each for elemental analysis along cross sections of membranes.

Membrane Morphology

To investigate the cross-section of the membrane and measure the thickness of selective layer of modified membrane, ion beam of the FEI Helios Nanolab Dual beam was used to cut out a small piece of the membrane. A small deposit of platinum with a thickness of around 60 nm was deposited over the area in order to protect the underlying surface during the process of cutting of cross-section by ion beam. A small cross section was cut out and lifted away from the rest of the membrane sample by welding a small bead of platinum to the platinum layer. This sample was then thinned out with a low power ion beam until the morphology of the mesoporous layer was visible using STEM mode in the Dual Beam. This sample was transferred into the JEOL 2010F (Peabody, MA, USA) for TEM imaging of the cross-section.

Flux Analysis

Flux profiles of PBI, PVA-alkyl modified PBI, inactive Aqp-SH modified PBI, and active Aqp-SH modified PBI membranes (called just Aqp-SH modified membranes) were obtained using dead end filtration in an Amicon filtration cell (Amicon Stirred Cell 8010—10 mL, Burlington, MA, USA) under a constant pressure of 0.48 MPa (4.83 bar) and continuous stirring. Flux values were calculated and plotted against the total permeate volume. Membrane samples were cut into circular pieces of area 4.1 cm^2 and supported by a Whatman^{TM} filter paper (125 mmø, Sigma Aldrich, St. Louis, MO, USA). Each membrane was precompacted with DI water for 1 h until a stable flux was reached.

Precompaction was followed by feed solutions of monovalent and divalent salt solutions in water under same conditions of pressure and stirring. Salt rejection was evaluated using five solutions of different concentrations of sodium chloride (NaCl, Sigma Aldrich, St. Louis, MO, USA) and calcium chloride (CaCl$_2$, Sigma Aldrich, St. Louis, MO, USA) in DI water: 3.4, 10, 20, 35 and 100 mM solutions. Salt concentrations were measured using conductivity meter. Solute rejections were calculated using Equation (1).

After each feed water filtration, reverse flow filtration using DI water was performed for 1 h to remove reversibly-attached foulants that were not adsorbed to the membrane and the filter paper support was changed. The flux recovery of the membrane was measured afterwards in order to study the effect of presence of aquaporins on removal of reversible fouling.

In order to analyze linearity of DI water flux through unmodified and Aqp-SH modified PBI membranes, flux values were measured using DI water as feed solution at four different pressure values: 1.38, 2.76, 4.14, and 5.52 bar.

Unmodified and modified membranes were subjected to dead end flow filtration using 0.5 M, 1 M, and 2 M NaCl and CaCl$_2$ solutions in order to compare the rejection trends of membranes under high salt concentration feed solutions. Inductively coupled plasma (ICP) analysis was used to measure the concentrations of permeates obtained from all membranes.

2.2.7. Estimation of Aquaporin Packing in Membrane Assembly:

Membrane porosity and double layer properties influence ion fluxes through the membrane. The flux values measured for Aqp-SH modified membranes exhibited weak sensitivity to ionic strength. These fluxes (j) can be estimated via an ion's concentration (c) gradient and its diffusion coefficient (D), as shown in Equation (2):

$$j = D\nabla c \tag{2}$$

assuming a concentration gradient was imposed perpendicular to a porous film. This concentration gradient was set by the ion concentrations in reservoirs to either side of the membrane as well as their separation. By relating the measured flux to the concentration gradient, an effective diffusion coefficient, De, could be determined. This allowed the inference of relative packing densities of aqp molecules incorporated in the active layer of membrane. This effective diffusion coefficient would be generally smaller than the ion's intrinsic diffusion rate in bulk media, and moreover, it would be proportional to the ratio of the accessible pores' surface area to the total surface area, assuming the channels were perfectly linear and aligned with the concentration gradient, e.g., $D_e = \frac{SA_{pore}}{SA_{total}} \times D$. According to the SEM imaging data of cross-sections of membranes published previously [40], it was further assumed that the PVA-alkyl and PBI were stacked in layers aligned perpendicular to the concentration gradient.

Based on these assumptions, a numerical partial differential equation was used to estimate how ionic fluxes were modulated by aquaporin surface densities, from which aquaporin packing densities compatible with experimentally-measured flux data could be determined. Namely, finite element simulations of the steady state Fickian diffusion Equation (3) were performed,

$$\frac{dc}{dt} = -\nabla j \tag{3}$$

subject to c(L) = 1 mM and c(R) = 0 mM, where c is the concentration of the ionic solution, D is the diffusion coefficient and L, R correspond to the left and right reservoir boundaries. From these simulations, an effective diffusion coefficient that reflected the impact of the channel geometry on transport was determined. This proceeded through recognizing the flux was related to the concentration gradient via Equation (4)

$$<j> = \frac{1}{A} \int D \nabla c \, dS \tag{4}$$

where A is the surface area of the film and S represents the surface. Flux could be expressed in terms of concentrations and De was given by Equation (5)

$$<j> A \sim D_e \frac{(c(L) - c(R))}{(x(L) - x(R))} \tag{5}$$

where c(i) is the concentration at boundary i (left and right) and x(i) is the position of the boundary. By numerically evaluating <j> at the film boundary, the equation was solved for De based on the concentrations imposed at the reservoir boundaries and their separation distance.

These equations were solved on three-dimensional finite element meshes [59,60], based on potential membrane and aquaporin configurations using the mesh generation tool GMSH [59,61] (See Figure 4). The meshes consisted of two reservoirs separated by a porous domain representing the film. Aqp or aggregates thereof were represented by cylinders of varying radii aligned parallel to the membrane. In principle, atomistic resolution surface geometries could have been used for the Aqp molecules [59,60], but since specific knowledge of the membrane structure at the solvent/membrane interface was not known, a simple cylindrical representation of the protein was used. These equations were solved, assuming Dirichlet conditions of c = 1.0 M and c = 0.0 M on the left and right reservoir boundaries [59,60] via the finite element method using FEniCS [59,62]. Thebulk diffusion coefficient was arbitrarily set to D = 1.0 m^2/s, since we present effective diffusion constants that are normalized with respect to the bulk value. Specifically, the weak form of these equations was solved using a piecewise linear Galerkin basis with FEniC's default direct linear solver and parameters. Concentration fluxes were determined by performing an 'assemble' call on an immersed boundary located at the middle and oriented parallel of the porous film. Details of the numerical procedure follow from previously published work [59,60]. To capture the behavior of monomeric AqpZ, the flux found at the boundary of a pore was normalized [59,60]. The packing fraction observed in the boundary layer then represented a boundary condition surrounding individual aquaporins. All code written in support of this study is publicly available [60]. Simulation input files and generated data are available upon request.

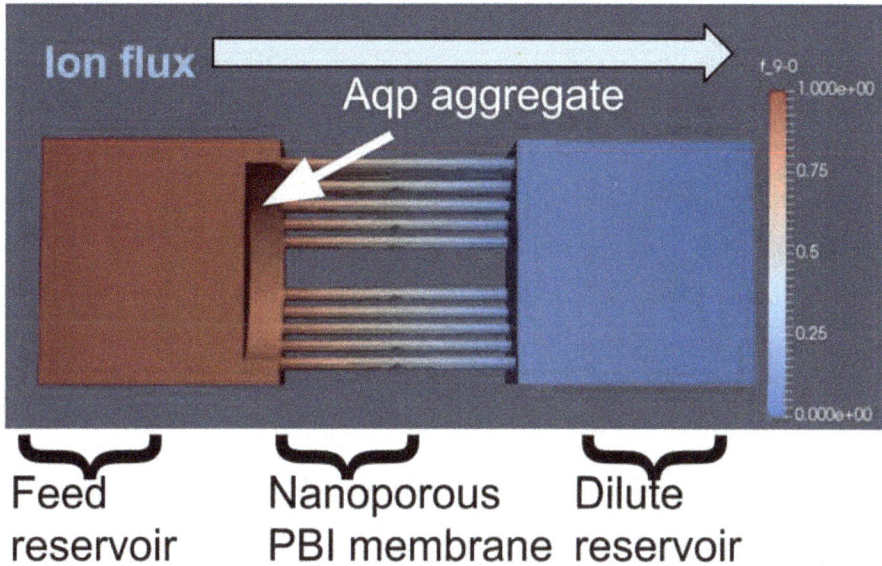

Figure 4. Representative simulation geometry of the membrane occluded by aquaporin (Aqp) aggregates. Here the left reservoir contained 1000 ppm NaCl solution versus pure solvent in the right reservoir (0 M NaCl). Aqp aggregate-size was simulated by cylinders of increasing diameter overlaid onto the membrane surface. Packing density was also tuned by changing the relative area of the PBI membrane. The effective ion diffusion rate was obtained by integrating the concentration gradient along the membrane surface, based on numerical simulation of the steady-state diffusion equation on this geometry.

3. Results

3.1. Dynamic Light Scattering

As shown in Figure 5, the particle size analysis of Aqp-SH solution obtained showed a sharp peak around 10 nm, which is the size range of aquaporin molecules and the bound detergent. In addition, the PDI obtained from the particle analyzer was 0.19. This showed that the proteins were not aggregated in the solution and the PDI of aquaporin solution was in the acceptable range [19].

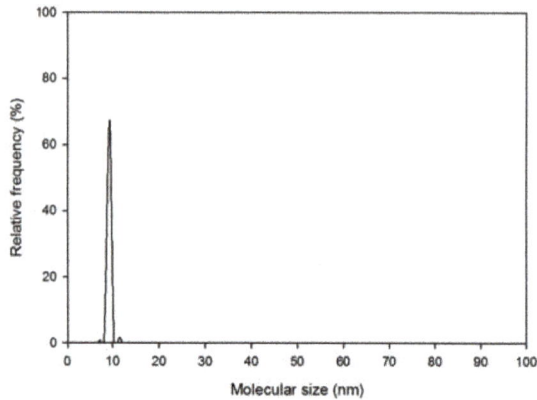

Figure 5. Size measurement of Aqp-SH by dynamic light scattering (DLS). DLS studies showed that aquaporins were not aggregated in the solution.

3.2. MWCO Analysis

Both unmodified PBI and PBI-CMBA membranes showed 90% rejection for PEG 1000 (Figure 6 and Table 2), which has a Stokes radius of 0.94 nm. This showed that the membranes were in the nanofiltration range. After modification with PVA-alkyl attachment to the membrane, the produced membranes showed a 90% rejection for PEG 600, which has a Stokes radius of 0.68 nm. This showed that PVA-alkyl modified membranes were also nanofiltration membranes but with smaller pores. Aqp-SH modified membranes showed rejections of 95.2% ± 3.7%, 97.2% ± 1.4%, 98.4% ± 0.4% for PEG 200, Sucrose, and PEG 400. In addition, these modified membranes showed complete rejections for PEG 600 and PEG 1000.

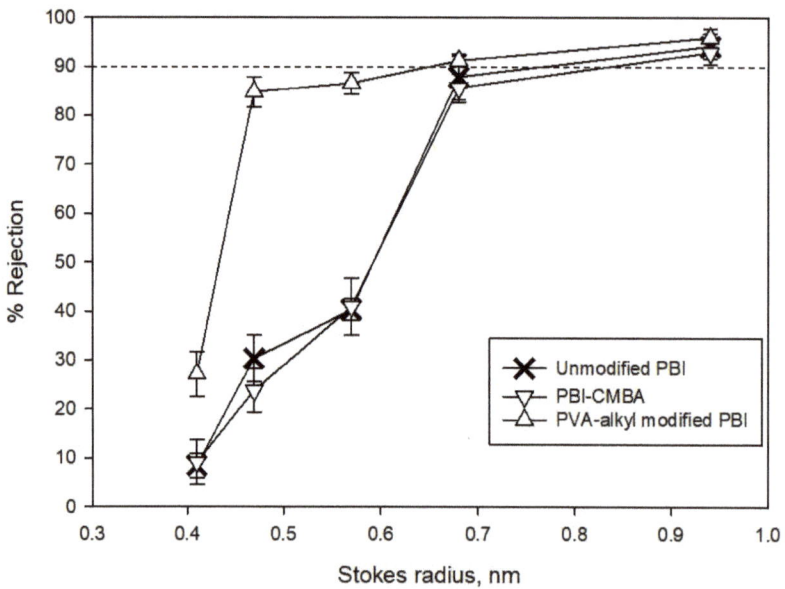

Figure 6. MWCO analysis of unmodified PBI, –COOH modified PBI, and PVA-alkyl modified PBI membranes.

Table 2. Rejections obtained for unmodified Polybenzimidazole (PBI), 4-chloromethyl benzoic acid (CMBA) modified PBI (PBI-CMBA) and aquaporin Z modified with cysteine (Aqp-SH) modified PBI membranes.

Membrane	Rejection > 90%
Unmodified PBI	0.94 nm (94.2% ± 2.5 %)
PBI-CMBA	0.94 nm (93.0% ± 2.4 %)
PVA-alkyl modified PBI	0.68 nm (91.3% ± 1 %)

3.3. Aquaporin Attachment Verification through Elemental Analysis

Depth profiling in XPS analysis was performed for both PBI-COOH and Aqp-SH modified PBI membranes in order to prove the change in sulfur concentration in the membranes after modification. Tables 3 and 4 show weight percentage of atoms of carbon, oxygen, nitrogen and sulfur present in the membrane samples. It can be seen from Table 3 that the amount of sulfur is negligible in PBI-COOH membrane, which was expected since the structure of –COOH modified PBI [40,43] does not contain any sulfur. A small amount of sulfur shown in unmodified membrane might be due to impurities present on the surface and polymer matrix of the membrane. Table 4 shows some amount of sulfur in Aqp-SH modified membrane. Each aquaporin monomer contained four cysteine groups including the one attached at the end groups. Considering the tetrameric form of aquaporins, there are 16 sulfur atoms present in one aquaporin molecule. Hence, for a point scan, the amount of sulfur present at a level in Aqp-SH modified membranes should be between 0.5% and 1%. Therefore, elemental analysis of both unmodified and modified PBI membranes showed the presence of sulfur in the Aqp-SH modified membrane.

Table 3. Elemental composition of elements in PBI-COOH membrane.

	Carbon	Nitrogen	Oxygen	Sulfur
Surface	85.22	10.7	4	0.07
Level 1	86.28	10.28	3.39	0.05
Level 2	86.2	10.39	3.33	0.09
Level 3	87.3	10.56	2.11	0.03

Table 4. Elemental composition of elements in Aqp-SH modified PBI membrane.

	Carbon	Nitrogen	Oxygen	Sulfur
Surface	92.13	4.07	3.3	0.5
Level 1	87.18	8.93	3.41	0.48
Level 2	87.58	8.73	2.99	0.7
Level 3	86.82	8.02	3.05	0.62

3.4. Hydrophobicity

Contact angle was used as a measure of hydrophobicity, and results are shown in Table 5. CMBA modified membranes were found to be more hydrophilic than PBI membranes [42,43]. This was most likely due to addition of a –COOH group in the modified molecule and its increased ability to form hydrogen bonds because of the presence of oxygen with a lone pair. After the addition of PVA-alkyl to the membranes, the contact angle decreased further showing a significant increase in the hydrophilicity of the membrane. This was most likely due to high hydrophilicity of PVA. It is hypothesized that hydrophobic part of PVA-alkyl was reoriented so that the alkyl chains were inside the membrane matrix whereas PVA was on the outside, thus making the membranes more hydrophilic [61].

After chemical attachment of Aqp-SH and PVA-alkyl, there was no significant difference in the contact angle showing that most of the surface of Aqp-SH membrane might be covered with PVA-alkyl, providing a protection to Aqp-SH. The middle portion of AqpZ is hydrophobic, but the ends are hydrophilic as these parts are exposed to the cytosol/periplasm. In case of aquaporins aligned to the feed direction and exposed to the surface, the hydrophilic ends would be facing up, and this would be responsible for an increase in contact angle if they were exposed on the surface of the membrane [4].

Table 5. Hydrophobicity via contact angle.

Membrane	Contact Angle
Unmodified PBI	75° ± 0.55
–COOH modified PBI	70.56° ± 1.04
PVA-alkyl modified PBI	60.5° ± 1.44
Aqp-SH modified PBI	57.5° ± 0.93

3.5. Zeta Potential and Surface Charge Analysis:

Aquaporins have histidine groups present at the pore opening which are positively charged [62]. In order to confirm that the aquaporins were not exposed on the surface of Aqp-SH modified membranes, zeta potential analysis was carried out of unmodified PBI, PBI-CMBA, and Aqp-SH modified PBI membranes over a pH range of 2–10 (Figure 7). There were no significant differences between the surface charge curves of the three membranes, with PBI-CMBA showing a more negative trend as compared to the others likely due to the additional of functional carboxylic end groups to the membrane surface. Since Aqp-SH membranes did not show more positive trends, and in fact showed no significant difference in zeta potential as compared to the unmodified PBI membranes, it is concluded that aquaporins were not exposed on the surface of the membranes.

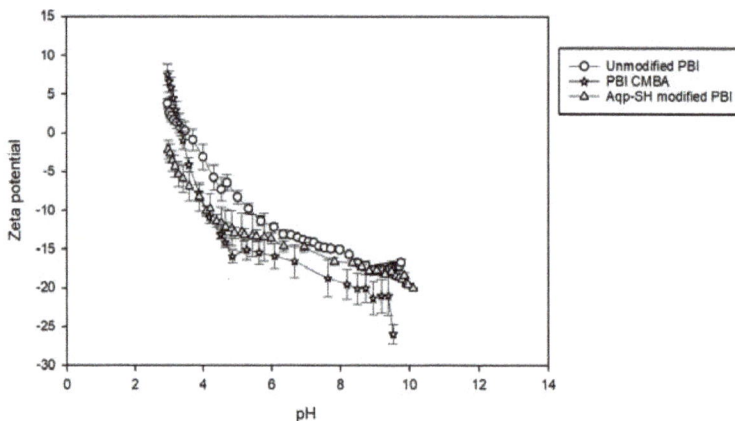

Figure 7. Zeta potential values of unmodified PBI, PBI-CMBA, and Aqp-SH modified PBI membranes over a pH range of 2–10.

3.6. Membrane Morphology

Membranes modified with Aqp-SH-PVA-alkyl showed a selective layer of approximately 50 nm. However, the cross-sectional images (Figure 8) did not provide any visual confirmation of Aquaporins present in the selective layer of the membrane. A selective layer as thick as 50 nm might be because of the surface modification of PBI membrane with PVA-alkyl. Lack of any visual confirmation of aquaporins in the selective layer might be because there were no vesicles in the system and aquaporins were present in the modification layer as individual molecules surrounded by hydrophobic and hydrophilic groups of PVA-alkyl.

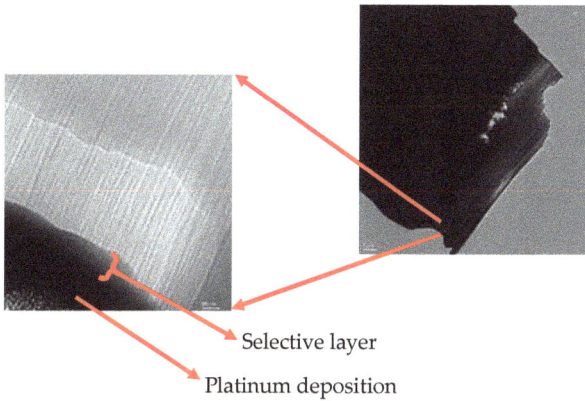

Selective layer

Platinum deposition

Figure 8. Cross sectional image of a cut out of the modified membrane obtained using TEM. The cutout was obtained using a focused ion beam instrument.

3.7. Flux Analysis

To investigate the ability of the Aqp-SH membranes to reject ions, filtration studies using different concentration solutions of NaCl and CaCl$_2$ in water were performed under a constant pressure of 4.83 bar. Experiments were conducted in parallel in order to study unmodified and modified membranes, and the flux values were plotted as a function of permeate volume for unmodified, PVA-alkyl modified PBI (to reflect the amphiphilic matrix without aquaporins), inactive Aqp-SH modified PBI (i.e., to be used as a negative control) and Aqp-SH membranes, as shown in Figure 9A–D, respectively.

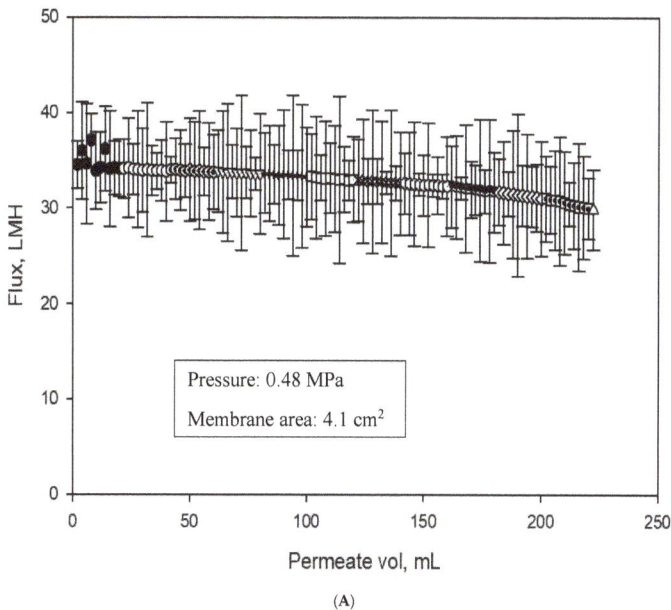

(A)

Figure 9. *Cont.*

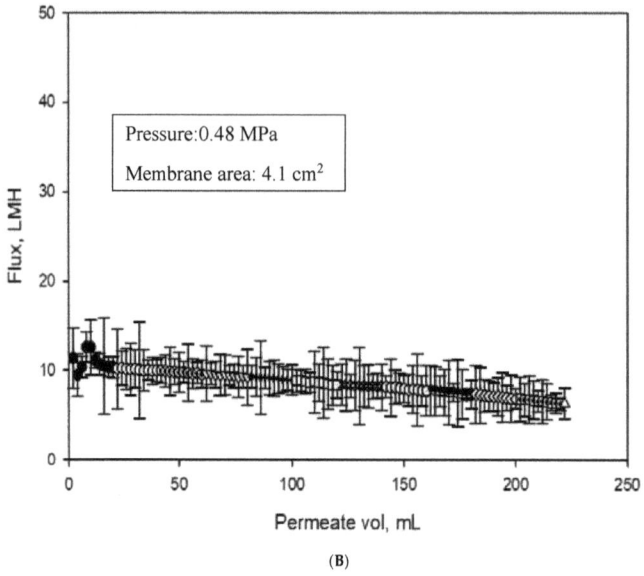

(B)

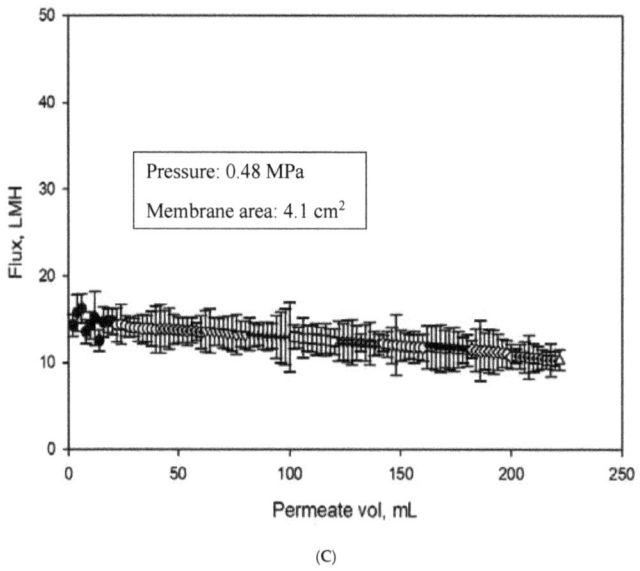

(C)

Figure 9. *Cont.*

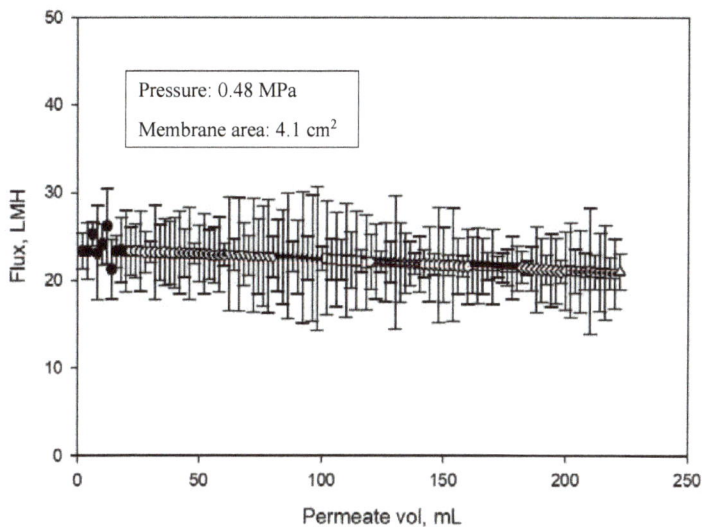

(D)

●	Precompaction	▽	NaCl, 20 mM	×	CaCl$_2$, 3.4 mM	◇	CaCl$_2$, 35 mM
○	NaCl, 3.4 mM	✦	NaCl, 35 mM	○	CaCl$_2$, 10 mM	▣	CaCl$_2$, 100 mM
⊕	NaCl, 10 mM	□	NaCl, 100 mM	✦	CaCl$_2$, 20 mM	△	Flux recovery

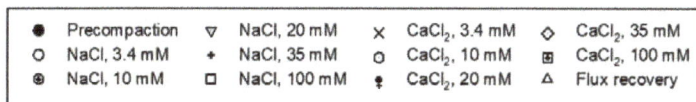

Figure 9. (A): Flux analysis of unmodified PBI membrane; **(B)**: Flux analysis of PVA-alkyl modified PBI membrane; **(C)**: Flux analysis of inactive Aqp-SH modified PBI membrane; and **(D)**: Flux analysis of Aqp-SH modified PBI. All flux analyses were carried out at constant pressure of 0.48 MPa (4.83 bar). Five concentrations each of NaCl and CaCl$_2$ solutions were used as feed solutions and reverse flow filtration was used after every solution filtration.

PVA-alkyl modified membranes showed the lowest initial flux, filtration flux and recovered flux among all membranes possibly because of added resistance to flow due to the addition of a dense layer to the surface and because of a decrease in pore size (Figure 6). Unmodified PBI membranes showed highest initial flux values, which might have been due to the absence of any layer adding resistance on the surface of membranes. The flux profile obtained for the inactive Aqp-SH membranes did not show any significant change when compared to that of PVA-alkyl modified membranes, possibly due to the lack of water permeability of inactive mutant of aquaporins (Aqp-SH R189A) [11,50]. The incorporation of aquaporins on PVA-alkyl modified membranes showed an increase in flux values as compared to PVA-alkyl membranes as well as the membranes modified with inactive mutant; however, the flux values of Aqp-SH membranes were still lower than those of unmodified PBI membranes. The addition of PVA-alkyl alone acted to both block pores and increased resistance to flow, and hence, decreased flux. The addition of functional aquaporins to these membranes provided them with flow channels, which increased the flux as compared to PVA-alkyl membranes. However, the flux was not as high as the modified membranes owing likely to the fact that aquaporin coverage was not complete over the surface of the PVA-alkyl, so there were still regions of minimal or no flow. Additional experiments were conducted in order to analyze the flux linearity of unmodified and modified membranes. Fluxes produced by all the membranes increased linearly with increment in pressure. Also, the incorporation of immobilized aquaporins and dense PVA-alkyl layer on the surface of PBI membrane did not affect the flux linearity of the membranes (Figure S1).

With respect to salt rejection (Figure 10), Aqp-SH membranes showed the highest rejections for the solutions as compared to unmodified PBI and PVA-alkyl modified PBI membranes. Unmodified PBI membranes showed $19 \pm 2.3\%$ rejection during filtration of the 3.4 mM NaCl solution, and as the NaCl concentration increased to 100 mM, the rejection decreased to $5.3 \pm 1.2\%$. PBI membranes modified with only PVA-alkyl showed a rejection of $37.24 \pm 2.5\%$ for a feed solution of 3.4 mM NaCl solution and $19.53 \pm 3.7\%$ rejection for 100 mM NaCl solution. PBI membranes modified with inactive mutant of Aqp (Aqp-SH R189A) showed $48.7 \pm 3.2\%$ rejection during filtration of the 3.4 mM NaCl solution, and as the NaCl concentration increased to 100 mM, the rejection decreased to $29.5 \pm 5.1\%$. On the other hand, Aqp-SH membranes showed a significantly higher rejection of $72.15 \pm 4.2\%$ for 3.4 mM feed solution of NaCl and $72.95 \pm 1.8\%$ for 100 mM NaCl. Similarly, unmodified PBI membranes showed $24.30 \pm 1.5\%$ rejection during filtration of the 3.4 mM $CaCl_2$ solution, and as the $CaCl_2$ concentration increased to 100 mM, the rejection decreased to $8 \pm 1.8\%$. PVA-alkyl modified PBI membranes showed $41.61 \pm 4\%$ rejection for 3.4 mM $CaCl_2$ and $25.82 \pm 4.5\%$ rejection for 100 mM $CaCl_2$. Aqp-SH R189A modified PBI membranes showed $53.4 \pm 3.2\%$ rejection for 3.4 mM $CaCl_2$ and $33.8 \pm 1.6\%$ rejection for 100 mM $CaCl_2$. On the other hand, Aqp-SH membranes showed a rejection of $73.01 \pm 3.7\%$ for 3.4 mM feed solution of $CaCl_2$ and $72.0.4 \pm 7.4\%$ for 100 mM $CaCl_2$. To demonstrate the effectiveness of the functionalizing aquaporins with cysteine end groups (i.e., Aqp-SH), results were compared to those using the exact same membranes (PVA-alkyl modified PBI membranes) with regular aquaporin Z (AqpZ) added to them [40]. As non-functionalized AqpZ does not have cysteine groups as anchors to attach chemically on membrane surface, the non-functionalized AqpZ were added via physical incorporation into PVA-alkyl, which acted as a surface modification layer on PBI membrane. Results showed that membranes modified with aquaporins displayed lower flux declines and higher flux recoveries after reverse flow filtration, along with improved rejection values for both protein and salt solutions as compared to PBI and PBI-PVA-alkyl membranes. However, there was leakage of ions between channels as observed by salt rejections decreasing as a function of feed concentration, from approximately 70% at 10 mM to less than 40% at 100 mM. On the other hand, membranes modified with functionalized aquaporins (Aqp-SH) showed consistently higher salt rejection values of ~70% irrespective of feed concentration, along with higher flux recoveries and lower flux declines.

All membranes were then subjected to high concentration feed solutions of NaCl and $CaCl_2$ (Figure 10). Unmodified PBI membranes showed $2.1 \pm 0.5\%$ rejection during filtration of the 0.5 M NaCl solution, and as the NaCl concentration increased to 2 M, the rejection decreased to $0.8 \pm 0.4\%$. PBI membranes modified with only PVA-alkyl showed a rejection of $15.21 \pm 5.1\%$ for a feed solution of 0.5 M NaCl solution and $2.13 \pm 1.7\%$ rejection for 2 M NaCl solution. Aqp-SH R189A modified PBI membranes showed $26.7 \pm 2.6\%$ rejection during filtration of the 0.5 M NaCl solution, and as the NaCl concentration increased to 2 M, the rejection decreased to $12.6 \pm 1.5\%$. On the other hand, Aqp-SH membranes showed a significantly higher rejection of $62.4 \pm 5.4\%$ for 0.5 M feed solution of NaCl and $49.3 \pm 7.5\%$ for 2 M NaCl. Similarly, unmodified PBI membranes showed $3.4 \pm 0.8\%$ rejection during filtration of the 0.5 M $CaCl_2$ solution, and as the $CaCl_2$ concentration increased to 2 M, the rejection decreased to $1.3 \pm 0.2\%$. PVA-alkyl modified PBI membranes showed $17.52 \pm 1.7\%$ rejection for 0.5 M $CaCl_2$ and $13.19 \pm 5.1\%$ rejection for 2 M $CaCl_2$. Aqp-SH R189A modified PBI membranes showed $28.7 \pm 3.2\%$ rejection during filtration of the 0.5 M $CaCl_2$ solution, and as the $CaCl_2$ concentration increased to 2 M, the rejection decreased to $16.7 \pm 1.0\%$ On the other hand, Aqp-SH membranes showed a rejection of $67.2 \pm 3.5\%$ for 0.5 M feed solution of $CaCl_2$ and $59.1 \pm 5.1\%$ for 2 M $CaCl_2$.

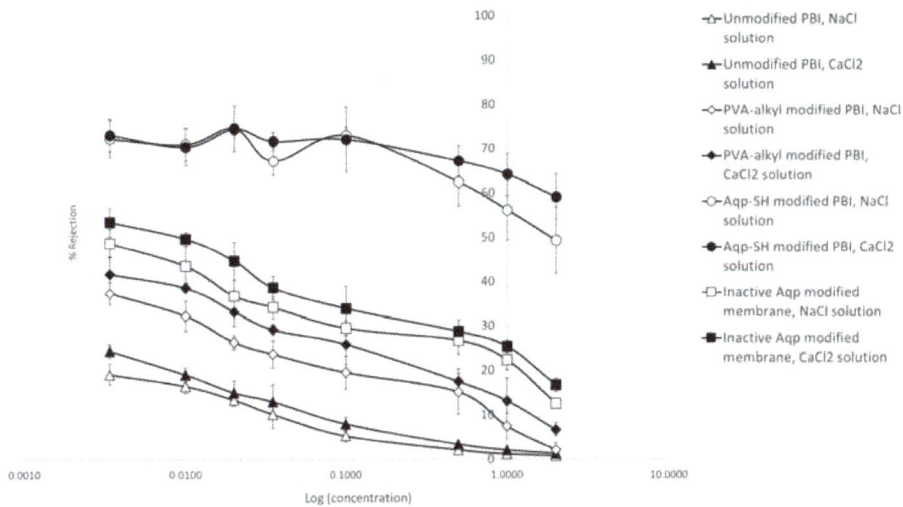

Figure 10. Rejection trends for Sodium chloride and calcium chloride filtration.

PVA-alkyl modified membranes showed higher salt rejection values as compared to unmodified PBI membranes possibly due to a decrease in MWCO, pore size of membranes, and charge interactions with ions. Membranes modified with inactive mutant of Aqp showed slightly higher rejections than PVA-alkyl modified PBI membranes. However, the difference between PVA-alkyl modified PBI membranes and Aqp-SH R189A modified PBI membranes was not significant. Two different samples were used in order to carry out flux analysis and salt rejection studies. The slight difference between salt rejections obtained from PVA-alkyl modified PBI and Aqp-SH R189A modified PBI might be due to the variation of concentration of PVA-alkyl on the surface of membrane sample. The rejection obtained for Aqp-SH modified PBI membranes for salt solutions were higher as compared to the unmodified PBI membranes. This might be due to the immobilized aquaporins on the membrane surface.

Rejection properties are in part determined by the electric double layer that arises from the electrostatic potential about charged surfaces in aqueous media; the magnitude of this potential and its rate of decay from the surface are determined by the surface charge and ionic strength, respectively [63]. As the ionic strength of feed solution increases, rejection decreases owing to a contraction of the electric double layer that enhances charge shielding [64,65] and reduces ion transport rates [66]. However, in the case of aquaporin modified membranes, membrane rejection remained fairly constant irrespective of the ionic strength of salt solutions (Figure 11). In the case of unmodified PBI membranes and membranes modified with PVA-alkyl, a decrease in salt rejection was observed as the ionic strength of feed salt solutions increased, which corresponds to double layer and charge shielding effects. The constant rejection of salt solution obtained with aquaporin modified membranes shows that immobilized aquaporins might be unaffected by charge interactions and provide the same rejection irrespective of the ionic strengths of the feed solutions. The reason might be because of the unique hourglass shaped structure of aquaporin pores and the electrostatic barrier able to reject all the charged entities present in feed other than water molecules [67,68], which means that any decreases in rejection would be due to leakage around the aquaporins. The free-energy profile for ion penetration through aquaporin modified membrane shows a significant difference between the overall barriers for ion and water penetration. The constant rejection observed with Aqp-SH modified membranes is another evidence of the presence of functionalized aquaporins that opened up more water channels, increased water flux through the membrane, and provided higher and constant rejection of feed salt solutions irrespective of their ionic strength. However, it is likely that aquaporins did not cover entire surface area of the membranes due to the presence of detergent or PVA-alkyl, so some of the feed solution might have gone around the aquaporins on the surface, providing a rejection less than the complete rejection that was expected with aquaporins.

3.8. Estimations of Aquaporin Packing in Membrane Assembly

Although the polymer layers could be resolved via electron microscopy, the distribution of surface-anchored Aqps were beyond limits of detection. Thus, to investigate the hypothesis that aquaporin surface deposition was incomplete, a computational model was used to measure ion fluxes across a membrane with aquaporins aggregates of varying sizes. In principle, complete coverage of the film surface with aquaporins should reduce electrolyte flux across the membrane to zero while permitting water flux owing to the high selectivity of aquaporins to water over ions. Although electrostatic interactions with charged surfaces can strongly influence ion conduction [69], the high ion concentrations at which the experiments were conducted strongly attenuated such effects.

To rationalize the flux values reported in Figure 9A–D that demonstrated significant variations in magnitude with respect to polymer membrane configuration, a computational geometry was developed. The 1 nm diameter pores were consistent with MWCO analysis; the pore spacing accounted for 28% surface coverage by nanopores. For this modeling, the PVA-alkyl porosity was assumed to be invariant across the characterized membrane configurations. Since ions do not permeate through the Aqp pore, the proteins were presumed to comprise a monolayer of cylindrical obstructions that resist flow through the PBI layer. Here it was assumed that Aqps capped the PBI-pores and the 64% reduction in flux reported in Figure 9B for PVA-alkyl-only relative to Aqp-SH modified membranes could be attributed to capping a commensurate percentage of available pores. However, these data are insufficient to density of the channels on the membrane surface.

Using a computational simulation of electrolyte diffusion through the Aqp-studded membranes (Figure S2; Table S1), we investigated the extent to which electrolyte fluxes at the membrane were influenced by the Aqps distribution: either as single proteins distributed uniformly or as aggregates. Hence, to determine whether the Aqp-SH behave as aggregates or uniformly distributed channels, we resorted to a three-dimensional, partial differential equation-based model that could account for a range of possible Aqp-SH distributions and packing densities. The model largely follows from our implementation described in previous studies [66,70] and we include implementation details in Methods Section 2.2.7.

Toward this end, the steady state diffusion equation was solved based on varying Aqp aggregate sizes. Figure 11 shows a disk of increasing radius that occluded the underlying pores as a representation of Aqp-aggregates of increasing size. In Figure 10, we demonstrate that the model predicted an increase in effective diffusion rate as the Aqp packing density approached 0, such that the faster diffusion observed experimentally for the PBI-only case was recovered. In other words, NaCl diffusion was not impeded by Aqp and diffused at rates typical of a PBI-only membrane. As the Aqp packing density approached 64%, the effective diffusion constant approached the experimental estimate for the Aqp-modified membranes, which we indicate in Figure 10 by the red vertical line. We additionally show in Figure 10 (model fit) the change in De with respect to packing density for a single Aqp monomer, by varying the surface area of the film for a fixed Aqp monomer size 0.4 nm × 0.4 nm. We simulated the effect of varying the monomeric Aqp packing fraction, by scaling the average concentration flux at the Aqp over a range of surface areas, as this flux will scale proportional to the occluded surface area. It was found that the effective diffusion rate scaled comparably to the aggregates, hence these two cases could not be discriminated based on diffusion alone.

It is important to note that Aqp monomers at a given packing density presented more exposed surface area compared to an aggregate of comparable density. In light of which, packing configurations could be discriminated under conditions that manifest strong surface/diffuser interactions. For instance, in the event that these experiments were performed under low ionic strength conditions, it was expected that ions could interact with the Aqp surface and thereby influence diffusion, either through weak electrostatic interactions or high affinity binding [69]. In addition, as shown in in Figure 9D, the initial flux obtained for Aqp-SH modified membranes could be due to the aggregation of some of the aquaporins on membrane surface. Using the experimental flux data obtained in Figure 9D, the packing fraction of dispersed aquaporins was found to be ~64%.

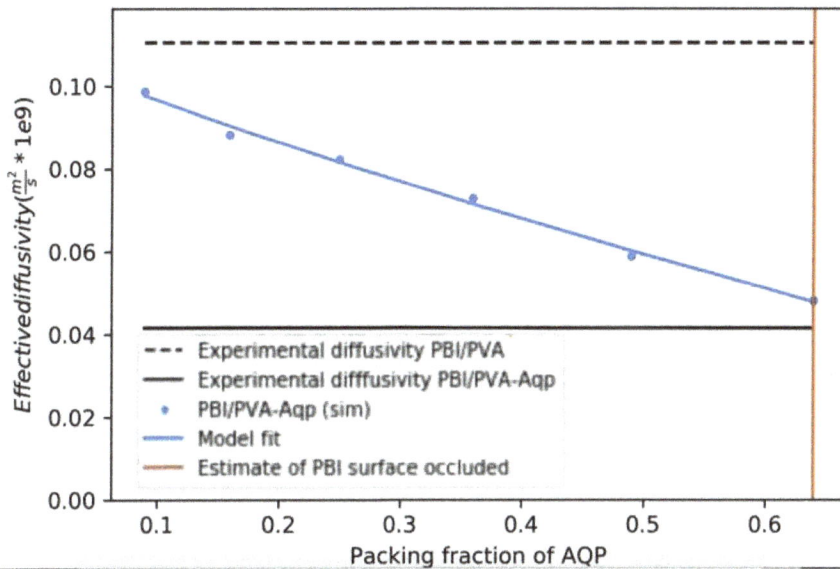

Figure 11. Predictions and experimentally measured effective diffusion coefficients, based on the geometry in Figure 4. Black lines correspond to experimental data found in Supplemental Table S1. Blue lines represent aggregate (solid) and monomeric (dots) Aqp models, respectively.

4. Concluding Remarks

Modification of aquaporins with a cysteine at the N-terminus and immobilization of these modified aquaporins on the membrane surface was successfully accomplished. Elemental analysis showed that aquaporins were immobilized on the membrane surface. It is proposed that four cysteine groups acting as anchors for aquaporin tetramer helped to align aquaporins on the surface of membranes. In agreement with pore size distributions, charge interactions, and added resistance to flow due to modification, PVA-alkyl modified PBI membranes showed lower flux values and slightly higher salt rejection as compared to unmodified PBI membranes. On the other hand, Aqp-SH modified membranes displayed lower flux values as compared to unmodified PBI but higher as compared to PVA-alkyl modified membranes. Membranes modified with an inactive Aqp-SH were used as a negative control to demonstrate the functionality of Aqp-SH incorporated into the membranes. Inactive Aqp-SH modified membranes did not show any improvement in flux values as compared to PVA-alkyl modified PBI membranes. Furthermore, owing to the presence of functional and immobilized aquaporins, Aqp-SH modified membranes displayed the highest salt rejection values among all membranes analyzed in the study. Aqp-SH modified membranes displayed a nearly constant salt rejection irrespective of the salt concentration for low feed concentration, while unmodified PBI and PVA-alkyl modified PBI membranes showed a decrease in rejections as feed salt concentration increased. However, due to the hindrance of detergent or PVA-alkyl in aquaporin solutions, the surface of the membrane was not completely covered with immobilized and aligned aquaporins, which in turn led to rejection values lower than 100%. Simulation studies showed that immobilized aquaporins with PVA-alkyl provided a diffusion rate equivalent to 64% coverage. This proved that aquaporins did not cover the entire surface area of the membranes, thereby providing a salt rejection of less than 100%. In addition, the flux obtained with Aqp-SH modified membranes was lower as compared to unmodified PBI membranes. This might be due to aggregation of some of the aquaporins added onto membrane surface. The packing fraction for dispersed aquaporins on membrane surface was calculated to be ~24%.

Supplementary Materials: The following are available online at http://www.mdpi.com/2227-9717/7/2/76/s1, Figure S1: Flux linearity and permeability consistency for unmodified PBI and Aqp-SH modified PBI membranes. Figure S2: Diffusion cell assembly with 1000 ppm NaCl and DI water in two compartments separated by membrane. Table S1: Salt concentrations measured every day for all three membranes in diffusion cell assembly.

Author Contributions: R.B. and P.M.K.-H. conceived and designed the computational experiments and wrote Sections 2.2.7 and 3.8. X.Z., P.R., and Y.W. carried out aquaporin expression and modification, wrote Sections 2.1.3 and 2.1.4. P.W. and I.C.E. conceived the idea of the manuscript, carried out membrane synthesis, modification, characterization experiments, and wrote remaining sections of the article.

Acknowledgments: This material is based upon work supported by the National Science Foundation under Cooperative Agreement No.1355438, and by the NSF KY EPSCoR Program and in addition the author acknowledges Center of Membrane Sciences at University of Kentucky. Research reported in this publication, release was supported by the Maximizing Investigators' Research Award (MIRA) (R35) from the National Institute of General Medical Sciences (NIGMS) of the National Institutes of Health (NIH) under grant number R35GM124977. We further acknowledge support from the American Chemical Society Petroleum Research Fund 58719-DN16. This work used the Extreme Science and Engineering Discovery Environment (XSEDE), which is supported by National Science Foundation grant number ACI-1548562. In addition, the authors want to acknowledge the sources of funding, NSF 1308095 and USGS 104(b): Ohio Water Development Authority GRT00028988/60040357 for funding this project. PBI Performance Products Inc. is thanked for supplying the PBI dope for the study.

Conflicts of Interest: The authors declare no conflict of interest.

References

1. Gena, P.; Pellegrini-Calace, M.; Biasco, A.; Svelto, M.; Calamita, G. Aquaporin Membrane Channels: Biophysics, Classification, Functions, and Possible Biotechnological Applications. *Food Biophys.* **2011**, *6*, 241–249. [CrossRef]
2. Wang, H.; Chung, T.S.; Tong, Y.W.; Jeyaseelan, K.; Armugam, A.; Chen, Z.; Hong, M.; Meier, W. Highly permeable and selective pore-spanning biomimetic membrane embedded with aquaporin Z. *Small* **2012**, *8*, 1185–1190. [CrossRef] [PubMed]
3. Taubert, A. Controlling water transport through artificial polymer/protein hybrid membranes. *Proc. Natl. Acad. Sci. USA* **2007**, *104*, 20643–20644. [CrossRef] [PubMed]
4. Agre, P. Aquaporin Water Channels. *Biosci. Rep.* **2005**, *24*, 127–163. [CrossRef]
5. Zhao, C.X.; Shao, H.B.; Chu, L.Y. Aquaporin structure-function relationships: Water flow through plant living cells. *Colloids Surf. B Biointerfaces* **2008**, *62*, 163–172. [CrossRef] [PubMed]
6. Zhong, P.S.; Chung, T.-S.; Jeyaseelan, K.; Armugam, A. Aquaporin-embedded biomimetic membranes for nanofiltration. *J. Membr. Sci.* **2012**, *407-408*, 27–33. [CrossRef]
7. Li, X.; Wang, R.; Tang, C.; Vararattanavech, A.; Zhao, Y.; Torres, J.; Fane, T. Preparation of supported lipid membranes for aquaporin Z incorporation. *Colloids Surf. B Biointerfaces* **2012**, *94*, 333–340. [CrossRef] [PubMed]
8. Li, X.S.; Wang, R.; Wicaksana, F.; Tang, C.Y.; Torres, J.; Fane, A.G. Preparation of high performance nanofiltration (NF) membranes incorporated with aquaporin Z. *J. Membr. Sci.* **2014**, *450*, 181–188. [CrossRef]
9. Tang, C.; Qiu, C.; Zhao, Y.; Shen, W.; Vararattanavech, A.; Wang, R.; Hu, X.; Torres, J.; Fane, A.G.; Hélix-Nielsen, C. Aquaporin Based Thin Film Composite Membranes. Patents WO/2013/043118, 28 March 2013.
10. Tang, C.Y.; Wang, Z.N.; Petrinic, I.; Fane, A.G.; Helix-Nielsen, C. Biomimetic aquaporin membranes coming of age. *Desalination* **2015**, *368*, 89–105. [CrossRef]
11. Zhao, Y.; Qiu, C.; Li, X.; Vararattanavech, A.; Shen, W.; Torres, J.; Hélix-Nielsen, C.; Wang, R.; Hu, X.; Fane, A.G.; et al. Synthesis of robust and high-performance aquaporin-based biomimetic membranes by interfacial polymerization-membrane preparation and RO performance characterization. *J. Membr. Sci.* **2012**, *423-424*, 422–428. [CrossRef]

12. Zhao, Y.; Vararattanavech, A.; Li, X.; Helixnielsen, C.; Vissing, T.; Torres, J.; Wang, R.; Fane, A.G.; Tang, C.Y. Effects of proteoliposome composition and draw solution types on separation performance of aquaporin-based proteoliposomes: Implications for seawater desalination using aquaporin-based biomimetic membranes. *Environ. Sci. Technol.* **2013**, *47*, 1496–1503. [CrossRef] [PubMed]

13. Sun, G.; Zhou, H.; Li, Y.; Jeyaseelan, K.; Armugam, A.; Chung, T.S. A novel method of AquaporinZ incorporation via binary-lipid Langmuir monolayers. *Colloids Surf. B Biointerfaces* **2012**, *89*, 283–288. [CrossRef] [PubMed]

14. Sun, G.F.; Chung, T.S.; Jeyaseelan, K.; Armugam, A. Stabilization and immobilization of aquaporin reconstituted lipid vesicles for water purification. *Colloids Surf. B Biointerfaces* **2013**, *102*, 466–471. [CrossRef] [PubMed]

15. Sun, G.F.; Chung, T.S.; Jeyaseelan, K.; Armugam, A. A layer-by-layer self-assembly approach to developing an aquaporin-embedded mixed matrix membrane. *RSC Adv.* **2013**, *3*, 473–481. [CrossRef]

16. Wang, H.; Chung, T.-S.; Tong, Y.W.; Meier, W.; Chen, Z.; Hong, M.; Jeyaseelan, K.; Armugam, A. Preparation and characterization of pore-suspending biomimetic membranes embedded with Aquaporin Z on carboxylated polyethylene glycol polymer cushion. *Soft Matter* **2011**, *7*, 7274–7280. [CrossRef]

17. Wang, H.L.; Chung, T.S.; Tong, Y.W.; Jeyaseelan, K.; Armugam, A.; Duong, H.H.P.; Fu, F.J.; Seah, H.; Yang, J.; Hong, M.H. Mechanically robust and highly permeable AquaporinZ biomimetic membranes. *J. Membr. Sci.* **2013**, *434*, 130–136. [CrossRef]

18. Xie, W.Y.; He, F.; Wang, B.F.; Chung, T.S.; Jeyaseelan, K.; Armugam, A.; Tong, Y.W. An aquaporin-based vesicle-embedded polymeric membrane for low energy water filtration. *J. Mater. Chem. A* **2013**, *1*, 7592–7600. [CrossRef]

19. Grzelakowski, M.; Cherenet, M.F.; Shen, Y.X.; Kumar, M. A framework for accurate evaluation of the promise of aquaporin based biomimetic membranes. *J. Membr. Sci.* **2015**, *479*, 223–231. [CrossRef]

20. Kumar, M.; Grzelakowski, M.; Zilles, J.; Clark, M.; Meier, W. Highly permeable polymeric membranes based on the incorporation of the functional water channel protein Aquaporin Z. *Proce. Natl. Acad. Sci. USA* **2007**, *104*, 20719–20724. [CrossRef]

21. Shen, Y.-X.; Saboe, P.O.; Sines, I.T.; Erbakan, M.; Kumar, M. Biomimetic membranes: A review. *J. Membr. Sci.* **2014**, *454*, 359–381. [CrossRef]

22. Habel, J.; Hansen, M.; Kynde, S.; Larsen, N.; Midtgaard, S.R.; Jensen, G.V.; Bomholt, J.; Ogbonna, A.; Almdal, K.; Schulz, A.; et al. Aquaporin-Based Biomimetic Polymeric Membranes: Approaches and Challenges. *Membranes* **2015**, *5*, 307–351. [CrossRef] [PubMed]

23. Nielsen, C.H. Biomimetic membranes for sensor and separation applications. *Anal. Bioanal. Chem.* **2009**, *395*, 697–718. [CrossRef] [PubMed]

24. Tang, C.Y.; Zhao, Y.; Wang, R.; Hélix-Nielsen, C.; Fane, A.G. Desalination by biomimetic aquaporin membranes: Review of status and prospects. *Desalination* **2013**, *308*, 34–40. [CrossRef]

25. Jensen, P.H.; Hansen, J.S.; Vissing, T.; Perry, M.E.; Nielsen, C.H. Biometric Membranes and Uses Thereof. Patents WO2010146365A1, 23 December 2010.

26. Jensen, P.H. Biomimetic Water Membrane Comprising Aquaporins Used in the Production of Salinity Power. Patents US20090007555A1, 8 January 2009.

27. Kaufman, Y.; Grinberg, S.; Linder, C.; Heldman, E.; Gilron, J.; Freger, V. Fusion of Bolaamphiphile Micelles: A Method to Prepare Stable Supported Biomimetic Membranes. *Langmuir* **2013**, *29*, 1152–1161. [CrossRef] [PubMed]

28. Zhang, X.; Tanner, P.; Graff, A.; Palivan, C.G.; Meier, W. Mimicking the cell membrane with block copolymer membranes. *J. Polym. Sci. Part A Polym. Chem.* **2012**, *50*, 2293–2318. [CrossRef]

29. Tanner, P.; Baumann, P.; Enea, R.; Onaca, O.; Palivan, C.; Meier, W. Polymeric Vesicles: From Drug Carriers to Nanoreactors and Artificial Organelles. *Acc. Chem. Res.* **2011**, *44*, 1039–1049. [CrossRef] [PubMed]

30. Choi, H.-J.; Montemagno, C.D. Artificial Organelle: ATP Synthesis from Cellular Mimetic Polymersomes. *Nano Lett.* **2005**, *5*, 2538–2542. [CrossRef]

31. Vriezema, D.M.; Garcia, P.M.L.; Sancho Oltra, N.; Hatzakis, N.S.; Kuiper, S.M.; Nolte, R.J.M.; Rowan, A.E.; van Hest, J.C.M. Positional Assembly of Enzymes in Polymersome Nanoreactors for Cascade Reactions. *Angew. Chem.* **2007**, *119*, 7522–7526. [CrossRef]

32. Stoenescu, R.; Graff, A.; Meier, W. Asymmetric ABC-Triblock Copolymer Membranes Induce a Directed Insertion of Membrane Proteins. *Macromol. Biosci.* **2004**, *4*, 930–935. [CrossRef]

33. Graff, A.; Fraysse-Ailhas, C.; Palivan, C.G.; Grzelakowski, M.; Friedrich, T.; Vebert, C.; Gescheidt, G.; Meier, W. Amphiphilic Copolymer Membranes Promote NADH:Ubiquinone Oxidoreductase Activity: Towards an Electron-Transfer Nanodevice. *Macromol. Chem. Phys.* **2010**, *211*, 229–238. [CrossRef]
34. Kumar, M.; Habel, J.E.O.; Shen, Y.-X.; Meier, W.P.; Walz, T. High-Density Reconstitution of Functional Water Channels into Vesicular and Planar Block Copolymer Membranes. *J. Am. Chem. Soc.* **2012**, *134*, 18631–18637. [CrossRef] [PubMed]
35. Wang, H.; Chung, T.-S.; Tong, Y.W. Study on water transport through a mechanically robust Aquaporin Z biomimetic membrane. *J. Membr. Sci.* **2013**, *445*, 47–52. [CrossRef]
36. Winterhalter, M.; Hilty, C.; Bezrukov, S.M.; Nardin, C.; Meier, W.; Fournier, D. Controlling membrane permeability with bacterial porins: application to encapsulated enzymes. *Talanta* **2001**, *55*, 965–971. [CrossRef]
37. Kumar, M. Biomimetic Membranes as New Materials for Applications in Environmental Engineering and Biology. Ph.D. Thesis, University of Illinois at Urbana-Champaign, Champaign, IL, USA, 2010.
38. Kaufman, Y.; Berman, A.; Freger, V. Supported Lipid Bilayer Membranes for Water Purification by Reverse Osmosis. *Langmuir* **2010**, *26*, 7388–7395. [CrossRef] [PubMed]
39. Wang, M.Q.; Wang, Z.N.; Wang, X.D.; Wang, S.Z.; Ding, W.D.; Gao, C.J. Layer-by-Layer Assembly of Aquaporin Z-Incorporated Biomimetic Membranes for Water Purification. *Environ. Sci. Technol.* **2015**, *49*, 3761–3768. [CrossRef] [PubMed]
40. Wagh, P.; Parungao, G.; Viola, R.E.; Escobar, I.C. A new technique to fabricate high-performance biologically inspired membranes for water treatment. *Sep. Purif. Technol.* **2015**, *156*, 754–765. [CrossRef]
41. Digman, B.; Escobar, I.C.; Hausman, R.; Coleman, M.; Chung, T.S. Surface functionalization of polybenzimidizole membranes to increase hydrophilicity and charge. *ACS Symp. Ser.* **2011**, *1078*, 303–321.
42. Hausman, R.; Digman, B.; Escobar, I.C.; Coleman, M.; Chung, T.-S. Functionalization of polybenzimidizole membranes to impart negative charge and hydrophilicity. *J. Membr. Sci.* **2010**, *363*, 195–203. [CrossRef]
43. Flanagan, M.F.; Escobar, I.C. Novel charged and hydrophilized polybenzimidazole (PBI) membranes for forward osmosis. *J. Membr. Sci.* **2013**, *434*, 85–92. [CrossRef]
44. Sawyer, L.C.; Jones, R.S. Observations on the structure of first generation polybenzimidazole reverse osmosis membranes. *J. Membr. Sci.* **1984**, *20*, 147–166. [CrossRef]
45. Zhu, W.-P.; Sun, S.-P.; Gao, J.; Fu, F.-J.; Chung, T.-S. Dual-layer polybenzimidazole/polyethersulfone (PBI/PES) nanofiltration (NF) hollow fiber membranes for heavy metals removal from wastewater. *J. Membr. Sci.* **2014**, *456*, 117–127. [CrossRef]
46. Inui, O.; Teramura, Y.; Iwata, H. Retention dynamics of amphiphilic polymers PEG-lipids and PVA-Alkyl on the cell surface. *ACS Appl. Mater. Interfaces* **2010**, *2*, 1514–1520. [CrossRef] [PubMed]
47. Teramura, Y.; Kaneda, Y.; Totani, T.; Iwata, H. Behavior of synthetic polymers immobilized on a cell membrane. *Biomaterials* **2008**, *29*, 1345–1355. [CrossRef] [PubMed]
48. Wang, Z.; Ye, C.; Zhang, X.; Wei, Y. Cysteine residue is not essential for CPM protein thermal-stability assay. *Anal. Bioanal. Chem.* **2015**, *407*, 3683–3691. [CrossRef] [PubMed]
49. Beitz, E.; Wu, B.; Holm, L.M.; Schultz, J.E.; Zeuthen, T. Point mutations in the aromatic/arginine region in aquaporin 1 allow passage of urea, glycerol, ammonia, and protons. *Proc. Natl. Acad. Sci. USA* **2006**, *103*, 269–274. [CrossRef] [PubMed]
50. Xin, L.; Helix-Nielsen, C.; Su, H.B.; Torres, J.; Tang, C.Y.; Wang, R.; Fane, A.G.; Mu, Y.G. Population Shift between the Open and Closed States Changes the Water Permeability of an Aquaporin Z Mutant. *Biophys. J.* **2012**, *103*, 212–218. [CrossRef] [PubMed]
51. Hosono, M.; Sugii, S.; Kitamaru, R.; Hong, Y.M.; Tsuji, W. Polyelectrolyte complex prepared from carboxymethylated and aminoacetalized derivatives of Poly(vinyl) alcohol. *J. Appl. Polym. Sci.* **1977**, *21*, 2125–2134. [CrossRef]
52. Totani, T.; Teramura, Y.; Iwata, H. Immobilization of urokinase on the islet surface by amphiphilic poly(vinyl alcohol) that carries alkyl side chains. *Biomaterials* **2008**, *29*, 2878–2883. [CrossRef]
53. Gung, B.W.; Dickson, H.D.; Seggerson, S.; Bluhm, K. A short synthesis of an acetylenic alcohol from the sponge Cribrochalina vasculum. *Synth. Commun.* **2002**, *32*, 2733–2740. [CrossRef]
54. Ling, K.; Jiang, H.; Zhang, Q. A colorimetric method for the molecular weight determination of polyethylene glycol using gold nanoparticles. *Nanoscale Res. Lett.* **2013**, *8*, 538. [CrossRef]

55. Dohmen, M.P.J.; Pereira, A.M.; Timmer, J.M.K.; Benes, N.E.; Keurentjes, J.T.F. Hydrodynamic Radii of Polyethylene Glycols in Different Solvents Determined from Viscosity Measurements. *J. Chem. Eng. Data* **2008**, *53*, 63–65. [CrossRef]
56. Lebrun, L.; Junter, G.A. Diffusion of sucrose and dextran through agar gel membranes. *Enzym. Microbial Technol.* **1993**, *15*, 1057–1062. [CrossRef]
57. Hernandez, S.; Porter, C.; Zhang, X.; Wei, Y.; Bhattacharyya, D. Layer-by-layer assembled membranes with immobilized porins. *RSC Adv.* **2017**, *7*, 56123–56136. [CrossRef]
58. Schultz, S.G.; Solomon, A.K. Determination of the Effective Hydrodynamic Radii of Small Molecules by Viscometry. *J. Gen. Physiol.* **1961**, *44*, 1189–1199. [CrossRef] [PubMed]
59. Lau, W.J.; Ismail, A.F.; Goh, P.S.; Hilal, N.; Ooi, B.S. Characterization Methods of Thin Film Composite Nanofiltration Membranes. *Sep. Purif. Rev.* **2015**, *44*, 135–156. [CrossRef]
60. Bit Bucket. Available online: https://bitbucket.org/pkhlab/pkh-lab-analyses (accessed on 30 January 2019).
61. Shang, Y.; Peng, Y. UF membrane of PVA modified with TDI. *Desalination* **2008**, *221*, 324–330. [CrossRef]
62. Gonen, T.; Walz, T. The structure of aquaporins. *Q. Rev. Biophys.* **2006**, *39*, 361–396. [CrossRef]
63. Israelachvili, J.N. *Intermolecular and Surface Forces*, 3th ed.; Elsevier: Amsterdam, The Netherlands, 2011; p. 710.
64. Braghetta, A.; DiGiano, F.A.; Ball, W.P. Nanofiltration of Natural Organic Matter: pH and Ionic Strength Effects. *J. Environ. Eng.* **1997**, *123*, 628–641. [CrossRef]
65. Xu, Y.; Lebrun, R.E. Investigation of the solute separation by charged nanofiltration membrane: effect of pH, ionic strength and solute type. *J. Membr. Sci.* **1999**, *158*, 93–104. [CrossRef]
66. Kekenes-Huskey, P.M.; Gillette, A.K.; McCammon, J.A. Predicting the influence of long-range molecular interactions on macroscopic-scale diffusion by homogenization of the Smoluchowski equation. *J. Chem. Phys.* **2014**, *140*, 174106. [CrossRef]
67. Burykin, A.; Warshel, A. What Really Prevents Proton Transport through Aquaporin? Charge Self-Energy versus Proton Wire Proposals. *Biophys. J.* **2003**, *85*, 3696–3706. [CrossRef]
68. Hu, G.; Chen, L.Y.; Wang, J. Insights into the mechanisms of the selectivity filter of Escherichia coli aquaporin Z. *J. Mol. Model.* **2012**, *18*, 3731–3741. [CrossRef] [PubMed]
69. Fields, J.B.; Németh-Cahalan, K.L.; Freites, F.A.; Vorontsova, I.; Hall, J.E.; Tobias, D.J. Calmodulin Gates Aquaporin 0 Permeability through a Positively Charged Cytoplasmic Loop. *J. Biol. Chem.* **2017**, *292*, 185–195. [CrossRef] [PubMed]
70. Kekenes-Huskey, P.M.; Scott, C.E.; Atalay, S. Quantifying the Influence of the Crowded Cytoplasm on Small Molecule Diffusion. *J. Phys. Chem. B* **2016**, *120*, 8696–8706. [CrossRef] [PubMed]

MDPI

St. Alban-Anlage 66

4052 Basel

Switzerland

Tel. +41 61 683 77 34

Fax +41 61 302 89 18

www.mdpi.com

Processes Editorial Office

E-mail: processes@mdpi.com

www.mdpi.com/journal/processes

www.ingramcontent.com/pod-product-compliance
Lightning Source LLC
Chambersburg PA
CBHW051853210326
41597CB00033B/5878